Compendium of Plant Genomes

Series Editor

Chittaranjan Kole, Raja Ramanna Fellow, Government of India, ICAR-National Research Center on Plant Biotechnology, Pusa, New Delhi, India

Whole-genome sequencing is at the cutting edge of life sciences in the new millennium. Since the first genome sequencing of the model plant Arabidopsis thaliana in 2000, whole genomes of about 100 plant species have been sequenced and genome sequences of several other plants are in the pipeline. Research publications on these genome initiatives are scattered on dedicated web sites and in journals with all too brief descriptions. The individual volumes elucidate the background history of the national and international genome initiatives; public and private partners involved; strategies and genomic resources and tools utilized; enumeration on the sequences and their assembly; repetitive sequences; gene annotation and genome duplication. In addition, synteny with other sequences, comparison of gene families and most importantly potential of the genome sequence information for gene pool characterization and genetic improvement of crop plants are described.

Interested in editing a volume on a crop or model plant?
Please contact Prof. C. Kole, Series Editor, at ckoleorg@gmail.com

More information about this series at http://www.springer.com/series/11805

Chittaranjan Kole
Editor

The Mango Genome

Springer

Editor
Chittaranjan Kole
ICAR-National Institute
for Plant Biotechnology
Raja Ramanna Fellow
Government of India
New Delhi, India

ISSN 2199-4781 ISSN 2199-479X (electronic)
Compendium of Plant Genomes
ISBN 978-3-030-47828-5 ISBN 978-3-030-47829-2 (eBook)
https://doi.org/10.1007/978-3-030-47829-2

This Springer imprint is published by the registered company Springer Nature Switzerland AG
The registered company address is: Gewerbestrasse 11, 6330 Cham, Switzerland

This book series is dedicated to my wife Phullara
and our children Sourav and Devleena

Chittaranjan Kole

Preface to the Series

Genome sequencing has emerged as the leading discipline in the plant sciences coinciding with the start of the new century. For much of the twentieth century, plant geneticists were only successful in delineating putative chromosomal location, function, and changes in genes indirectly through the use of a number of "markers" physically linked to them. These included visible or morphological, cytological, protein, and molecular or DNA markers. Among them, the first DNA marker, the RFLPs, introduced a revolutionary change in plant genetics and breeding in the mid-1980s, mainly because of their infinite number and thus potential to cover maximum chromosomal regions, phenotypic neutrality, absence of epistasis, and codominant nature. An array of other hybridization-based markers; PCR-based markers; and markers based on both facilitated construction of genetic linkage maps, mapping of genes controlling simply inherited traits, and even gene clusters (QTLs) controlling polygenic traits in a large number of model and crop plants. During this period, a number of new mapping populations beyond F2 were utilized and a number of computer programs were developed for map construction, mapping of genes, and for mapping of polygenic clusters or QTLs. Molecular markers were also used in the studies of evolution and phylogenetic relationship, genetic diversity, DNA fingerprinting, and map-based cloning. Markers tightly linked to the genes were used in crop improvement employing the so-called marker-assisted selection. These strategies of molecular genetic mapping and molecular breeding made a spectacular impact during the last one and a half decades of the twentieth century. But still they remained "indirect" approaches for elucidation and utilization of plant genomes since much of the chromosomes remained unknown and the complete chemical depiction of them was yet to be unraveled.

Physical mapping of genomes was the obvious consequence that facilitated the development of the "genomic resources" including BAC and YAC libraries to develop physical maps in some plant genomes. Subsequently, integrated genetic–physical maps were also developed in many plants. This led to the concept of structural genomics. Later on, emphasis was laid on EST and transcriptome analysis to decipher the function of the active gene sequences leading to another concept defined as functional genomics. The advent of techniques of bacteriophage gene and DNA sequencing in the 1970s was extended to facilitate sequencing of these genomic resources in the last decade of the twentieth century.

As expected, sequencing of chromosomal regions would have led to too much data to store, characterize, and utilize with the then-available computer software could handle. But the development of information technology made the life of biologists easier by leading to a swift and sweet marriage of biology and informatics, and a new subject was born—bioinformatics.

Thus, the evolution of the concepts, strategies, and tools of sequencing and bioinformatics reinforced the subject of genomics—structural and functional. Today, genome sequencing has traveled much beyond biology and involves biophysics, biochemistry, and bioinformatics!

Thanks to the efforts of both public and private agencies, genome sequencing strategies are evolving very fast, leading to cheaper, quicker, and automated techniques right from clone-by-clone and whole genome shotgun approaches to a succession of second-generation sequencing methods. The development of software of different generations facilitated this genome sequencing. At the same time, newer concepts and strategies were emerging to handle sequencing of the complex genomes, particularly the polyploids.

It became a reality to chemically—and so directly—define plant genomes, popularly called whole genome sequencing or simply genome sequencing.

The history of plant genome sequencing will always cite the sequencing of the genome of the model plant Arabidopsis thaliana in 2000 that was followed by sequencing the genome of the crop and model plant rice in 2002. Since then, the number of sequenced genomes of higher plants has been increasing exponentially, mainly due to the development of cheaper and quicker genomic techniques and, most importantly, the development of collaborative platforms such as national and international consortia involving partners from public and/or private agencies.

As I write this preface for the first volume of the new series "Compendium of Plant Genomes," a net search tells me that complete or nearly complete whole genome sequencing of 45 crop plants, 8 crop and model plants, 8 model plants, 15 crop progenitors and relatives, and 3 basal plants is accomplished, majority of which are in the public domain. This means that we nowadays know many of our model and crop plants chemically, i.e., directly, and we may depict them and utilize them precisely better than ever. Genome sequencing has covered all groups of crop plants. Hence, information on the precise depiction of plant genomes and the scope of their utilization are growing rapidly every day. However, the information is scattered in research articles and review papers in journals and dedicated Web pages of the consortia and databases. There is no compilation of plant genomes and the opportunity of using the information in sequence-assisted breeding or further genomic studies. This is the underlying rationale for starting this book series, with each volume dedicated to a particular plant.

Plant genome science has emerged as an important subject in academia, and the present compendium of plant genomes will be highly useful to both students and teaching faculties. Most importantly, research scientists involved in genomics research will have access to systematic deliberations on the plant genomes of their interest. Elucidation of plant genomes is of interest not only for the geneticists and breeders, but also for practitioners of an array of plant science disciplines, such as taxonomy, evolution, cytology, physiology, pathology, entomology, nematology, crop production, biochemistry, and obviously bioinformatics. It must be mentioned that information regarding each plant genome is ever-growing. The contents of the volumes of this compendium are, therefore, focusing on the basic aspects of the genomes and their utility. They include information on the academic and/or economic importance of the plants, description of their genomes from a molecular genetic and cytogenetic point of view, and the genomic resources developed. Detailed deliberations focus on the background history of the national and international genome initiatives, public and private partners involved, strategies and genomic resources and tools utilized, enumeration on the sequences and their assembly, repetitive sequences, gene annotation, and genome duplication. In addition, synteny with other sequences, comparison of gene families, and, most importantly, the potential of the genome sequence information for gene pool characterization through genotyping by sequencing (GBS) and genetic improvement of crop plants have been described. As expected, there is a lot of variation of these topics in the volumes based on the information available on the crop, model, or reference plants.

I must confess that as the series editor, it has been a daunting task for me to work on such a huge and broad knowledge base that spans so many diverse plant species. However, pioneering scientists with lifetime experience and expertise on the particular crops did excellent jobs editing the respective volumes. I myself have been a small science worker on plant genomes since the mid-1980s and that provided me the opportunity to personally know several stalwarts of plant genomics from all over the globe. Most, if not all, of the volume editors are my longtime friends and colleagues. It has been highly comfortable and enriching for me to work with them on this book series. To be honest, while working on this series I have been and will remain a student first, a science worker second, and a series editor last. And I must express my gratitude to the volume editors and the chapter authors for providing me the opportunity to work with them on this compendium.

I also wish to mention here my thanks and gratitude to the Springer staff, particularly Dr. Christina Eckey and Dr. Jutta Lindenborn for the earlier set of volumes and presently Ing. Zuzana Bernhart for all their timely help and support.

I always had to set aside additional hours to edit books beside my professional and personal commitments—hours I could and should have given to my wife, Phullara, and our kids, Sourav and Devleena. I must mention that they not only allowed me the freedom to take away those hours from them but also offered their support in the editing job itself. I am really not sure whether my dedication of this compendium to them will suffice to do justice to their sacrifices for the interest of science and the science community.

New Delhi, India Chittaranjan Kole

Foreword

Mango (*Mangifera indica* L.) is one of the most ancient fruits of the Indian sub-continent grown since as early as 2000 BC and is designated as the "King of Fruits" in the Tropical world owing to its unique quality and high nutritive value and vast diversity in color, size, and aroma. Besides, several value-added and novel processed products form the part of popular mango industry. It is the fifth most important fruit commodity traded worldwide, along with bananas, apples, grapes, and oranges. At present, the world produces over 42 M tons of fresh mango fruits in over 100 countries, while India, China, Thailand, Mexico, Indonesia, Pakistan, Brazil, Egypt, and Bangladesh are the major growers. Its cultivation has spread to almost all the continents of the globe except Antarctica. The global production of mango is projected to reach 65 million ton by 2028, with annual increment of 2.1 percent for the next one decade. While the production from the leading exporting countries from Latin America and the Caribbean—critically Ecuador, Brazil, Guatemala, Colombia, Costa Rica, and Mexico—is expected to reach 34 million ton, due to expected import demand from the major importing countries.

Mango is an allotetraploid (2n = 40), highly heterozygous, and cross-pollinated fruit tree with a small genome size of 450 Mb. It is reported to have over 70 species scattered in the region extending from Northeastern India to Southeast Asia. Both conventional and non-conventional crop improvement attempts have led to the development of over three dozen hybrids globally, though the existence of enormous genetic diversity in *M. indica* has led to the identification of over 1200 registered chance seedlings, superior clones, hybrids, and rootstocks.

The development of molecular tools in mango is extremely limited, and thus its genes, genetics, and genomics remain scantly understood. The whole genome sequencing and the development of genetic maps of this species are important components in marker-assisted breeding and thereby targeted and effective genetic improvement. The high heterozygosity in mango is a big challenge in terms of whole genome sequencing-based SNP/polymorphism discovery. Accordingly, there are several scientific groups/consortia, which are involved in deciphering its complete genome information. Hence, complete information on genetic resources including the development of large numbers of SNP molecular genetic markers; development of genetic map of mango; association of phenotypic traits to the genetic map to identify useful unique markers for breeding; assessment of the genetic diversity in mango germplasm collections; and sequencing, assembly, and annotation of the mango genome are still to be achieved. The robust genetic resources will facilitate identification of genetic components associated with useful traits for hastening breeding efforts with precision. In this direction, next-generation genomics tools, bioinformatics, and omics approaches could be used in tandem to uncover large amount of genetic information. Advanced techniques like RADSeq can be used for rapid marker discovery and genotyping in crops like mango, which are highly heterozygous and out-breeding fruit species. Further, allelic database of mango for varietal differentiation has been reported though limited number of markers could not differentiate genotypes. Such leads achieved globally are though limited in the form of genomic resources in public domain which would thus be logically used as research tool for mapping, QTL studies, and genome completion in near future.

Considering the importance of the crop, i.e., mango the above book with focus on genome and omics is a timely effort undertaken to compile and present the up-to-date information in genomic science using state-of-the-art tools on different aspects pertaining to mango. It has a wide diversity of aspects like botany, germplasm and genetic resources, alternate flowering, propagation, classical genetics and traditional breeding, *in vitro* regeneration and genetic transformation, molecular mapping and breeding, genome sequencing, organelle genome, functional genomics, genomic resources, databases, etc. The effort made would go a long way in upgrading and updating the available information, which is being added vigorously in mango by the global scientific community. This is a monumental effort made by Prof. Chittaranjan Kole on Mango—a crop very close to my heart. Prof. Kole is an internationally reputed scientist with original contributions on crop genomics and breeding including several horticultural crops made through his researches in India and USA on fruit trees, vegetables, and medicinal and aromatic plants. Out of his 126 already published books, 55 are directly on horticultural crops and trees. "The Mango Genome" will remain as another valuable contribution from him for the academic community. The information compiled in this book would serve as a valuable reference source for researchers, students, plant breeders, and other stakeholders alike interested in high-end mango research. I personally congratulate the authors and the entire team of dedicated scientists who have contributed their reviews, research findings, and above all compiled up-to-date information in this publication.

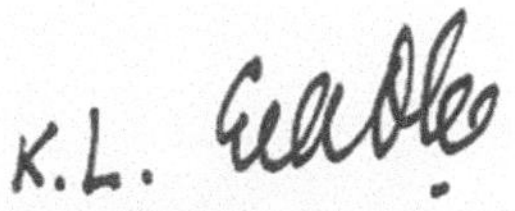

New Delhi, India
April 2020

K. L. Chadha
Former Deputy Director General (Hort.)
Indian Council of Agricultural Research
Former ICAR National Professor (Hort.)
President, Indian Academy of Horticultural Sciences

Preface

Mango is one of the most ancient fruit trees of the world. This tree and its parts, mainly leaves and fruits, have been cited profusely in the Sanskrit literature. Its leaves and fruits were used in Hindu ceremonies and worships, and the fruits had been a popular food for the masses. The mango fruit trees are grown in India for at least the last 4000 years.

Scientific studies and mentions in literature by travelers suggest the Indo-Burma region in the Indian Sub-continent to be the center of origin of this crop. Presently, mango is grown as a popular fruit tree in the tropical or subtropical regions of Asia and Africa. However, it was later introduced to other continents and is grown now in pockets of some countries in North and South America, Europe, and Australia. It is reportedly cultivated in 120 countries of the world; however, India, China, Thailand, Indonesia, and Mexico contribute to the major share of production.

The mango fruit is not only a source of delicious food but also very rich in the content of nutritional components and nutraceutical bioactives. Considering all these aspects, the mango fruit is rightly called as "the King of fruits"! The mango tree caters to the needs of fuel for a large number of families specifically in the rural and tribal areas of the developing countries.

In spite of such importance of mango tree, there is no book available that presents a compilation on fundamental aspects such as basic botany, genetic diversity, classical genetics, traditional breeding, and advanced research areas including molecular genetics and breeding, genetic transformation, and genomics. This book on "The Mango Genome" attempted to fill precisely such a gap with 13 chapters. Chapter 1 presents a brief introduction on the fruit tree with information on its ancient history, distribution, cultural significance, nutritional and nutraceutical importance, and various usage. Chapter 2 deliberates on the origin and distribution, taxonomy, morphology, variability, phylogenetic relationships, and other botanical aspects such as pollination, self-incompatibility, and polyembryony. Both sexual and asexual means may be used for propagation in mango. However, the mode of propagation depends on the existence of polyembryonic and monoembryonic mango plants. Various vegetative propagation methods including cuttings, air layering, and graftings are widely used in mango. Micropropagation has also been practiced profusely. All these methods have been discussed in Chap. 3. Genetic resources are highly useful in genetic improvement in an economic plant species. Mango, *Mangifera indica*, has 69 wild allied species that serve as the source of genes conferring mainly biotic and abiotic stresses. Some of these allied species are also used as rootstocks for the same purpose. Chapter 4 depicts these characteristics of the wild gene pool useful for wide hybridization. In addition, this chapter describes different varieties and cultivars, and presents details on collection, characterization, conservation, and utilization of genetic resources of mango. The extent of genetic diversity in a crop helps in planning the breeding strategies. Chapter 5 enumerates the methods used in characterization of genetic diversity in mango. These includes morphological observations, biochemical characterization, and molecular analysis employing molecular markers. Chapter 6 deliberates on alternate bearing in mango, a phenomenon in perennial fruit crops that invokes for understanding of the underlying mechanisms for planning profitable orcharding. Chapter 7 presents the classical genetic studies including mode of inheritance of qualitative and quantitative traits, characterization and cataloguing of varieties, and sources of resistance or tolerance of stresses in mango. This chapter also enumerates the breeding objectives, different methods of breeding practiced, and the achievements made in this fruit crop. Various *in vitro* culture techniques including *in vitro* shoot establishment and micropropagation, *in vitro* regeneration through organogenesis and somatic embryogenesis, encapsulation and cryopreservation, and embryo culture practiced in mango have been discussed in Chap. 8. In addition, it presented the outcome of the researches on genetic transformation in this crop. Molecular mapping of genes and QTLs followed by marker-assisted breeding are useful techniques in recent plant genetic improvement. These techniques are much more important in mango because traditional breeding is constrained with long juvenile phase, high heterozygosity, alternate bearing, polyembryony, heavy fruitlet drop, etc. Molecular mapping itself is also constrained because of heterozygosity and long juvenile period. However, a number of molecular genetic maps have been constructed for mango using several molecular markers such as RFLP, AFLP, SLAP, SNP but mainly for F1 populations. A number of genes and QTLs have also been mapped in mango. These information have facilitated marker-assisted selection, association mapping, and genome-wide association studies. Chapter 9 elucidates these techniques and achievements of molecular mapping and breeding in mango. Chapter 10 presents mango genome sequences of four varieties reported by four groups including Amrapali from India, "Tomy Atkins" from USA, "Kensington Pride" from Australia, and "Hong Jian Ha" and "Alphonso" from China. The chapter also enumerated sequences of mango transcriptome and chloroplast genome. Chapter 11 depicts the first draft genome of mango chloroplast of the "Langra" cultivar and enumerates many new functional genes in mango. Chapter 12 presents the status of functional genomics resources generated through studies on transcriptomics, proteomics, and metabolomics in mango and their applications in trait mapping as well as in breeding for specific target traits. The last chapter, Chap. 13, deals with all available, viz., linkage map-based, whole genome-based, transcriptome-based, and SNP marker-based genomic resources and database available which can be further used for variety of development programs as well as intellectual property protection.

The chapters of this book have been contributed by 47 authors precisely 39 scientists from 7 countries including Brazil, China, India, Malaysia, Spain, Thailand, and USA. I express my thanks to them for their lucid and resourceful deliberations. I am also grateful to Dr. K. L. Chadha, former Deputy Director General (Horticulture), Indian Council of Agricultural Research (ICAR), former ICAR National Professor (Horticulture) and presently the President of the Indian Academy of Horticultural Sciences for writing the Foreword of this book. He is an authority on one of the leading fruit crops of the tropics and subtropics.

I thank Dr. Zuzana Bernhart, the Executive Editor, Life Sciences of Springer Nature and Mr. Praveen Anand Sachidanandam of the Production Division of Springer Nature for their cooperation right from inception till completion of this book project.

Hope this book becomes useful for the students, teachers, and scientists interested in this fruit tree and the concepts and techniques dealt with.

New Delhi, India
July 2020

Chittaranjan Kole

Contents

Contributors

K. Abirami ICAR-Central Island Agricultural Research Institute, Port Blair, India

UB Angadi Centre for Agricultural Bioinformatics, ICAR-Indian Agricultural Statistics Research Institute, New Delhi, India

Anju Bajpai Division of Crop Improvement and Biotechnology, ICAR-Central Institute for Subtropical Horticulture, Lucknow, India

Julapark Chunwongse Department of Horticulture, Faculty of Agriculture at KamphaengSaen, Kasetsart University, Nakhon Pathom, Thailand

Alberto Carlos de Queiroz Pinto Consultant on Tropical Fruits, Eng. Agr. Ph.D. Researcher from Embrapa (Retired), Visitor Professor, University of Brasilia-DF, Brasilia-DF, Brazil

M. R. Dinesh ICAR-Indian Institute of Horticultural Research, Bengaluru, Karnataka, India

Janejira Duangjit Department of Horticulture, Faculty of Agriculture, Kasetsart University, Bangkok, Thailand

Fabio Gelape Faleiro Eng. Agr. Dr. Researcher of Embrapa Cerrados, Brasilia-DF, Brazil

Francisco Ricardo Ferreira Eng. Agr. Dr. Researcher of Embrapa Genetic Resources and Biotechnology, Brasilia-DF, Brazil

Victor Galán Saúco Consultant on Tropical Fruits, Ing. Agr, Ph.D, Research Professor (Retired), Instituto Canario de Investigaciones Agrarias, Canary Islands, Spain

Xin Hua He State Key Laboratory for Conservation and Utilization of Subtropical Agro-Bioresources, College of Agriculture, Guangxi University, Nanning, People's Republic of China

José I. Hormaza Departamento de Fruticultura Subtropical y Mediterránea, Instituto de Hortofruticultura Subtropical y Mediterránea La Mayora (IHSM La Mayora—UMA-CSIC), Algarrobo-Costa, Málaga, Spain

Mir Asif Iquebal Centre for Agricultural Bioinformatics, ICAR-Indian Agricultural Statistics Research Institute, New Delhi, India

Sarika Jaiswal Centre for Agricultural Bioinformatics, ICAR-Indian Agricultural Statistics Research Institute, New Delhi, India

Pawan K. Jayaswal ICAR-National Institute for Plant Biotechnology, New Delhi, India

Dinesh Kumar Centre for Agricultural Bioinformatics, ICAR-Indian Agricultural Statistics Research Institute, New Delhi, India

Sandeep Kumar Division of Crop Improvement and Biotechnology, ICAR-Central Institute for Subtropical Horticulture, Lucknow, India

Sunil Kumar Centre for Agricultural Bioinformatics, ICAR-Indian Agricultural Statistics Research Institute, New Delhi, India

Richard E. Litz Department of Horticultural Sciences, Tropical Research and Education Center, University of Florida, Homestead, FL, USA

Cong Luo State Key Laboratory for Conservation and Utilization of Subtropical Agro-Bioresources, College of Agriculture, Guangxi University, Nanning, People's Republic of China

Ajay K. Mahato ICAR-National Institute for Plant Biotechnology, New Delhi, India

Sisir Kumar Mitra Board Member of International Society for Horticultural Science (ISHS), Brasilia, Brazil

C. Murugan Botanical Survey of India, Regional Office, Coimbatore, India

M. Muthukumar Division of Crop Improvement and Biotechnology, ICAR-Central Institute for Subtropical Horticulture, Lucknow, India

César Petri Departamento de Fruticultura Subtropical y Mediterránea, Instituto de Hortofruticultura Subtropical y Mediterránea La Mayora (IHSM La Mayora—UMA-CSIC), Algarrobo-Costa, Málaga, Spain

Anil Rai Centre for Agricultural Bioinformatics, ICAR-Indian Agricultural Statistics Research Institute, New Delhi, India

Shailendra Rajan ICAR- Central Institute for Subtropical Horticulture, Lucknow, Uttar Pradesh, India

Hutchappa Ravishankar ICAR-Central Institute for Subtropical Horticulture, Lucknow, UP, India

K. V. Ravishankar ICAR-Indian Institute of Horticultural Research, Bengaluru, Karnataka, India

Shahril Ab Razak Biotechnology and Nanotechnology Research Centre, MARDI Headquarters, Selangor, Malaysia

Heiplanmi Rymbai Scientist, Division of Horticulture, ICAR Research Complex for NEH Region, Umiam, Meghalaya, India

M. Sankaran ICAR-Indian Institute of Horticultural Research, Bengaluru, India

Nimisha Sharma Division of Fruits and Horticultural Technology, ICAR-Indian Agricultural Research Institute, New Delhi, India

Nagendra K. Singh ICAR-National Institute for Plant Biotechnology, New Delhi, India

Sanjay K. Singh Division of Fruits and Horticultural Technology, Indian Agricultural Research Institute, New Delhi, India

V. K. Singh ICAR-Central Institute for Subtropical Horticulture, Lucknow, UP, India

Manish Srivastav Division of Fruits and Horticultural Technology, ICAR-Indian Agricultural Research Institute, New Delhi, India

Pumipat Tongyoo Center for Agricultural Biotechnology, Kasetsart University, Nakhon Pathom, Thailand

Abbreviations

2,4-D	2,4-dichlorophenoxyacetic acid
2D-GE	Two-dimensional gel electrophoresis
2iP	N6-(2-isopentenyl)adenine
AB	Alternate bearing
ABA	Abscisic acid
ABI	Alternate bearing index/intensity
ACC	Aminocyclopropane-1-carboxylic acid
ACO	Aminocyclopropane-1-carboxylate oxidase
ADB	Asian Development Bank
AES (PARIA)	Agricultural Experiment Station, Paria (Gujarat, India)
AFL2	*Apple floricula/lfy* (gene)
AFLP	Amplified fragment length polymorphism
AGP	Alpha-1,4 glucan phosphorylase
AM	Association mapping
ANMG	Australian National Mango Genebank
AP	*Apetala* (gene)
APAU	Andhra Pradesh Agricultural University (India)
B5	Gamborg B5 (medium)
BA	6-benzyladenine
BBI	Biennial bearing index
BFT	Brothers of FT
BP	Barba and Pateña
BSSKVV	Balasaheb Sawant Konkan Krishi Vishwavidyalaya (India)
Cas9	CRISPR associated protein 9
cDNA	Complementary-DNA
CH	Carbohydrate
CiFT	*Citrus flowering locus T* (gene)
CISH	Central Institute for Subtropical Horticulture (ICAR, India)
cM	CentiMorgan
CO	*Constans* (gene)
cpDNA	Chloroplast DNA
cpgenome	Chloroplast genome
CPLLs	Combinatorial peptide ligand libraries
CR	Chromosomal rearrangement
CRISPR	Clustered regularly interspaced short palindromic repeats

CW	Coconut water
ddRAD	Double digest restriction-site-associated DNA
DEGs	Differentially expressed genes
DUS	Distinctness, uniformity, and stability
EMS	Ethylmethyl sulfonate
ERF	ET-responsive transcription factor
EST	Expressed sequence tag
ETS	External transcribed spacer
FAO	Food and Agriculture Organization
FAOSTAT	FAO Corporate Statistical Database
FAP	Full air porosity
FBD	Fruit bud differentiation
FI	Flower induction
FLC	*Flowering locus c* (gene)
FRS	Fruit Research Station, Sangareddy (Andhra Pradesh, India)
FT	*FLOWERING LOCUS T* (gene)
FUL	Fruitful
GAs	Gibberellins
GAU	Gujarat Agricultural University (India)
GC	Gas chromatography
GD	Genetic distance
GDJ	Jaccard's coefficient
GDMR	Modified Rogers' distance
GDNL	Nei and Li's coefficient
GDSM	Simple matching coefficient
GEA	Global Environment Agency
GI	Geographical indication
GLM	Generalized linear model
GS	Genetic similarity
GST	Glutathione S-transferase
GUs	Growth units
GWAS	Genome-wide association studies
HB-1	Healthy bud stage
HB-2	Healthy panicle stage
HPLC	High-performance liquid chromatography
HPP	High-pressure processing
HSSR	Highly variable SSR
IAA	Indole-3-acetic acid
IARI	Indian Agricultural Research Institute (ICAR, India)
IBA	Indole-3 butyric acid
ICAR	Indian Council of Agricultural Research
IIHR	Indian Institute of Horticultural Research (ICAR, India)
INSDC	International Nucleotide Sequence Database Collaboration
IP	Intellectual property
IPGRI	International Plant Genetic Resources Institute
ISSR	Inter-simple sequence repeat

ITS	Internal transcribed spacer
IU	International unit
IUCN	International Union for the Conservation of Nature
KEGG	Kyoto Encyclopedia of Genes and Genomes
LB	Luriabertani
LC	Liquid chromatography
LFY	*Leafy* (gene)
LINEs	Long interspersed nuclear elements
lncRNA	Long non-coding RNA
LRR	Leucine rich repeat
LTR	Long terminal repeat
MABI	Modified alternate bearing index
MAF	Minor allele frequency
MALDI	Matrix-assisted laser desorption/ionization
MAS	Marker-assisted selection
miRNA	Micro-RNA
MLM	Mixed linear model
MMD	Mango malformation disease
MPKV	Mahatma Phule Krishi Vishwavidyalaya (India)
mRNA	Messenger-RNA
MS	Mass-spectrometry/Murashige and Skoog (medium)
MTA	Material Transfers Agreement
NAA	Naphthalene-acetic acid
NCBI	National Center for Biotechnology Information
ncRNA	Non-coding RNA
NGS	Next-generation sequencing
NMU	*N*-nitroso-*N*-methyl urea
*npt*II	Neomycin phosphotransferase
PAGE	Polyacrylamide gel electrophoresis
PAL	Phenylalanine ammonia-lyase
PCA	Principal coordinate analysis
PCR	Polymerase chain reaction
PDB	Protein Data Bank
PEE	Pulsed electric field
PEM	Pro-embryonic mass
PGRC	Plant Genetic Resources Center
PPO	Polyphenol oxidase
PRJNA#	BioProject accession
PVP	Polyvinyl pyrrolidone
qPCR	Quantitative PCR
qRT-PCR	Quantitative reverse transcription PCR
QTL	Quantitative trait locus
QTLs	Quantitative trait loci
RACE	Rapid amplification of cDNA ends
RAPD	Random amplified polymorphic DNA
RDA	Recommended daily allowance
RE	Repetitive element
RFLP	Restriction fragment length polymorphism

RNA-Seq	RNA-sequencing
ROS	Reactive oxygen species
RT-PCR	Reverse transcription PCR
RTS	Ready-to-serve
SAM	S-adenosyl methionine/shoot apical meristem
SCoT	Start codon targeted
SDS	Sodium dodecyl sulfate
SINEs	Short interspersed nuclear elements
SLAF	Specific locus-amplified fragment
SMRT	Single molecule real time
SNP	Single nucleotide polymorphism
SoC1	Suppressor of over expression of constants
SPL5	Squamosa promoter binding
SRA	Sequence Read Archive
SRAP	Sequence-related amplified polymorphism
SSR	Simple sequence repeat
SVP	Short vegetative phase
TDF	Transcript derived fragment
TE	Transposable element
TFL1	Terminal flower1 (gene)
TNAU	Tamil Nadu Agricultural University (India)
TOF	Time of flight
TPI	Triosephosphate isomerase
TSS	Total soluble solids
UPGMA	Unweighted pair group method with arithmetic averages
VAD	Vitamin A deficiency
VNTRs	Variable number tandem repeats
WGD	Whole genome duplication
WPM	Woody plant medium

Mango: The King of Fruits

1

Shailendra Rajan

Abstract

Mango, the "King of Fruits," is an economically important fruit in various parts of the world. In addition to its excellent tropical flavor, mangoes embody nutrition and make eating healthy and delightful sensory experience. Though mango cultivation is reported from more than 120 countries, merely 15 countries have share of more than 1% in its world production. More than 60% of world mango production comes from India, China, Thailand, Indonesia, and Mexico. Major portion of mango production is consumed as fresh mangoes; however, dozens of its processed products are also being commercially produced. Mangoes are low in sodium while there is ample quantity of potassium, phosphorous, and calcium in the fruit. Apart from nutritional benefits for human health, mangiferin and lupeol, native bioactive compounds in mango, have exhibited several health-promoting characteristics including antitumor-promoting activity.

1.1 Introduction

Mango is one of the most important fruits grown in the tropics and subtropics of the world. The Mango tree is not merely revered as a religious object but also is valued owing to its enormous economic potential as all its parts are considered usable and are used for various purposes (Singh 1960). It is a source of food, fuel, and fodder, mostly in the traditional Asian countries. The Mango fruit is a livelihood for millions of people across the globe banks. Apart from its numerous usages, it is a rich source of nutrients and several rare bioactive compounds. Mango acts as a source of vitamin A for millions of people living in different subtropical, semi-arid, and tropical regions of Africa and Asia. Mango also finds a place in the high-tech commercial cultivation in Western hemisphere and is fast emerging as a super food fruit. The ripe mango fruit has a distinctive combination of taste and flavor. Realizing as a super fruit, the consumption and area under its cultivation are increasing incessantly. But the king of fruits is also facing several challenges due to perennially changing climate and emerging biotic stresses under varied agroecologies. It is becoming an export crop commodity for several countries resulting in international export market competition.

S. Rajan (✉)
ICAR- Central Institute for Subtropical Horticulture, Rehmankhera, Kakori, Lucknow 226101, Uttar Pradesh, India
e-mail: srajanlko@gmail.com; shailendra.rajan@icar.gov.in

C. Kole (ed.), *The Mango Genome*, Compendium of Plant Genomes,
https://doi.org/10.1007/978-3-030-47829-2_1

1.2 Voyages to Different Continents

The mango has a fascinating history of domestication and distribution as the sea transportation was exceedingly important before the early twentieth century. The common mango (*Mangifera indica*) is believed to have been originated in the Indo-Burma region. Its wild types are still existing at many places in the Indian subcontinent (Yadav and Rajan 1993; Rajan and Hudedamani 2019). As an important consequence of its origin and domestication in Indian subcontinent, mango has become a common fruit plant in India (Mukherjee 1951, 1953, 1972). Old historical works and its descriptions made by ancient travelers took place earlier in Southeast Asia than Western hemisphere. The origin in the Northeastern India indicated by many workers is explained by the prevalence of related *Mangifera* species in the Indo-Burma region (Mukherjee 1997). The belief that mango might have been naturalized in India (Hooker 1876) has also been substantiated by opinion of several researchers. "On the basis of taxonomic and recent molecular studies results, there is now evidence that mango has developed in a vast area, including northwestern Myanmar, Bangladesh and northeast India" (Mukherjee 1997).

The mango has been known in India since time immemorial. De Candolle (1884) estimates that it has been cultivated for the last 4000 years, a figure based on repeated reports of botanists; however, Hill (1952) views it to be 6000 years. On the basis of 60-million-year-old fossil impressions of carbonized mango leaves in the Palaeocene sediments Meghalaya, Srivastava and Mehrotra (2013) proposed the origin of *Mangifera* in Indian subcontinent. This also supports the lineage of *Mangifera* in the Indian subcontinent often debated due to existing dozens of wild relatives in Malay Archipelago (Ram and Rajan 2003).

Traders, travelers, and rulers were fascinated with Indian mangoes and their efforts during the last 2,500 years established mango as a crop in tropics and subtropics. Perhaps the efforts of Buddhist monks during the fourth-fifth centuries BC introduced mango to the Malaya Peninsula and East Asia. Chinese explorer Hwen T'sang introduced mango into China after his return from India to his native country during the seventh century AD (Singh 1960). During the tenth century AD, the Persians brought mango to East Africa (Purseglove 1969) whereas Portuguese took them to West Africa and Brazil during the sixteenth century AD (Singh 1960).

In Americas, at the outset, mango was introduced and established in Brazil. Thereafter, it was bought in West Indies, first in Barbados (in mid of the eighteenth century) and later came to Jamaica in around 1782. In the early nineteenth century, it arrived in Mexico from Philippines and the West Indies. During 1862, mango came to Miami around the same time. About 40 types of mangoes were transported from India and planted in North Queensland part of Australia in 1875 (Morton 1987).

Buddhists and "Tamils" played an important role in introducing mango from India to Southeast Asian countries. At an early stage, it was brought to Malay Peninsula by the "Tamils." Introduction from Indochina Peninsula resulted in spread of polyembryonic varieties in the Philippines. Perhaps Carabao and Pico varieties are the outcome of these introductions. Valmayor (1962) documented the introduction of mango in Eastern and Western hemispheres. In Eastern hemisphere, it was introduced between thirteenth and twentieth centuries. Knight and Schnell (1992) documented the details of introduction in the USA.

Its travel from the Eastern hemisphere to the Western hemisphere and acceptability as an important fruit ensued its cultivation in many countries and expanded newer area under its cultivation. In Western hemisphere, in some of the developed countries, modern ways of cultivation were adapted, mostly based on temperate fruit production. Newer production technologies made it more important in the international market and established it as a commercial export crop in Latin American countries. Still, mango is traveling to different parts of the world where it was not cultivated earlier and presently being grown in about 140 countries.

1.2.1 Deeply Associated with Asian Culture

Prominence of mango in Indian mythology and religious ceremonies forms part of the long history in the country. Ancient Hindus valued mango highly not merely because of religious or sentimental considerations but also they fully realized its importance in the economic and cultural life of the society (Singh 1960). The Sanskrit word "Amra" (mango) implies that it is the fruit of masses. The Hindi word "Aam" means the masses or the populace. It has been intimately connected with Hindu religion and highly valued as fabled transformation of Prajapati (God of creation). Inflorescence and leaves are used by Hindus at various ceremonies and for the worship of "Saraswati," Goddess of Wisdom and Arts. The "pancha-pallava" or the bunch of five sprigs from the mango tree is used by Hindus at various ceremonies (Singh 1960).

The influence of mango on Indian culture is well known as in ancient India, man and women linked their names with mango. The lady who presented mango grove to Lord Buddha was herself known as "Amradarika." In the epic Ramayana, Valmiki mentions gardens and even forests of mango trees. Name of *ragas* (scores of music) as *Amra-takeswara* and *Amra-panchma* are based on this fruit. It clearly states that in India, mango cultivation is nearly as old as Indian civilization. Indian art also recognizes the supreme merit of mango. "Barhut Stupa" of 110 BC, "Ajanta," and "Elora" bear testimony of its antiquity as being excellent examples of sculptures of that time (Singh 1960).

During the fifteenth century, traders and Buddhist monks transported superior seedlings to Southeast Asia. Important mango varieties of Thailand, Malaysia, Indonesia, Cambodia, Vietnam, and Philippines are polyembryonic in nature. This indicates the common origin disseminated through seedlings to these countries. In Malaysia and Indonesia, the Sanskrit or Tamil term for mango suggests that it was introduced from India to Peninsular Malaysia and eastern Asia. Huien T'Snag (632-45), probably the first to document mango, puts the fruit to people's attention outside India. Ibn Hankel (902-68), Ibn Batuta (1325-40), and Jordanus (1330) were important foreigners who described mango in their works. Praise of this fruit reached its climax when a Turkman saint and poet, Amir Khusroo (c.1330), admired mango in his verse. The Ain-i Akbari, an encyclopedic treatise of the Moghul emperor Akbar (c.1500-1600), gives a lengthy account of mango and its varieties. Akbar is famous for his creation of an enormous plantation of one lakh mango trees, the "Lakh Bhag" in Darbhanga, of which traces are still surviving (Singh 1960). There are also numerous places where mangoes have been mentioned in cultural and theological writings.

1.2.2 Gaining Importance in the New World

Until the middle of the twentieth century, mango was considered as an important commercial crop in the Asian countries. Later on mango cultivation expanded and still continues to extend in newer subtropical and tropical environments. It has become an important export commodity in non-traditional mango-growing countries where the considerable proportion of their fruit production is exported. Analysis of FAOSTAT, 2014 data indicates that Mexico is the highest exporter of mango with about 17% of the country production. Although Peru produces < 1% of the world production but its share in the export is larger (27.5%). The production of Floridian varieties of mangoes is adapted to wide range of agro-ecologies and responsible for the expansion of newer production areas. Brazil is also an important exporter of mango (8.08% of the country's production). Indubitably, when it comes to mango production, Asia has a major share. The top five countries producing mangoes are from Asia. Additionally, Bangladesh and Philippines are in the list of top 10 countries. America's tropical countries, Nigeria, Brazil, and

Mexico are representative of the top 10 mango producers. Mango pulp export is one of the important export commodities for India. Pulp export of about 100 m US$ from India takes place to more than 50 countries and Saudi Arab, Yemen Republic, Netherland, Kuwait, UK, and USA are the major importers (Anonymous 2017).

The mango has become a crop grown in many warm subtropical areas and throughout the tropics. Widely adaptable varieties have led to widespread mango cultivation in differing agroecologies in different countries. It is remarkable in Australia's tropical and subtropical northern areas. The major development regions are Queensland, the North and Western Australia (Bally et al. 2000). Superior Indian cultivars have become common in Bangladesh particularly in Rajshahi, Dinajpur, Kushtia, and new Satkhira (Abedin and Quddus 1990) (Fig. 1.1).

Mango has gained a lot of importance in most parts of the world due to its taste and economic value where it might not be grown as a commercial crop but its importance is well recognized. It is a source of livelihood for millions and has immense importance in the Asian continent and a few of them like India, Philippines, and Pakistan have declared it as the national fruit. It is the national tree of Bangladesh. Many countries have issued postage stamps on mangoes on account of its importance and wide adaptability among the people. Great Britain issued the world's first postage stamp "Penny Black" officially on May 6, 1840. Later on, the U.S. Post Office Department issued its first postage stamps on July 1, 1847. Not only Asian countries, where mango is important for centuries, Western hemisphere has also used stamps. Many Caribbean countries, like Anguilla, Aruba, British Virgin Island, Curacao, Montserrat, Netherlands Antilles, and Saint Maarten, have issued stamps on mangoes. With the increasing importance of the fruit, Angola, Cameroon, Central African Republic, Comoros, Congo Republic, Gabon, Gambia, Ghana, Guinea, Madagascar, Madeira Island, Mali, Mayotte, Senegal, South Africa/Venda, Sudan, Togo, Upper Volta (Burkina Faso), and Kenya have also issued such stamps.

1.3 Changing Face of Cultivars in International Markets

Red-peel cultivars dominate the international mango export market because of their eye-appeal but in Asian countries consumers select the fruits on the basis of overall understanding of the variety. Naturally, this is because of populous developed preferences for specific local yellow- and green-colored varieties. The outstanding local varieties are sold at a premium price but

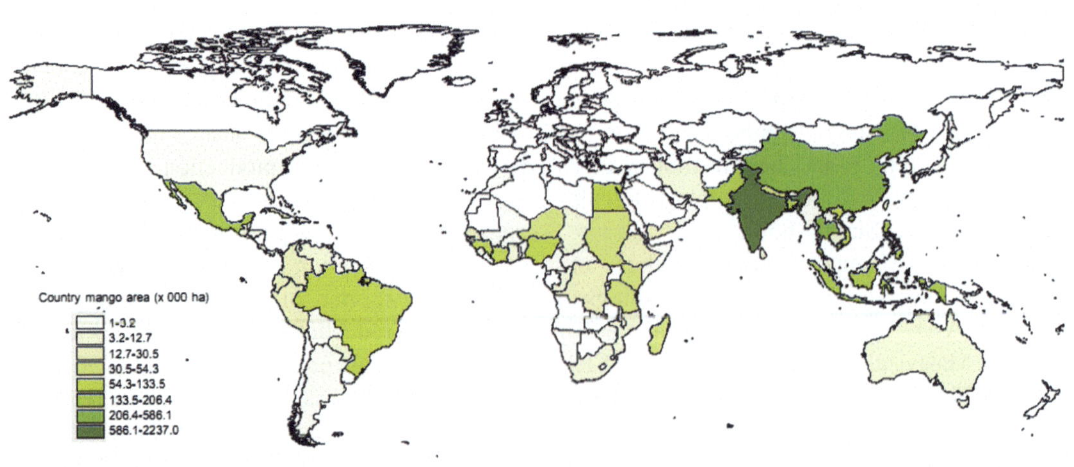

Fig. 1.1 Area under cultivation in different countries

they don't find place in the international market because of their yellow or yellow green color even at the ripe stage. Gradually, the scenario is changing and yellow skin mangoes are also recognized in the US market because of their quality. "Ataulfo," a yellow-colored variety has gained increased interest of consumers in the US (Ledesma and Campbell 2019). At least this variety has played a role in breaking red-peel variety preference in the market. This change may help in opening way for the excellent Indian varieties in the United States, not much preferred because of their yellow or greenish peel. The scenario is also changing in India where people are now interested in the red-colored varieties. Many of the red-peel varieties might not be even sweet enough to match expectations of the Indian consumers but have firm pulp and longer shelf life. Breeding programs for developing new mango varieties in India have red-peel color also, as one of the important objectives. Thus many of the newly developed varieties are red in color. Although "Carabao," "Manila," "Nam Doc Mai," "Alphonso," "Kesar," "Totapuri," "Edward," "Osteen," and "Palmer" are cultivars of minor importance in US but now there is a growing interest for these varieties. Production of "Kent" has shown a significant increase in the recent past due to its use as fresh fruit and products. Acreage under "Tommy Atkins" and "Haden," the leading international cultivars, is declining in the Western hemisphere due to less demand (Campbell and Ledesma 2013).

The consumer in the conventional United States does not fully apprehend the use of mature-green mangos, but in recent past there has been considerable growth in demand in ethnic markets for mature-green fruit (Campbell and Ledesma 2013). Indian varieties like "Alphonso," "Kesar," and "Totapuri" have received considerable attention in the United States as a result of air shipment from India in the recent past. Commercial cultivation of Totapuri in the Western hemisphere had little success. Kesar is a new cultivar for masses and not much tried for cultivation in United States. Excellent Indian varieties like Alphonso have limitation due to narrow adaptability (Ram and Rajan 2003).

1.4 Nutritionally Rich Fruit

Mango fruit is rich in carbohydrates, minerals, dietary fiber (pectin) (Bello-Pérez et al. 2007; USDA 2020), vitamin C, and vitamin A (β-carotene) (Saleem-Dar et al. 2016). Mango pulp has an energy value between 60 and 190 kcal (250-795 kJ) for 100 g of the pulp (Tharanathan et al. 2006). Several other phytochemicals of mango help to sustain good health (Masibo and He 2009).

Ripened mango fruit is a significant source of sugars (glucose, fructose, and sucrose) along with other carbohydrates including starch and pectins (Bello-Pérez et al. 2007). From a nutritional and flavor standpoint, these compounds are essential for human health.

The nutritional content of mango varies with variety, location, and maturity stage. The content of major constituents of pulp is hereunder.

Food value per 100 g of ripe mango pulp	
Calories	62.1–63.7 Cal
Moisture	78.9–82.8 g
Protein	0.36–0.40 g
Fat	0.30–0.53 g
Carbohydrates	16.20–17.18 g
Fiber	0.85–1.06 g
Ash	0.34–0.52 g
Calcium	6.1–12.8 mg
Phosphorus	5.5–17.9 mg
Iron	0.20–0.63 mg
Vit A (carotene)	0.135–1.872 mg
Thiamine	0.020–0.073 mg
Riboflavin	0.025–0.068 mg
Niacin	0.025–0.707 mg
Ascorbic Acid	7.8–172.0 mg
Trptophane	3–6 mg
Methionine	4 mg
Lysine	32–37 mg

Values based on Indian mango varieties (Gopalan et al. 1977)

Values based on analyses of Floridan cultivars (Tommy Atkins, Keitt, Kent, and/or Haden cultivars)

Name	Amount	Unit	Min	Max
Water	83.46	G	80.84	88.1
Energy	60	Kcal		
Energy	250	kJ		
Protein	0.82	G	0.44	1.14
Total lipid (fat)	0.38	G	0.21	0.5
Ash	0.36	G	0.25	0.42
Carbohydrate, by difference	14.98	G		
Fiber, total dietary	1.6	G	1.2	2.3
Sugars, total including NLEA	13.66	G		
Sucrose	6.97	G	0.07	9.37
Glucose (dextrose)	2.01	G	0.66	5.97
Fructose	4.68	G	3.8	7.99
Calcium, Ca	11	Mg	7	16
Iron, Fe	0.16	Mg	0.09	0.41
Magnesium, Mg	10	Mg	8	19
Phosphorus, P	14	Mg	10	18
Potassium, K	168	Mg	120	211
Sodium, Na	1	Mg	0	3
Vitamin C, total ascorbic acid	36.4	Mg	13.2	92.8
Thiamin	0.028	Mg	0.014	0.041
Riboflavin	0.038	Mg	0.02	0.072
Niacin	0.669	Mg	0.198	1.31
Pantothenic acid	0.197	Mg	0.161	0.24
Vitamin B-6	0.119	Mg	0.053	0.164
Folate, total	43	µg	20	69
Folate, food	43	µg	20	69
Folate, DFE	43	µg		
Choline, total	7.6	Mg		
Vitamin A, RAE	54	µg		
Carotene, beta	640	µg	185	1680
Carotene, alpha	9	µg	0	40
Cryptoxanthin, beta	10	µg	0	32
Vitamin A, IU	1082	IU		
Lycopene	3	µg	0	6
Lutein + zeaxanthin	23	µg	6	63
Vitamin E (alpha-tocopherol)	0.9	Mg	0.79	1.02

(continued)

Source USDA https://fdc.nal.usda.gov/fdc-app.html#/food-details/169910/nutrients

During the preclimateric process, fructose is the main monosaccharide (Bernardes et al. 2008), while, in ripe mango fruit, sucrose is the main sugar (Saleem-Dar et al. 2016). Significant differences for carbohydrates (10.5 percent and 32.4 percent) were found in different African mango cultivars (Othman and Mbogo 2009). Starch is the most significant ingredient of unripe mango, which is hydrolyzed at a mature stage to glucose (Derese et al. 2017). Pectin, a structural carbohydrate, is abundantly present at unripe stage in mango pulp and the molecular weight decreases during ripening (Bello-Pérez et al. 2007; Saleem-Dar et al. 2016) because of its hydrolysis (Prasanna et al. 2004).

Alike several other fruits, mango has low content of protein and fat in the pulp. In Indian cultivars, the total protein in the pulp has been reported to be 0.5-1% (Saleem-Dar et al. 2016). Fruit maturity and cultivars influence the amino acid content. Mango contains considerable amount of alanine, arginine, leucine, glycine, isoleucine, and serine at ripe stage while others are in trace amount (Tharanathan et al. 2006).

Vitamin A is an essential nutrient in limited quantities necessary for good human health (FAO/WHO 2004). Mango acts as a source of vitamin A for millions living in the world's various subtropical, semi-arid, and tropical regions, particularly in countries where vitamin A deficiency (VAD) is prevalent (Thomas and Oke 1983) as carotene is an essential precursor of vitamin A. In many countries, mango is distributed in abundance in rural areas of developed nations. Dried mango pulp is also a good source of carotenoids and can be stored for 4-6 months, when fruits are out of season (Drammeh et al. 2002). Vitamin A content of mango fruit ranges from 1,000 to 6,000 IU (Matheyambath et al.

2016). Extraordinarily, high plastoglobule content in mango promotes high carotenoid bioaccessibility (Jeffery et al. 2012; Mayne 1996).

Mango fruit pulp has very less lipid content and can be considered as fat free. Seeds are rich in lipids and fatty acids' composition is comparable with cocoa butter. Those fatty acids might play a role in the cosmetics, medicinal, and food sectors (Sonwai et al. 2012; Jahurul et al. 2015). Prominent fatty acids in the mango kernel include lignoceric, arachidic, linolenic, and behenic acids (Jahurul et al. 2015). Association between the aroma intensity and taste depends on the ratio of palmitic–palmitoleic acid in the maturing mango. It also defines the aroma and flavor of mangoes (Gholap and Bandyopadhyay 1977). Such compounds play important part in mango flavoring biochemistry (Desphande et al. 2016).

Acidity in mango fruit is mainly due to the citric and malic acid (Matheyambath et al. 2016). The content of pyruvic, citric, succinic, oxalic, and malic acids is significant and the content varies according to the variety (Shashirekha and Patwardhan 1976; Kumar et al. 1993). As per the United States National Research Council (1989) recommended Daily Allowance (RDA), mango is a rich source of essential minerals such as calcium, iron, magnesium, phosphorus, potassium, zinc, copper, manganese, and selenium necessary for human health. The fruits are low in sodium (Njiru et al. 2014).

1.4.1 Bioactive Compounds

Mango has been linked to many pharmacological effects, including antidiabetic (Mustapha et al 2014), antioxidant, antiviral (Amin et al. 2019), and anti-inflammatory. Various effects have also been documented, including antibacterial, antifungal, anthelmintic, antiparasitic, anti-HIV, antibone resorption, antispasmodic, antipyretic, antidiarrheal, immunomodulation, hypolipidemic, antimicrobial, hepatoprotective, and gastroprotective (Shah et al. 2010).

Components of mango are recurrently used as a traditional medicine to cure numerous ailments. Active constituents are present in stem bark, leaves, heartwood, roots, and fruit, and have antioxidant, anti-inflammatory, radioprotective, antitumor, immune-modulatory (Garcia et al. 2003), anti-allergic, antidiabetic (Mustapha et al 2014), antibone resorption, mono-amine oxidase inhibiting, antiviral (Guha et al. 1996; Amin et al. 2019), antifungal (Kanwal et al. 2010), antibacterial (Engels et al. 2011), antispasmodic, antidiarrheal, antiparasitic as well as lipolytic (Yoshikawa et al. 2002), hepatoprotective, and gastroprotective (Shah et al. 2010) properties. In spite of considerable progress in phytochemical and medicinal analysis of mango, more efforts are needed to explore its active components and their application in pharmaceutical industries.

Mango fruits have various phenolics required for maintaining good human health. The pulp contains several polyphenolic compounds, such as mangiferin, gallotannins, quercetin, gallic acids, isoquercetin, ellagic acid, and β-glucogallin (Schieber et al. 2000). The peel also possesses mangiferin, quercetin, ellagic acid, kaempferol, and rhamnetin (Berardini et al. 2005a, b).

The compound, mangiferin (1,3,6,7-tetrahydroxyxanthone-C2-β-D-glucoside), a predominant bioactive ingredient isolated from different parts of mango tree has been extensively studied both in vivo and in vitro for pharmacological effects like antioxidant activity, antidiabetic, antitumor, lipometabolism regulating, cardioprotective, antihyperuricemic, neuroprotective, antioxidant, anti-inflammatory, antipyretic, analgesic, antibacterial, antiviral, and immunomodulatory effects (Du et al. 2018). It has been shown that Zvnamite®, a mango leaf extract rich in the natural polyphenol and mangiferin, enhances sprint exercise performance when given in combination with luteolin or quercetin (Gelabert-Rebato et al. 2019a, b).

Lupeol, a natural pentacyclic triterpene present in mango fruit, has gained attention in recent past which has shown strong anti-inflammatory, anti-arthritic, antimutagenic, and antimalarial action. Lupeol has also been found to have antitumor-promoting activity in mouse skin tumorigenesis (Saleem et al. 2004; Saleem 2009) and an effect on the oxidative stress caused by dimethylolbutanoic acid in liver (Prasad et al. 2008).

1.5 Usage

Quality, nutrients, and bioactive compounds in mango fruit are affected by genotype and ripening stage at harvest and growing environment. Mango is widely used in food, cosmetics, and pharmaceutical industries, right from root to the top of the tree (Singh 1960). The fruit is used in all stages of development. Premature fruit is used for extraction of tannins and in the preparation of other astringent products, viz., chutneys, curries, cold drinks, pickles, etc. while ripe fruits are utilized for puree, squash, nectar, drinks, mango leather, fruit bars, canned and frozen mango slices, and candies (Ram and Rajan 2003). Mango stones are used for raising seedlings, oil, fiber, and as animal feed. The kernel is a rich source of carbohydrates, calcium, and fat, and starch can be produced for industrial purposes. Goats and cattle are fond of eating mango leaves while branches and wood are utilized for fuel, timber besides use of bark for extraction of tannins and gums (Singh 1960).

1.5.1 Processed Mango Products

In terms of the variety of processed items, mango is a special crop, used for processing at raw and mature crop stage. At an early stage of fruit, sweet or sour chutney (mango sauce) and whole fruit pickle are prepared. While when reached the stone hardening stage, products like "amchoor," pickles, and green mango beverages prepared. The mature fruit is distinguished by a distinctive flavor and taste. The technique has been developed for various products, viz., canned and frozen slices, pulp, jelly, squash, tea, nectar and ready-to-serve (RTS) beverages, mango cereal flakes, concentrates, dehydrated products, mango fruit bars, and mango toffee.

There are a large number of traditional mango products processed at home or cottage industry level. Indian National Mango Database details about 230 recipes based on mango (Rajan et al. 2020). Mangoes are used for several hundred dishes and are used with pulses, buttermilk, yogurts, soups, meat, fish curry, spices, etc. Large variation in mango pickles can be attributed to the recipes used in different parts of the world. In India, several varieties, specific to region, are known for making specialized pickles. Appimedi pickle famous from Western Ghats of Karnataka is made of tender mango mainly grown in forests and also on the riversides. Mango pulp and pickles are the most exported commodities from India (Vasugi et al. 2012).

With the advancement of processing technology, research on innovative processing of mango has taken place in the recent years. Research on high-pressure processing (HPP), pulsed electric field (PEF) processing, Holmic heating, microwave heating, radio frequency heating, ultraviolet (UV) light, ionizing radiation, pulse light technology, ultrasound, and ozone treatments has been undertaken (Ahmed and Ozadali 2012; Ahemad 2017). Non-thermal processing techniques for retaining nutritional and sensory properties have been used. These innovative technologies are at development stages and commercialization in near future will ensure better utilization of the mangoes.

1.6 Concluding Remark and Future Prospects

Tropical and subtropical mango production continues to face both abiotic and biotic threats. Rising demands of mango fresh fruit and value-added products are related to customer's choice. Being a climate stress-sensitive plant, the task for the breeders is to create robust plants as per the need of the fresh fruit and finished product industries. With increasing export and changing consumer demand, the cultivar need of different regions is going to change. Traditional mango breeding is time taking and may become difficult due to the requirement of large human and other resources for developing sufficient breeding population. Mango is vulnerable to many new disease, insect-pest, and climate change problems, and wild relatives, as a source of genes for biotic and abiotic stress resistance, may be utilized as gene donors. Useful mango germplasm identified for many horticultural traits and identification of favorable alleles and quantitative trait loci (QTLs) for marker-assisted selection will help in reducing the time taken in breeding varieties. Molecular biology has added to the scope of mango breeding providing an option to manipulate plant expressions in efficient manner. Researchers will have to strive to combine the best conventional and modern molecular approaches to improve mango germplasm to keep the "king of fruits" an economically viable global crop. This is possible with a multidisciplinary approach of handling long-term integrated breeding programs.

References

Abedin, MZ, Quddus MA (1990) Household fuel situation, home garden and agrofor- estry practices at six agro-ecologically different locations of Bangladesh. In: Abedin MZ , Chun KL, Omar MA (eds.) Homestead plantation and agroforestry in Bangladesh, BARI/RWEDP/FAO/Winrock International, Joydebpur, Bangakok/Dhaka, pp. 19–53

Ahmed J, Ozadali F (2012) Novel processing technologies for fruits. In: Siddiq M (ed) Tropical and subtropical fruits: postharvest physiology, processing and packaging. Wiley, Ames, IA, pp 71–96

Ahemad SK (2017) Evaluation of antihyperglycemic activity of aqueous extract of *Carica papaya* linn. leaves in Alloxan induced diabetic Albino rats. J Med Sci Clin Res 5. https://doi.org/10.18535/jmscr/v5i11.187

Amin ASR, Hajir SHDLAL, Marwa AARAL (2019) Antiviral activity of *Mangifera* extract on influenza virus cultivated in different cell cultures. J Pure Appl Microbiol 13(1):455–458

Anonymous (2017) Agri Export Statistics. Agriculture and processed food products export development authority. Ministry of Commerce and Industry, New Delhi, India

Bally ISE, Johnson PR, Kulkarni VJ (2000) Mango production in Australia. Acta Hort 509:59–68

Bello-Pérez LA, García-Suárez F, Agama-Acevedo E (2007) Mango carbohydrates. Food 1(1):36–40

Berardini N, Knoedler M, Schieber A, Carle R (2005a) Utilization of mango peels as a source of pectin and polyphenolics. Inn Food Sci Emerg Technol 6:442–452

Berardini N, Schieber A, Klaiber I, Beifuss U, Carle R, Conrad J (2005b) 7-O-methylcyanidin 3-O-β-D-galactopyranoside, a novel anthocyanin from mango (*Mangifera indica* L.) cv. 'Tommy Atkins' peels. Chem Sci 60(7): 801–4

Bernardes SAP, do Nascimento JRO, Lajolo FM, Cordenunsi BR (2008) Starch mobilization and sucrose accumulation in the pulp of keitt mangoes during postharvest ripening. J Food Biochem 32(3):384–395. https://doi.org/10.1111/j.1745-4514.2008.00175.x

Campbell RJ, Ledesma N (2013) The changing face of cultivars for the Western Hemisphere. Acta Hort 992:55–58

De Candolle A (1884) Origin of cultivated plants. Hanfer, Kengan Paul, Trench, London, UK

Derese S, Guantai EM, Souaibou Y, Kuete V (2017) Chapter 21 *Mangifera indica* L (Anacardiacea). In: Kuete V (ed) Medicinal spices and vegetables from Africa: therapeutic potential against metabolic, inflammatory, infectious and systemic diseases. Elsevier, London, pp 451–483. https://doi.org/10.1016/B978-0-12-809286-6.00021-2

Desphande AB, Chidley HG, Oak PS, Pujari KH, Giri AP, Gupta VS (2016) Data on changes in the fatty acid composition during fruit development and ripening of three mango cultivars (Alphonso, Pairi and Kent) varying in lactone content. Data Brief 9:480–491. https://doi.org/10.1016/j.dib.2016.09.018

Drammeh BS, Marquis GS, Funkhouser E, Bates C, Eto I, Stephensen CB (2002) A randomized, 4-month mango and fat supplementation trial improved vitamin A status among young Gambian children. J Nutr 132:3693–3699

Du S, Liu H, Lei T, Xie X, Wang H, He X, Tong R, Wang Y (2018) Mangiferin: An effective therapeutic agent against several disorders (Review). Mol Med Rep 18(6):4775–4786. https://doi.org/10.3892/mmr.2018.9529

Engels C, Schieber A, Ganzle MG (2011) Studies on the inhibitory spectrum and mode of antimicrobial action of gallotannins from mango kernels (*Mangifera indica* L.). Appl Environ Microbiol 77:2215–2223

FAO (2014) FAOSTAT database collections. Food and Agriculture Organization of United Nations, Rome, Italy. http://faostat.fao.org

Garcia D, Escalante M, Delgado R, Ubeira FM, Leiro J (2003) Anthelminthic and antiallergic activities of *Mangifera indica* L. stem bark components Vimang and mangiferin. Phytother Res 17:1203–1208

Gelabert-Rebato M, Martin-Rincon M, Galvan-Alvarez V, Gallego-Selles A, Martinez-Canton M, Vega-Morales T, Wiebe JC, Fernandez-Del Castillo C, Castilla-Hernandez E, Diaz-Tiberio O et al (2019a) A single dose of the mango leaf extract Zvnamite® in combination with quercetin enhances peak power output during repeated sprint exercise in men and women. Nutrients 11:2592

Gelabert-Rebato M, Wiebe JC, Martin-Rincon M, Galvan-Alvarez V, Curtelin D, Perez-Valera M, Habib JJ, Pérez-López A, Vega T, Morales-Alamo D, Calbet JAL (2019b) Enhancement of exercise performance by 48 hours, and 15-day supplementation with mangiferin and luteolin in men. Nutrients 11:1–24

Gholap AS, Bandyopadhyay C (1977) Characterisation of green aroma of raw mango (*Mangifera indica* L.). J Sci Food Agri 28:885–888
Gopalan C, Rama Shastri BV, Balasubramanian SC (1977) Nutritive value of Indian foods. National Institute of Nutrition, Hyderabad, India
Guha S, Ghosal S, Chattopadhay U (1996) Antitumor, immuno-modulatory and anti-HIV effect of mangiferin, a naturally occurring glucosylxanthone. Chemotherapy 42:443–451
Hill AF (1952) Economic botany. A textbook of useful plants and plant products. McGraw Hill Book Co, New York, Toronto, London
Hooker JJW (1876) The flora of British India 2. Reeve, London, UK
Jahurul MHA, Zaidul ISM, Ghafoor K, Al-Juhaimi FY, Nyam KL, Norulaini NAN, Sahena F, Mohd Omar AK (2015) Mango (*Mangifera indica* L.) by-products and their valuable components: a review. Food Chem 183:173–180
Jeffery JL, Turner ND, King SR (2012) Carotenoid bioaccessibility from nine raw carotenoid-storing fruits and vegetables using an in vitro model. J Sci Food Agri 92:2603–2610. https://doi.org/10.1002/jsfa.5768
Kanwal Q, Hussain I, Siddiqui LH, Javaid A (2010) Antifungal activity of flavonoids isolated from mango (*Mangifera indica* L.) leaves. Nat Prod Res 24:1907–1914
Knight and Schnell (1994) Mango introduction in Florida and the "haden" cultivar's significance to the modern industry. Economic Botany, 48(2), 139–145. https://doi.org/10.1007/BF02908201
Kumar S, Das DK, Singh AK, Prasad US (1993) Changes in non-volatile organic acid composition and pH during maturation and ripening of two mango cultivars. Indian J Plant Physiol 36:107–111
Ledesma N, Campbell RJ (2019) The status of mango cultivars, market perspectives and mango cultivar improvement for the future. Acta Hort 1244:23–27
Masibo M, He Q (2009) Mango bioat place?ctive compounds and related nutraceutical properties – a review. Food Rev Int 25:346–370
Masibo M, He Qian (2008) Mango polyphenols and their potential significance to human health. Compr Rev Food Sci Food Saf 7:309–319
Matheyambath AC, Subramanian J, Paliyath G (2016) Mangoes reference module in food science. In: Paul BC, Told FF (eds) Encyclopedia of Food and Health. Elsevier, Switzerland, pp 641–645
Mayne ST (1996) Beta-carotene, carotenoids, and disease prevention in humans. FASEB J 10:690–701
Morton J (1987) Mango. In: Julia F (ed) Fruits of warm climates. Morton, Miami, FL, USA, pp 221–239
Mukherjee SK (1951) Origin of mango. Indian J Genet 11:49–56
Mukherjee SK (1953) The mango - Its botany, cultivation, uses and future improvement, especially as observed in India. Econ Bot 7:130–162
Mukherjee SK (1972) Origin of Mango (*Mangifera indica*). Econ Bot 26:260–264
Mukherjee SK (1997) Introduction: botany and importance. In: Litz RE (ed) The Mango: botany, production and uses. CAB International, Wallingford, UK, pp 1–19
Mustapha AA, Enemali MO, Olose M, Owuna G, Ogaji JO, Idris MM, Aboh VO (2014) Phytoconstituents and antibacterial efficacy of mango (*Mangifera indica*) leave extracts. J Med Plants Stud 2:19–23
Njiru MK, Nawiri PM, Wanjau R, Odundo JO (2014) Residues of *Mangifera indica* L. as alternative animal feed in Embu county, Kenya. Green Chem 16:1–10
Othman OC, Mbogo GP (2009) Physicochemical characteristics of storage-ripened mango (*Mangifera indica* L.) fruits varieties of Eastern Tanzania. Tanzania J Sci 35:57–65
Prasad S, Kalra N, Singh M, Shukla Y (2008) Protective effects of lupeol and mango extract against androgen induced oxidative stress in Swiss albino mice. Asian J Androl 10(2):313–318
Prasanna V, Prabha TN, Tharanathan RN (2004) Pectic polysaccharides of mango (*Mangifera indica* L.): structural studies. J Sci Food Agric 84:1731–1735. https://doi.org/10.1002/jsfa.1874
Purseglove JW (1969) Some aspects of Mango culture in the western tropics. In: Proceedings of the International symposium on mango and mango culture, Indian Agriculture Research Institute, New Delhi, India
Rajan S, Hudedamani U (2019) Genetic resources of mango: Status, threats, and future prospects. Conserv Utiliz Hort Genet Resour 1:217–249
Rajan S, Mishra PK, Srivastav V, Aditya K, Sagar P, Tripathi PK (2020) A study on the visitor preference for different modules of the National Mango Database. J Appl Hort 22(2). https://doi.org/10.37855/jah.2020.v22i02.06
Ram S, Rajan S (2003) Status report on genetic resources of mango in Asia Pacific Region International Plant Genetic Resource Institute. New Delhi, India, p 196
Saleem M (2009) Lupeol, a novel anti-inflammatory and anti-cancer dietary triterpene. Cancer Lett 285(2):109–115
Saleem M, Afaq F, Adhami VM, Mukhtar H (2004) Lupeol modulates NF-kappaB and PI3K/Akt pathways and inhibits skin cancer in CD-1 mice. Oncogene 23:5203–5214
Saleem-Dar M, Oak P, Chidley H, Deshpande A, Giri A, Gupta V (2016) Chapter 19. Nutrient and flavor content of mango (*Mangifera indica* L.) cultivars: an appurtenance to the list of staple foods. In: Simmonds MSJ, Preedy VR (eds) Nutritional composition of fruit cultivars. Elsevier, Amsterdam, Netherlands, pp 445–468
Schieber A, Ullrich W, Carle R (2000) Characterization of polyphenols in mango puree concentrate by HPLC with diode array and mass spectrometric detection. Inn Food Sci Emerg Technol 1:161–166

Shah KA, Patel MB, Patel RJ, Parmar PK (2010) *Mangifera indica* (mango). Pharmacogn Rev 4:42–48

Shashirekha MS, Patwardhan MV (1976) Changes in amino acids, sugars and nonvolatile organic acids in ripening mango fruit (*Mangifera indica* L. Badami variety), Lebensm. Wiss Technol 9:369–370

Singh LB (1960) The Mango: botany, cultivation and utilization. Leonard Hill, London, UK

Sonwai S, Kaphueakngam P, Flood A (2012) Blending of mango kernel fat and palm oil mid-fraction to obtain cocoa butter equivalent. J Food Sci Technol 51:2357–2369. https://doi.org/10.1007/s13197-012-0808-7

Srivastava G, Mehrotra RC (2013) Endemism due to climate change: Evidence from Poeciloneuron Bedd. (Clusiaceae) leaf fossil from Assam. India. J Earth Syst Sci 122:283–288. https://doi.org/10.1007/s12040-013-0277-z

Tharanathan RN, Yashoda HM, Prabha TN (2006) Mango (*Mangifera indica* L.), the king of fruits—an overview. Food Rev Int 22:95–123

Thomas P, Oke MS (1983) Improvement in quality and storage of "Alphonso" mango by cold adaptation. Sci Hort 19:257–262

USDA (2020) National Nutrient Database for Standard Reference, SR-28, Full Report (All Nutrients): 09176, Mangos, Raw National Agricultural Library. USDA. https://fdc.nal.usda.gov/fdc-app.html#/?query=ndb Number:9176

Valmayor RV (1962) The mango, its botany and production. University of the Philippines, College of Agriculture, College, Laguna, p 90

Vasugi C, Dinesh MR, Sekar K, Shivashankara KS, Padmakar B, Ravishankar KV (2012) Genetic diversity in unique indigenous mango accessions (Appemidi) of the western ghats for certain fruit characteristics. Curr Sci 103:199–207. https://doi.org/10.1007/s13592-011-0065-1.24

Yadav IS, Rajan S (1993) Genetic resources of mango. In: Chadha KL, Pareek OP (eds) Advances in horticulture, vol 1. Malhotra Publishing. New Delhi, India, pp 77–93

Yoshikawa M, Ninomiya K, Shimoda H, Nishida N, Matsuda H (2002) Hepatoprotective and antioxidative properties of *Salacia reticulata*: preventive effects of phenolic constituents on CCl4-induced liver injury in mice. Biol Pharm Bull 25:72–76

Botany of Mango

2

M. Sankaran, M. R. Dinesh, K. Abirami, and C. Murugan

Abstract

Mango (*Mangifera indica* L.), the king of fruits, belongs to the family Anacardiaceae. Myanmar, Thailand, Indo-China, and Malaya during the Eocene or an earlier period in the Cretaceous were assumed to be the origin of genus *Mangifera*. Later, the genus spread to India, to Sri Lanka in the west, to Eastern Malaysia and to the Philippines in the East. Southeast Asia is said to be the centre of diversity of genus Mangifera with existence of maximum diversity in Peninsular Malaysia. Presently, maximum number of *Mangifera* species is found in Borneo, Sumatra, Java and Malay Peninsula. 69 species of *Mangifera* exist worldwide based on morphological characterization which are classified into two sub-genera *Mangifera* and *Limus* with several sections. Out of 69 species, *M.andamanica, M. griffithii, M.khasiana, M.camptosperma, M.sylvatica, M.odorata, M.zeylanica* and *M. nicobarica* are reported to occur in India. Huge variability has been recorded in *M. indica* due to its inherent characteristics such as cross-pollination, heterozygosity, incompatibility and polyembryony. The molecular-based phylogenetic tree generated among the different species of *Mangifera* showed their genetic relatedness based on amplification of chloroplast DNA (cpDNA) and ITS genomic regions.

2.1 Introduction

Mango (*Mangifera indica* L.), popularly known as 'King of fruits', is the most important among the tropical fruits. The fruit is rich in vitamins, minerals and other phytochemical compounds. Since time immemorial, mango is cultivated and consumed in India. This is evidenced by ancient Indian literature and scripts. Mango is highly associated with socio-cultural and economic life of the people. Due to invasions, mango wealth is enriched in India by introduction of superior quality germplasm. Among the different invasions, Muslim kings, Nawabs were mainly involved in conservation of best varieties of mango. The fruits of mango play a significant role in both nutritional and livelihood security of the society by providing essential nutrients to consumers. A drink made from boiled unripe mango with salt is a wonderful remedy for heat stroke.

M. Sankaran (✉) · M. R. Dinesh
ICAR-Indian Institute of Horticultural Research, Hesaraghatta Lakepost, Bengaluru 560089, India
e-mail: kmsankaran@gmail.com; m.sankaran@icar.gov.in

K. Abirami
ICAR-Central Island Agricultural Research Institute, Garacharma Post, Port Blair 744101, India

C. Murugan
Botanical Survey of India, Regional Office, TNAU Campus, Coimbatore 641003, India

C. Kole (ed.), *The Mango Genome*, Compendium of Plant Genomes,
https://doi.org/10.1007/978-3-030-47829-2_2

Mango juice is a restorative tonic; it should be taken throughout the season to stay healthy. Mango bark is used in the treatment of Jaundice. Dried leaf powder is used in dental care. Stones of mango are also used with leaf powder in powdered form. Oil extracted from stones is used for industrial purposes. Traditional users use it as hair tonic. It stops premature greying. Fresh mango bark with warm water is recommended for the patients having Gonorrhea. Mangoes are also a good source of antioxidants. The phenolic compounds ranged from 48.40 mg/100 g in the cv. Haden to 208.70 mg/100 g in the cv. Ubá. The beta-carotene content in the cv. Ubá was found to be 2220 mg/100 g and 77.71 mg/100 g of ascorbic acid was also observed (Rebeiro et al. 2007).

2.2 Origin and Distribution

Eastern India, Assam to Burma or the Malayan region is believed to be the origin of the cultivated mango (*Mangifera indica* L.). Hooker (1876) reported that mango might have originated naturally in India as wild. Originally, the cultivated mango was an alloploid (Popenoe 1920). However, Indo-Burma was also reported as the centre of origin of mango (Vavilov 1926). The historical records, related species and genus distribution, fossil remains and the presence of innumerable varieties and many indigenous wild species show enormous evidence for India to be the geographical origin for mango. Based on the fossil remains of leaf imprints, it was established that Assam was origin of *M. pentandra* (Seward 1912). With respect to origin of mango, there were many reports by different researchers. The origin of genus *Mangifera* was probably Burma, Siam, Indo-China and the Malay Peninsula according to Mukherjee (1951). However, he also reported that cultivated mango originated from Assam–Burma region rather than the Malay and during Quaternary period. In controversy, De Candolle in 1884 stated that 'if mango is not originated from South Asia or the Malay Archipelago, it is impossible to find the wide variability among the cultivars found in these regions, evidenced by local names, Sanskrit names, taxonomy and recent molecular findings. After many controversies, the recent findings firmly state that both origin and maximum diversity of the genus *Mangifera* exist in Southeast Asia (Mukherjee 1997; Bompard and Schnell 1997). As per Indian History, mango originated in India and from there it was spread to other regions like Southeast Asia, New World and Africa, as Northeastern India is at the northernmost edge for distribution of the *Mangifera* species (Mukherjee 1997). The Sudanic origin of the genus *Mangifera* was reported by Bompard and Schnell (1997) based on the existed diversity in Malay Peninsula, Borneo and Sumatra. Apart from this, region between Myanmar and Indo-China is one of the main centres of diversification, as wide range of species belonging to the section Euantherae (viz. *M. caloneura, M. cochinchinensis and M. pentandra*) exists here. In addition, *M. indica* apparently originated in region on Western border of the secondary centre of diversification mentioned above. Wildly grown common mango trees have been reported in Bangladesh (Chittagong Hills), Northeastern India (Assam Valley), Andaman and Nicobar Islands and Myanmar. The forests of Indian subcontinent are enriched with semi-wild trees of mango. Maximum species diversity of the genus *Mangifera* exists in Northeast India, in peninsular Malaya, Borneo and Sumatra (Ara et al. 2005). Seedling variations of primitive characters like polyembryony and dwarf growth habit with species richness are found in forests of Borneo, Malay Peninsula, Indonesian archipelago, Thailand, Indo-China and Philippines. It is now evidently clear that the genus *Mangifera* evolves in a large geographical area of Burma, Siam, Indo-China and the Malay Peninsula as the centre of origin, with maximum diversity in Southeast Asia (Popenoe 1920; Mukherjee 1951; Iyer and Subramanian 1989). The maximum concentrations of different *Mangifera species in* major geographical regions are in Malay Peninsula (19), Sunda Islands (16), Eastern Peninsula (14) including *M. indica* and its allied species such as *M. sylvatica*, *M. pentandra*, *M. caloneura*, *M. caesia*, *M. foetida* and *M. odorata*

(Mukherjee 1949a). Wild *M. indica* which are botanically similar to *M. sylvatica, M. caloneura, M. zeylanica,* and *M. pentendra* and other *Mangifera* species studied so have chromosome number of $2n = 2\,x = 40$ (Mukherjee 1950; Roy and Visweswariya 1951; Sharma 1987; Mustaffa 1993; Mitra 2001).

2.3 Taxonomy

Mango belongs to the genus *Mangifera* which comes under the family Anacardiaceae and order Sapindales. The family Anacardiaceae comprises 73 genera and 850 species mostly tropical origin, which also include some crops of temperate regions. The other relatives of *Mangifera* are cashew (*Anacardium occidentale*), gandaria (*Bouea gandaria*), pistachio (*Pistacia vera*), marula (*Sclerocarya birrea*), ambarella (*Spondias cytherea*), yellow mombin (*Spondias mombin*), red mombin (*Spondias purpurea*), imbu (*Spondias tuberosa*), dragon plums (*Dracontomelum* spp.) kaffir plum (*Harpepbyllum caffrum*), etc. which are mainly distributed in Malay Peninsula stretching from south of the Kangar-Pattani line to the Bismarck archipelago east of New Guinea (Whitemore 1975). The family Anacardiaceae consists of species that yield various products like wood, gums and resins, wax and varnishes and tanning materials other than edible fruits. The resinous sap of some members of the family Anacardiaceae causes allergic reactions like dermal irritation including some *Mangifera* spp.

***Mangifera* species**

Cultivated mango comes under the family Anacardiaceae and order Sapindales. The botanical classification of mango is as follows:

Division: Magnoliophyta
Class: Magnoliopsida
SubClass: Rosidae
Order: Sapindales
Family: Anacardiaceae
Genus: *Mangifera*
Species: *indica*

The genus *Mangifera* is mainly distributed in the Tropical Asia with maximum species diversity in Malay Peninsula, Sumatra, Java and Borneo (Western Malaysia). The primary centre of origin of the genus *Mangifera* is Myanmar (Burma)-Siam-Indochina or Malay Archipelago, whereas Sunda Islands (Java, Sumatra, Borneo), the Philippines and Celebes-Banda-Timor group is considered to be the secondary centre of origin of the genus Mukherjee (1985).

The species diversity of the *Mangifera* spreads from Malaysia to other parts of the world. One species spreads to Pacific area, nine species to India and three species to Srilanka (Kostermans and Bompard 1993). Four species were found endemic in Andaman and Nicobar Islands. The indigenous species of mango in Philippines are *M. merrilli* and *M. monandra.* The species *M. altissima* is specifically found distributed in Celebes Island (Kostermans and Bompard 1993).

The genus *Mangifera* L. is divided into two sub-genera, namely, *Mangifera* and *Limus* (Marchand) Kosterm. These two sub-genera do not have common base and reported to be originated from two different ancestors. Though this fact is established, the ancestors of the two sub-genera and species of *Mangifera* are still a misery (Kostermans and Bompard 1993). The mango species distributed in India are *M.indica* L. *M. sylvatica* Roxb., *M. khasiana, M. andamanica* and *M. camptosperma, M. griffithi and M. nicobarica.* The cultivated mango, *M. indica* L., is related to *M. longipes* Griff. However, wide variability exists in distributed populations of *M. sylvatica* Roxb. These species have adaptability to specific regions of precise climatic requirements. Therefore, it is very essential to conserve these species either in situ or ex situ, where similar climate prevails.

Mangifera indica L. belonging to the sub-genus *Mangifera have four sections, namely,* (i) *Marchandora* Pierre with types *M. gedebe* Miq; (ii) *Euantherae* Pierre with types *M. duperreana* Pierre (*M. caloneura* Kurz.) and species like *M. cochinchinesis* Engler and *M. pentandra* Hooker f.; (iii) *Rawa Kosterm* with types *M. griffithii* Hooker f. and species like *M. andamanica* King,

M. gracilipes Hooker f., *M. parvifolia*, *M. merrillii* Mukherjee, *M. microphylla, M. nicobarica Kosterm* and *M. minutifolia* and (iv) *Mangifera* with types *M. indica* L. and 12 species are included in this section. Sub-genus Limus (Marchand) Kosterm of genus *Mangifera* L. consists of six sections, viz. (i) *Marchand* Revis. with types *M. foetida* Lour.; (ii) *Manga* Marchnad with types *M. leschenaultii*; (iii) *Eudiscus* Pierre with types *M. superba* Hooker f.; (iv) *Microdiscus* Pierre with types *M. caesia* Jack; (v) *Deciduae* Kosterm with types *M. caesia* Jack including species like *M. kemanga, M. pajang, M. superba* and *M. caesia* Jack and (vi) Perini's Kosterm., sect. nov. with types *M. foetida* Lour including species like *M. foetida* Lour., *M. leschenaultii, M. macrocarpa, M. odorata* and perhaps *M. lagenifera*. The ancestral origin of these two sub-genera is different (Kostermans and Bompard 1993). Wide diversity is reported in terms of intra-specific variability in the species *M. zeylanica,* which is known in the local name Etamba at Sri Lanka.

A total of 69 species of *Mangifera* are reported worldwide. The diversity of the *Mangifera species* is found naturally distributed between 27° latitude North to Caroline Islands at East (Bompard and Schnell 1997). Wild mango species are distributed in different parts of the world like India, Sri Lanka, Bangladesh, Myanmar, Thailand, Kampuchea, Vietnam, Laos, Southern China, Malaysia, Singapore, Indonesia, Brunei, the Philippines, Papua New Guinea, and the Solomon and Caroline Islands. 28 species of the genus *Mangifera* are known to occur in Malayan region and reported as maximum diversity. The altitudinal distribution of the *Mangifera* mostly occurs below 300 m MSL. However, there are also reports that few *Mangifera* species may be distributed at 600–1900 m above MSL. The tropical lowland rain forests enrich more diversity of *Mangifera.* Areas with well-drained soils show maximum distribution of *Mangifera* species (44), about 9 species in flooded region. *Mangifera* species like *M. gedebe*, *M. griffithii* and *M. parvifolia* are found distributed in particular type of swamp forest regions. Someao *Mangifera* species distributed in submontane forests include *Mangifera bompardii*, *M. dongnaiensis* and *M. orophila that exists* 1000 m above MSL and rarely up to 1500–1700 m above mean sea level. *M. caloneura*, *M. collina*, *M. timorensis* and *M. zeylanica* are adapted to grow in seasonal dry climatic region like deciduous or semideciduous forests. Wild types of *M. indica and M. sylvatica* are distributed at altitudes 600–1900 m MSL in specific regions of South China and Northeast India.

Out of 69 reported species, only few of them are edible and rest of them are non-edible (Table 2.1). According to Iyer (1991), the species must fulfill two criteria such as it should have edible fruits and must possess resistant/tolerant to biotic/abiotic stresses to utilize in the breeding programme to introgress certain trait/traits in the cultivated variety background (Tables 2.2 and 2.3).

Table 2.1 *Mangifera* species with edible fruits

M. altissima	*M. longipes*
M. caesia	*M.macrocarpa*
M. cochinchinensis	*M. odorata*
M. foedita	*M. pajang*
M. griffithii	*M. pentandra*
M. indica	*M. sylvatica*
M. lagenifera	*M. zeylanica*

Iyer (1991)

Table 2.2 Species close to cultivated *Mangifea indica*

1. *M. indica* complex	11. *M. odorata* complex
2. *M. andamanica* King	12. *M.caesia* Jack
3. *M. caloneura* Kurz	13. *M. kemanga* Jack *casesia* var.*kemanga*
4. *M. camptosperma* Pierre	14. *M. foetida* Lour.
5. *M. indica* L.	15. *M. lagenifera* Griff.
6. *M. longipes* Griff.	16. *M. macrocarpa* Blume
7. *M. minor* Blume	17. *M. odorata* Griff.
8. *M. siamensis* Warb. ex Craib	18. *M. pajang* Kosterm.
9. *M. sylvatica* Roxb.	19. *M. superba* Hook.f.
10. *M. zeylanica* (Blume) Hook.f.	

Mukherjee (1985)

Table 2.3 *Mangifera* species as donors

Sl.No	Species	Traits	Region of occurrence
1	*M. caesia* Jack.	Pulp is aromatic, sweet and white in colour	East Borneo, Karangan reserve
2	*M. decandra* Ding Hou	Survives in water-logged condition. Hence may be exploited as rootstock	Sumatra, Pakambarue
3	*M. gedebe* Miq.	Survives in water-logged condition. Hence may be exploited as rootstock	C.E. Borneo
4	*M. indica* var. *mekongensis*	Multi-season fruiting	Saigon
5	*M. inocarpoides* Merr. & L.M. Perry	Survives in water-logged condition. Hence may be exploited as rootstock	Papua, Morehead
6	*M. pajang* Kosterm.	Fruits may be peeled like banana	Sarawak, Kapit District
7	*M. similis* Blume	Free stone mangoes	Borneo

Iyer (1991)

Though there is a growing awareness about conservation and sustainable utilization of wild relatives of cultivated species, yet little efforts have been made to conserve *Mangifera* species except *M. odorata, M. laurina and M. zeylanica.* In fact, the basic data on cytology, crossability with the cultivated and prevalence of apomixis in the wild species are lacking to initiate any meaningful breeding programme.

Though enormous diversity exists in the genus *Mangifera, threats also exists in the distribution of different species.* Mukherjee (1985) has reported mango species under threat which are as follows and they are also listed in the red data book published by IUCN (International Union for conservation of Nature and Natural Resources).

1. **Endangered**:	*M. cochinchinensis*
	M. flava
	M. lagenifera
	M. pentandra
	M. reba
	M. superba
2. **Vulnerable**:	*M. duperreana*
	M. inocarpoides
	M. monandra
	M. timorensis
	M. zeylanica

(continued)

3. **Rare**:	*M. andamanica*
	M. camptosperma
	M. gedebe

Efforts must be made among the Southeast Asian countries to conserve and utilize the *Mangifera* species. The incorporation certain traits these wild relatives are very important considering the impact of global climate change phenomenon.

The genetic diversity and utilization of wild species in perennial crops play a major role in both rootstock and scion breeding. In the genera *Mangifera,* wide species diversity exists with various utilities. However, very less work is done to collect, characterize and the possibility of crossing wild species with cultivated species. Pest and disease occurrence is a major threat for commercial cultivation of mango that results in yield loss. So, exploiting the wild species for biotic resistance is a major need. However, the main issue is the conservation and regeneration of wild species in regions other than their original habitat. This may be due to difficulty in adaptation and acclimatization of the species in other habitats. Hence, it is very much essential for in situ conservation of wild and threatened species of mango in their original habitats or in regions where similar climatic condition prevails.

2.3.1 Description of the *Mangifera* Spp

The genus *Mangifera* has evergreen trees/shrubs with aerid or resinous juice. Leaves are simple, alternate, rarely entire, petiolate; stipules 0. Inflorescence is terminal or axillary panicles, often crowded at the apices of twigs. Flowers are small, male and bisexual (polygamous), 4–5 merous; bracts caducous. Calyx is 4–5 lobed with imbricate. Petals are 4–5 in numbers, free and imbricate. Disc is fleshy, usually 4–5 lobed, sometimes obsolete in male flowers. Stamens are usually 5, rarely 10–12, usually 1–2 fertile, the others imperfect or sterile, much smaller, rarely all 5 fertile. Ovary sessile, superior, 1 loculed, 1 ovuled; style excentric or lateral; stigma simple. Fruit is usually a fleshy drupe.

KEY to the Six Indian Species

1a	Flowers 4 merous	**2**
B	Flowers 5 merous	**3**
2a	leaves up to 27 × 3 cm,; Sepals ovate; petals broadly ovate-lanceolate	***M. andamanica***
B	leaves up to 15 × 15 cm; sepals triangulate; petals spathulate	***M.nicobarica***
3a	Drupes distinctly flat	***M.camptosperma***
B	Drupes only slightly compressed	**4**
4a	Drupes acute or obtuse at tip	***M.indica***
b	Drupes acuminate, with protruded tip	**5**
5a	Leaves 25–30 × 5–7 cm; panicles peduncled, pyramidal, 20–30 cm or more; petals lanceolate, tip reflexed	***M.sylvatica***
b	Leaves 7–13 × 2–3 cm; petals ovate-oblong, tips not reflexed panicles sessile with fascicles of branches at the base, 5–10 cm long	***M.khasiana***

2.3.2 Botanical Descriptions of *Mangifera* Species in India

Based on the IPGRI descriptors, some *Mangifera* sp. such as ***Mangifera Odorata, Mangifera, Camptosperma, Mangifera zeylanica, Mangifera griffithii and Mangifera andamanica*** are characterized morphologically and the results showed wide variation among the species with respect to leaf, fruit, tree and stone (Sankaran et al. 2018).

(i) ***Mangifera andamanica*** King (J. As. Soc. Bengal 65: 470. 1896; Parkinson, For. Fl. Andaman IsIs. 139. 1923; Chandra and Mukherjee 2000).

Trees are 9–12 m tall. Leaves are elliptic-oblong, 9–23 × 4–6.5 cm, thinly coriaceous, base cuneate, apex notched and petioles are 1–2 cm long. Panicles are 10–12 cm long, terminal. Bisexual flowers are 4 mm across; pedicel 3 mm long. Sepals are 4, ca 3 × 1 mm and petals are 4, ca 4 × 1.8 mm, with 4–5 prominent ridges, confluent at base. Disc is four-lobed, cupular, fleshy, 1.5 mm in diameter. Stamens are 4, 1 fertile, 2.5 mm long, rest 3 short, teeth-like, and sterile. Ovary is 1 mm in diameter, 1 loculed, and 1 ovuled. Drupes ovate, 2.5 × 1.7 cm, apex subacuate and is in *distributed* in India (Assam and Andaman Islands).

(ii) ***Mangifera camptosperma*** Pierre (Fl. Cochinch. t. 363A. 1897. Thothathri & Balakr. in Bull. Bot. Surv. India 24: 175. 1982; D Chandra & S.K. Mukherjee in Fl. India 5: 466.2000)

Large trees up to 30 m tall. Leaves are oblong, 14–45 × 4–8 cm, glabrous, cuneate at base, entire at margin, apex acuminate to caudate; lateral nerves are 15–20 pairs; petioles are 1.5–2.8 cm, glabrous, pulvinate. Panicles up to 30 cm long, pubescent. Fruits are flat, elliptic, 6–9 × 6–8 cm, green when unripe and yellow when ripe; epicarp is thin; mesocarp is fibrous, 0.8–3 cm thick and endocarp is woody. Seeds reniform, 5 × 3 cm, filling most of the cavity of the stone. Distributed in India (Andaman and Nicobar Islands), Myanmar, Vietnam and Thailand.

(iii) ***Mangifera indica*** L. (Sp. Pl. 1: 200. 1753; Hook.f., Fl. Brit. India 2: 13. 1876; D Chandra & S.K. Mukherjee in Fl. India 5: 466.2000)

Variability exists in tree height ranging from 20 m to 45 m. All parts of the tree are glabrous except the inflorescence. Shape of the leaf is oblong-lanceolate, 15 to 30 × 3.5–6.5 cm, thinly coriaceous, base cuneate, margin entire, acute-acuminate apex; petioles are 1.5–6 cm long. Length of panicles is 15–35 cm, terminal, rarely axillary. Hermaphrodite and male flowers are 5 mm in width and length of pedicel is 1.5–3 mm. Calyx is 5 lobed; lobes ca 2 × 1 mm, tomentose. Petals are elliptic, 5, ca 3 × 1.2 mm, with branched ridges and reflexed tips. Stamens are 5 in number with 1 fertile stamen and length 2.5 mm. Rest of the stamens are sterile, small, teeth like; Disc ca 1.5 mm in diam., distinctly 5 lobed. Ovary ca 1 mm in diam and obliquely ovoid, 1 loculed, 1 ovuled; styles ca 2 mm long, lateral. The shape of drupes is oblong or sub-reniform with sweet juicy pulp. *Distributed* in India (cultivated almost in all states), Nepal, Bangladesh, Myanmar and Malesia.

(iv) ***Mangifera khasiana*** Pierre (Fl. Cochinch. t. 364C. 1897. Mukherjee in Lloydia 12: 97. 1949; D Chandra & S.K. Mukherjee in Fl. India 5: 468.2000)

Leaves of the tress are lanceolate, 7.5–13 × 1.8–3 cm, thinly coriaceous, apex acuminate; petioles are 0.8–2 cm long and slender. Panicles are 5–10 cm long, terminal and glabrous. Flowers are ca 4 mm across; pedicel 2–3 mm long. Calyx 5 lobed; lobes ovate, ca. 2 × 1 mm glabrous. Petals are 5, ca. 3.5 × 1.3 mm, concave, with 3 prominent swellings, non-reflexive tips. Stamens are 5 in number with 1 fertile stamen of length 3 mm. Rest of the stamens are teeth-like staminodes. Disc is fleshy, erect, ca 0.7 × 1 mm, with 5 ridges and furrows. Ovary ca 1 mm in diameter, conical; style is lateral, ca 2 mm long. Fruits not seen. *Distributed* in India (Sikkim and Meghalaya). Tree is endemic.

(v) ***Mangifera nicobarica*** Kosterm. (in Kosterm. & Bompard, The Mangoes 52. 1993; D Chandra & S.K. Mukherjee in Fl. India 5: 469.2000)

Trees are up to 20 m tall; branches reddish-brown. Leaves are oblong or elliptic-oblanceolate, *ca* 15 × 5 cm, gradually tapering and decurrent at base apex acute-shortly acuminate; lateral nerves are 8–12 pairs; petioles up to 2.5 cm long. Inflorescence is pseudo-terminal panicles, fascicled; peduncles are pseudo-racemoid, 9–15 cm long. Flowers are white or yellowish, 4 merous. Sepals are triangulate, up to 1.25 mm long. Petals are spathulate, *ca* 1.5 mm long. *Distributed* in India (South Nicobar). Tree is endemic. The species is closely allied to *M. andamanica* and *M. quadrifida*, but differs from them in leaf characteristics, and in case of latter in floral details as well.

(vi) ***Mangifera sylvatica*** Roxb. (Fl. Ind. 2: 644. 1824; D Chandra & S.K. Mukherjee in Fl. India 5: 469.2000. (Common name: Asm: *Ban-am*, *Lakshmi am*; Chittagong: *Basam*, *Kosam*; Lep: *Katur*; Nep: *Chuchi-am*; Sans.: *Kosamra*; Tipp.: *Haibamin)*

Trees are 30–40 m tall. Leaves are lanceolate, 25–30 × 5–7 cm, thinly coriaceous. Apex is acuminate; petioles are 5 cm or more. Panicles are 30–40 cm, terminal, glabrous, laxly flowered. Flowers are ca 4 mm across; pedicel is ca 3 mm long. Calyx is 5 lobed, ca 1.5 × 0.5 mm, glabrous. Petals are 5, linear-lanceolate, ca 4 × 0.8 mm, with 3 longitudinal swellings. Stamens are 5, 1 fertile, 3 mm long and rest teeth-like staminodes. Disc is ca 1 × 0.8 mm, erect, with 5 ridges and furrows. Ovary globose, ca 2 mm in diam., style sub-terminal. Drupes are

Fig. 2.1 Fruits of different Mangifera species

4.5–6 × 2.5–3 cm, obliquely acuminate, beak much drawn out; flesh is thin, white, slimy, without any fibres. It is distributed in India (Sikkim, Assam, Arunachal Pradesh, Meghalaya, and Andaman and Nicobar Islands), Nepal, Bangladesh and Myanmar (Fig. 2.1).

2.3.3 Detailed Description of *Mangifera Indica* L

2.3.3.1 Tree Canopy

The trees from seedling origin are of vigorous types with strong tap root and straight bole. The seedling origin tree shows sympodial growth which is known as Scarrone's model. However, grafted trees are comparatively dwarf with spreading branches. The canopy shape depends on the availability of space, edaphic factor and climatic influence. Naturally grown isolated seedling trees may get enough canopy space and hence exhibits vigorous growth. In contrast to this, the orchard grown trees will be of less vigour due to the reduced canopy space. There are also few cultivars like Latra and Creeping which are of spreading in growth habit and can be trained as creeper without utilization of any canopy space unlike other mango varieties. The Indian origin mango cultivars also show wide variation in canopy characteristics like compactness of the canopy, branching pattern and leaf component. The variation in canopy characteristics was observed based on ecogeographical distribution of mango cultivars. The longevity of seedling origin trees is more than 100 years. However, the grafted mango trees have comparatively lower longevity and may survive up to 80 years or less. In India, it is reported that a largest mango tree is observed in Chandigarh with a wide trunk diameter of 3.5 m and branches of 75 cm diameter. The canopy space occupied by the tree is about 2250 $m^{2.}$ The yield of the tree is about 16000 fruits in peak bearing years even after the tree age of 100 years (Singh 1960a). The largest trees are also found in various parts of the world. One such seedling tree is reported (Popenoe 1920) from Brazil which has a canopy spread of 125 ft and a girth of 25 ft..

In general, the trees of *Mangifera indica* are medium to large in stature with height ranging from 10 to 40 m. The trees are evergreen with round canopy. The canopy shape varies among the different cultivars ranging rom low and dense to upright and open. The bark of the tree is usually rough and dark grey brown to black coloured, cracked outside or sometimes inconspicuously fissured. Presence of resinous canals is the characteristic feature of all members of Anacardiaceae and the bark exudate is dark yellowish brown in colour and comprises resin mixed with gum. On drying, the exudate changes colour to brown. Tannic acid, 78% resin and 15% gum are present in the bark.

Lanceolate acute bud scales envelop the small terminal which appears during the initial stage of flowering. The twigs on the tree branches are thick, glabrous and dark green in colour with smooth and glossy appearance. The root system of the mango tree is strong with both tap root and feeder roots. The tap root is long which grow deeper into the soil and is unbranched. The length of the tap root ranges from 6 to 8 m or more. The feeder roots are highly dense and are developed at the base of the trunk or little deeper. The feeder roots are very important as anchor roots are produced by these feeder roots. The presence of good water table enhances the production of feeder roots and more feeder roots are produced above the water table. The root system of an 18-year-old tree grows to a depth of 1.2 m and spread to an area of 7.5 m laterally (Bojappa and Singh 1974).

2.3.3.2 Leaves

Wide variation exists in the appearance of mango leaves like oval-lanceolate, lanceolate, oblong, linear-oblong, ovate, obovate-lanceolate or roundish-oblong (Singh 1960b). Margin of the leaf also shows variations like wavy, entire, undulated, twisted or folded. Leaf apex is acuminate or rounded. The leaves are simple, exstipulate with alternate arrangement and groove on upper side. The petiole has a swollen base and the length varies from 1 to 12 cm. The leaves appear whorled as they are very closely arranged at the tips. Based on the growing condition and region, leaf size shows variation in both length and breadth. The range of length varies from 12 to 45 cm and breadth ranges from 2 to 12 cm. Fibres are present in leaves, which becomes brittle after drying and crackles when crushed. Prominent secondary veins ranges from 18 to 30 pairs. The leaf's lower surface is glabrous and light green in colour whereas the upper surface is shiny and dark green in colour. The new leaves emerge in flushes and are flaccid and pendulous while young. Varietal difference exists in the colour of young leaves and is tan-red, pink or yellow brown in colour depending on the variety. The colour of the leaf changes to dark green when fully matured. In some varieties, turpentine smell is observed and few varieties are without any odour. Xanthones, mainly the mangiferin, are present in leaves. If cattle are excessively fed with mango leaves, they die due to xanthones.

2.3.3.3 Inflorescence

Both the new leaf sprout and inflorescence emerge from the bud and they are pseudo-terminal. However, there are a few cultivars which have laterally positioned inflorescence. The shape of inflorescence is narrow at the tip and broad at the base with a conical shape and the length is about 45 cm; however, the length of the inflorescence varies depending on the cultivar. Lanceolate acute bud scales envelop the small terminal which appears during the initial stage of flowering. Many cultivars have inflorescence with bracts and a very few are ebracteate. The shape of the bract if present is elliptical, concave and leafy. The inflorescence has panicles with yellowish green colour or light green with crimson-coloured patches. The inflorescence is pubescent but rarely glabrous in few cultivars. Tertiary and sometimes quaternary

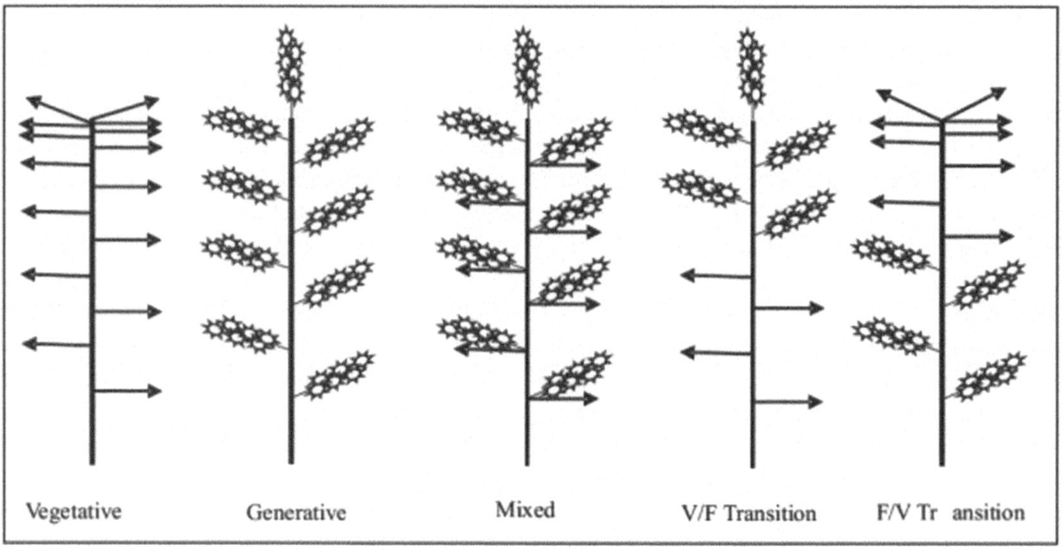

Fig. 2.2 Types of shoots and inflorescence in mango (Davenport 2007)

branching of inflorescence called cymose exists in the species (Singh 1960a, b). The number of flowers per panicle varies between cultivars and usually ranges from 500 to 6000 flowers. 70% of flowers are bisexual and remaining 30% are male flowers. However, the ratio of bisexual and male flowers depends on the varieties and also vary due to the temperature variations existing during floral development (Fig. 2.2).

2.3.3.4 Flower

Both male and hermaphrodite flowers exist in the same panicle with more proportion of male flowers when compared with hermaphrodite flowers. The male and hermaphrodite flowers are small in size and ranges from 6 to 8 mm in diameter. Flowers are aromatic, subsessile and pedicellate rarely. Pedicels are usually absent and very short if present. Panicle branch diameter is same as pedicel and is of articulate nature, and hence sometimes assumed as pedicel (Barfod 1988). Calyx has five sepals with ovate-oblong and concave lobes. The corolla is double the length of calyx with five petals (rarely four to eight), which are yellow coloured and three to five ridges on the ventral side. At bud stage, petals are imbricated and are contorted. The petals are free at base with irregular and non-reflexed upper half that are thin. Slightly dark ridges are found in petals. The petal colour changes to pink on drying. Five-lobed disc is found in between corolla and androecium (Singh 1960a, b; Kostermans and Bompard 1993). Together, five staminodes and stamens are found in androecium, of which one would be fertile and rarely two. Rest of them are sterile. The fertile stamens have longer length when compared to staminodes and equal to the length of the pistil. Few variations observed in the structure of androecium like cultivar Pico have three fertile stamens (Juliano and Cuevas 1932). A rare variation is found in few other members of the genus, where ten stamens in the form of primordia only are present (Maheshwari 1934). The stamens are attached on the inner margin of the disc. Stamen and pistil are distributed and parallel or oblique position to each other (Naik and Gangolly 1950). Generally, the anthers are pink coloured and turn into purple colour during flower shed. The ovary is slightly compressed laterally and is sessile, oblique shaped. The disc has the ovary. In few genotypes, three carpels develop into a flower. The ovule shows one-sided growth, which is anatropous and pendulous nature. The style ends in stigma which arises from the edge of the ovary (Singh 1960a, b). The

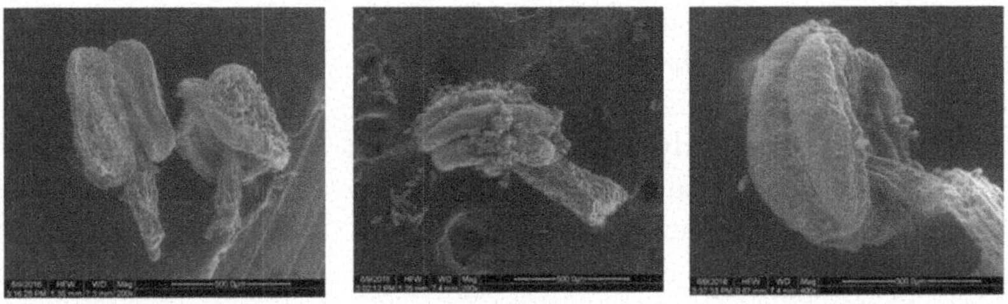

M. camptosperma M. odorata M. indica

Fig. 2.3 Filament attachment in the anther sacs of different *Mangifera* spp

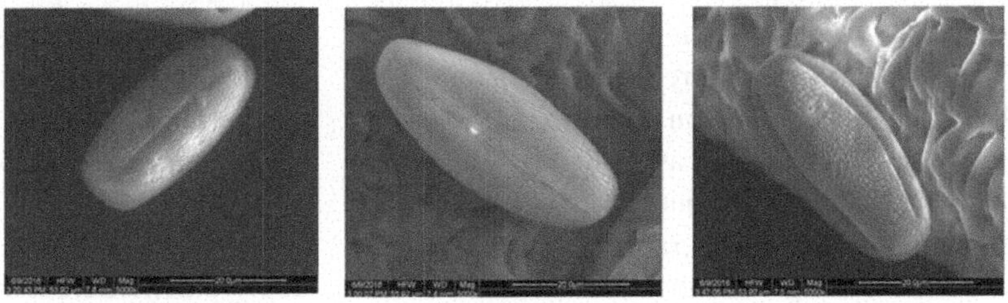

M. camptosperma M. odorata M. indica

Fig. 2.4 Pollen grain size and shape of different *Mangifera* spp

size of pollen grains varies among different varieties and ranges from 20 to 35 micron (Mukherjee 1950; Singh 1954a). The pollen grains are in variable shapes like spheroidal to prolate spheroidal, radially symmetrical, subangular in polar view, isopolar, and have few big triploid grains with size up to 50 micron. *Mangifera indica* usually produces pollen grain in small quantity and they are 3-monocolporate and goniotreme. The sides of pollen grain are convex-subprolate with equidistant apertures and the external aperture (colpus) extends like a slit between the two poles.

2.3.4 Pollen Characteristics

The shape of the pollen grains in all species is elliptic to oblong. The pollen of *Mangifera indica* is elliptic to oblong shaped with poles slightly flattened. The pollen of *Mangifera indica* has the presence of three colpi along the polar axis with an inner porate aperture. The structure of exine is semi-tectate and the tectum is perforate with *striato*-microreticulate sculpturing. The length of the polar axis (P) is 35–40 μm and breadth is 15–20 μm with prolate shape. The pollen of *M. odorata* is elliptic in shape and is tapering towards the two poles. The pollen of the species *M. odorata* is zono-tricolporate shaped with endoaperturaten pores arranged in 3 colpi. *M. odorata* has semi tectate exine with striato-reticulate ornamentation. The length of polar axis is 40–45 μm and breadth is 15–20 μm with perprolate shape. The pollen grains are oblong shaped in the species *M. camptosperma.* Endoaperture is absent in *M. camptosperma* pollen and hence is called as zonotricolpate type. The structure of exine is semiectate type with reticulate sculpturing. The length of polar axis is 25–30 μm and breadth is 5–20 μm with perprolate shape (Sankaran et al. 2018) (Figs. 2.3 and 2.4).

2.3.4.1 Floral Biology

The inflorescence is generally terminal in all the species of the genus *Mangifera*; however, multiple and axillary panicle from axillary buds is also observed. Both male and hermaphrodite flowers are seen in same panicle. The number of flowers per panicle varies among different species and varieties and the range is usually 1,000–6,000 (Mukherjee 1953). The fruit set at initial stage is determined by the proportion of bisexual flowers (Iyer et al. 1989). Hence, for economic fruit set, the distribution of perfect flowers is an important factor and when the percentage of perfect flowers is less than 1%, fruit set will be critical. Variation is observed among the different species of *Mangifera* for anthesis. Flower usually starts opening in the early morning hours in the genus *Mangifera* and continues up to noon in sunny situations. Between 9 and 10 a.m. maximum flower opening is observed in many varieties of the genus *Mangifera.* After anthesis, stigma shows maximum receptivity up to 6 h; however, it extends up to 72 h in some varieties. There are reports that in few varieties' stigma receptivity initiates before flower opening (Singh 1960a, b). The pollination and fruit set depend on both the factors such as pollen viability and stigma receptivity. About 1.5 h is required for germination of pollen grain in *Mangifera* (Sen et al. 1946: Singh 1954b; Spencer and Kennard 1955). Under proper controlled condition, the pollen viability could be retained for a long period of time. It is reported that pollen grains of *Mangifera* are viable even after 11 months of storage with 98% viability under controlled storage conditions at 7 °C and 25% RH. At decreased storage temperature, the viability of pollen grains can still be prolonged and stored up to 24 months at $0^{\circ}C$ and 25% RH (Singh and Singh 1952).

2.3.4.2 Fruit Characters

Botanically the fruit of *Mangifera* is fleshy drupe with compressed nature. Variability occurs in shape, size, colour, aroma, fibre content, flavour, taste and biochemical composition among the different varieties and species of *Mangifera*. The presence of beak (a small conical projection develops at proximal end of fruit) is also an important characteristic feature of the fruits of *Mangifera.* However, variation occurs in the appearance of beak among different genotypes. Above the beak, a sinus is present in the fruits. The base of the fruit is elevated or depressed or intermediate to both depending on the variety. At the base of beak, *nak* is present which is the pistillate area of the fruit. The fruit shape is not uniform and it varies among different species of the genus *Mangifera* and also among different varieties of *Mangifera indica*. The fruit shape varies from round to ovate-oblong among different varieties with length ranging from 2.5 cm to 30 cm. The peel colour of the fruit at maturity also shows wide variation and colours exhibited green, yellow and red shades. The peel is smooth or rough with dotted glands. Both the fruits and fruit stalk have some acrid juice with turpentine smell due to the presence of myrcene and ocimene in some varieties and is called as *chenp* in Hindi. The acrid juice causes irritation to the tongue and is due to the presence of chemical called urushiol, 5-heptadecenylreorcinol.

2.4 Pollination in Mango

The genus *Mangifera* is highly heterozygous and cross-pollinated. However, there are also reports that self-pollination occurs in mango (Dijkman and Soule 1951). Wide range of pollinators are reported as pollinating agents including wind, gravity and insects. Among the insects, honey bees, house fly and thrips are major pollinating agents (Popenoe 1917; Maheshwari 1934; Malik 1951). Less than 50% of flowers get pollinated when there is poor distribution of pollinating agents.

2.4.1 Incompatibility

Self incompatibility is reported in the genus *Mangifera* by several workers (Singh et al. 1962a, b; Mukherjee et al 1968 and Sharma et al. 1972). Sporophytic self-incompatibility is reported in mango. However, cross-incompatibility also occurs in mango (Ram et al. 1976). The existence of self-sterility was also doubted in genus *Mangifera* (Dijkman and Soule 1951). Based on the embryological studies, it was observed that 15 days after pollination, degeneration of endosperms occurs after self-pollination in spite of normal fertilization (Mukherjee et al. 1968).

2.4.2 Polyembryony

Both monoembryonic and polyembryonic modes of reproduction are reported in the genus *Mangifera*. The seedling per seed is single in monoembryonic type of germination and more than one seedling per stone are produced in polyembryonic types. The polyembryonic seedlings arise from both the zygotic tissue and the nucellar tissue which are surrounding the embryo sac. The seedlings originated from the nucellar tissue are true to type. However, the monoembryonic seedlings are variable as they are of zygotic origin. In polyembryonic mango varieties, the nucellar seedlings will be vigorous and the zygotic seedlings will degenerate. One exception is observed in the variety Pico from Philippines where both nucellar and zygotic seedings remain after seed germination. The embryological study by Maheshwari and Rangaswamy (1958) showed the presence of thick and dense cytoplasm and starch content in adventitious embryos originated from nucellar cells. The adventitious embryos thus formed in the nucellar region are slowly pushed into the cavity of embryo sac and they further divide and differentiate into embryos. The occurrence of polyembryony in mango is determined genetically. Polyembryony in mango is governed by recessive genes and is controlled by single pair of genes (Leroy 1947 and Sturrock 1968). However, an exception in this is that an Indian monoembryonic cultivar Malgoa (Mulgoba) showed polyembryonic nature when grown in Florida (USA). Popular polyembryonic mango cultivars reported are Bappakai, Chandrakaran, Kurukkan, Mazagaon, Mylepalium, Muvandan, Nekkare, Nileswar Dwarf, Olour and Peach, from India. Polyembryonic genotypes are reported to be introduced in India from Southeast Asia (Ravishankar et al. 2004). Cambodiana, Carabao, Corazon, Edward, Palo, 13-1, Pahutan, Pico, Senora, Starch, Strawberry and Florigon are the polyembryonic cultivars reported from other countries. Occurrence of polyembryony is reported more from coastal regions of India. The polyembryonic mango cultivars are highly fibrous and are of inferior quality with less preference as edible varieties. Hence, the monoembryonic cultivars are of commercial importance in India, Florida, South America and Africa due to the superior fruit quality parameters. The polyembryonic mango cultivars are valued in countries like Southeast Asia, Central America, Haiti, Hawaii, Australia and South Africa due to their utility in crop improvement (Singh 1960; Shukla et al. 2004). Classification of mango cultivars based on their embryo type and geographical origin was done by RAPD molecular characterization (Lopez-Valenzuela et al. 1997). The genetic diversity analysis of both polyembryonic and monoembryonic varieties by RAPD markers, total genomic DNA and chloroplast DNA RFLP (Ravishankar et al. 2004; Abirami et al. 2008) showed that polyembryonic and monoembryonic varieties have different genetic base.

2.5 Variability in *Mangifera indica*

The total number of cultivars reported in *Mangifera indica* was 1595 in 1998 by Pandey. However, the list was revised as 1663 in 2010 (Pandey and Dinesh 2010) and updated checklist was reported as 1682. The recently updated cultivars/varieties are Ambika, Arunika, Pusa Pratibha, Pusa Shreshth, Pusa Lalima, Pusa Peetamber, Arka Neelachal Kesari, Arka Ravi, Pant Chandra, Pant Sinduri, Mankurad Aldona, Paiyur-1, Vanlaxmi, Rajrani, Parish, Suvarna, Sai Sugandha, Martmanbhog and Khirma. There is a huge variability, which occurs among the cultivated and wild genotypes of mango (Pandey and Dinesh 2010). Seven hot spots of mango diversity reported in India are (i) humid subtropical region (Manipur, Tripura, Mizoram and south Assam); (ii) Chhota Nagpur Plateau (trijunction of Madhya Pradesh, Odisha and Bihar); (iii) Santhal Paragana; (iv) Southern Madhya Pradesh (tribal area) adjoining Odisha and Andhra Pradesh; (v) Dhar Plateau of Madhya Pradesh adjoining South Rajasthan and Gujarat, (vi) humid tropical southern peninsular India and (vii) Andaman and Nicobar group of islands (Yadav and Rajan 1993). In addition to this biodiversity hot spots of mango, Gangetic Plains, eastern peninsular region, Eastern and Western Ghats and Deccan plateau of India also show diversity richness. The genes controlling specific traits like dwarfness, fruit size, red-peel colour, high pulp content, high content of total soluble solids, long shelf life of fruit, regularity in fruit bearing, earliness and lateness in fruit maturity and good processing quality are enriched in the mango cultivars reported worldwide (Pandey and Dinesh 2010). Some of the cultivars are also used as a source of genes for both biotic and abiotic resistances. The commercial and export quality mango varieties listed worldwide are good source for high-yield and good-quality fruits (Pandey and Dinesh 2010). Characterization based on IPGRI descriptors for 151 mango cultivars is catalogued (Dinesh and Vasugi 2002). The classification of nomenclature of South Indian mangoes was earlier reported by Naik and Gangolly (1950) and Pandey (1984) in monograph of mango. Recently, DNA barcodes are developed for 223 mango cultivars based on molecular characterization and biodiversity international descriptors (Dinesh et al. 2012). Characterization of polyembryonic and pickle-making varieties from Western Ghats showed wide variation in physico-chemical attributes (Dinesh et al. 2012). Based on the size of fruits, the mango varieties are classified into different groups as very small (99 g and below), small (100–149 g), medium large (150–299 g), large (300–500 g) and very large (more than 500 g) for ease in grading and marketing (Pandey and Dinesh 2010). The commercial and most popular mango varieties like Dusehri Aman or Dashehari, Alphonso Bombay, Pairi, Langra, Pusa Arunima, Himsagar, Totapari, Sukul, Rumani, Neeleshwari, Rajapuri, Neelum, Suvarnarekha, Ambika, Zardalu, Bombay Green, Arka Puneet, Arka Anmol of India, Tommy Atkins, Eldon and Sensation of Florida, Aroumanus and Golek of Indonesia, Kyo Savoy and Okrong of Thailand, Vallenato of Colombia and Momi-K (Momi Kinney) of Hawaii belong to the category of medium large-sized fruits. Varieties like Arka Aruna, Mallika, Alfazli, Manjeera, Mulgoa, Samarbehist Chowsa from India; Amelie of West Africa; Bourbon of Brazil; Golden Brooks, Florigon, Davis Haden, Zrifin, Glenn and Van Dyke of Florida; Kensington of Australia; Nam Doc Mai, Nuwun Chan, Namtal and Pak-Kra Bok of Thailand; and Tahar and Naomi of Israel come under the classification category of large fruit size group. Among the Indian varieties, very large fruit-bearing cultivars are Sufed Mulgoa (1813 g), Himam Pasand (536 g), Anardana (717 g), Balakondapari (647 g), Tenneru (944 g), Cowasji Patel (713 g), Sora (1166 g), Gaddemar (713 g), Hathijhool, Pansera and Fazli. The cultivars which are categorized in very large-sized fruits from other countries are Manzanillo Nunez of Mexico, Osteen, Keitt, Anderson and Haden of Florida and R2E2 of Australia. Other very large fruited varieties are Nymath (1328 g) and Rebello (627 g). The Indian varieties which are classified under fruit size of very small group are Bhowani (50 g) and Chowras (70 g). Indian cultivars like Amrapali (130 g) and Rataul (107 g) are classified under small fruit size group (Anon. 2014; Pandey 1984). In USA, 19 varieties among the reported 61 varieties are classified into large and very large fruit size group (Brooks and Olmo 1972). Though thousands of mango varieties are reported worldwide, the consumer acceptance is based mainly on fruit quality. A wide variation is observed in the physico-chemical attributes of mango varieties and hybrids based on their geographical location and genetic base. Hence, there are about only less than 100 varieties which are commercially acceptable and grown worldwide. Among the noticeable commercial mango varieties, export quality varieties prevalent worldwide are Alphonso, Dashehari, Banganpalli and Kesar of India; Bourbon, Carlota and Haden of Brazil; Mabrouka of Egypt; Madame Frances of Haiti; Haden and Maya of Israel; Apple, Borino and Ngowe of Kenya; Haden, Keittoand Manila of Mexico; Anwar Rataul, Samerbehist chowsa and Fazli Khan of Pakistan; Carabao of the Philippines; Haden, Keitt and Zill of South Africa; Hindi Be Sennar and Kitchener of Sudan; Okrong of Thailand; and Julie of Trinidad and Venezuela (Pandey and Dinesh 2010). Based on the quality attributes, the mango varieties can be utilized for various purposes like table purpose and processing into various value-added products. Based on the recovery and quality attributes the Indian cultivars have been classified for utilization in processing of mango into different value-added products. Ramkela, Pulian, Chandrakaran, Karanjio, Aswina and Sandersha are grouped at pickling cultivars. Amrapali, Gautjit and Jauhari Safeda are highly suitable for processing into beverages. Quality mango nectar is produced from Amrapali, Dashehari and Saheb Pasand. Processing of mango into pulp is one of the major sectors for commercialization as the fruit is made available throughout the year. Bangalora or Totapuri is the variety which is mainly used for production of mango pulp on commercial scale as the production per tree is high with high pulp content and the cost of the fruits are also less. Mango varieties identified for juice making are Sukul, Mithuwa Ghazipur, Neelphanso and Safeda Lucknow. Mallika, syrup of varieties of Dashehari, Banganpalli and Nazuk Pasand are highly suited to produce mango slices on commercial scale. Due to the presence of variation in quality of the mango varieties, the juice of Amrapali and Nisar Pasand is blended with fruits of inferior quality to produce mango beverages. Canned mangoes are one of the major commercial value-added products in mango and the varieties suitable for this are Alampur Baneshan, Mallika, Nazuk Pasand, Dashehari and Alphonso.

2.6 Phylogenetic Analysis of *Mangifera* Species

Molecular markers like genomic restriction fragment length polymorphisms (RFLPs) and amplification of chloroplast DNA (cpDNA) were used to study the genetic relationship among the different species of the genus *Mangifera* and the phylogenetic trees were constructed (Eiadthong et al. 1999). The phylogeny of different species of Mangifera was also studied using the sequence analysis of External Transcribed Spacer (ETS) (Vavilov et al. 1996). Non-coding chloroplast regions are successfully employed for studying the phylogenetic relationships at genus and species level (Jasper et al. 2008; Shaw et al. 2005). Among the different molecular techniques, the non-coding chloroplast regions are more informative in generating the phylogenetic relationships among the different species (Shaw et al. 2013). The phylogenetic relationship among the wild mango species distributed in Andaman and Nicobar Islands and Northeast India, viz. *M. indica*, *M. griffithii*, *M. camptosperma*, *M. odorata* and *M. andamanica,* was studied using chloroplast markers like petB-petD intergenic spacer, rps16 gene, trnL-trnF intergenic spacer and nuclear marker—External Transcribed Spacer (ETS). The results showed that the cultivated mango, *Mangifera indica,* was grouped with *M. griffithii and M. camptosperma* showing their closed relatedness and all these three species belong to the sub-genus *Mangifera*. The wild species, *M. odorata,* was grouped separately along with *M. andamanica* and it belonged to the sub-genus Limus. Also, the study indicated that the endemic species of Andaman and Nicobar Islands, *M. andamanica.* is closely related with *Bouea oppositifolia* as analysed from both ITS and *psbAtrnH* rDNA analyses and therefore the taxonomic position of *M. andamanica* may be reconsidered (Sankaran et al. 2018).

2.7 Conclusion

Study of taxonomy of *Mangifera* species is essential to know the inter- and intra-species diversity. The wide diversity of species existing in the genus *Mangifera* has more utility in breeding programme of cultivated mango. The species diversity can be utilized in two different ways: (i) utilizing the better fruit quality parameters for improvement of cultivated mango and (ii) may act as donor for specific traits like resistant to insect pest and diseases (Iyer 1991). The different species of Mangifera are vital gene source for biotic and abiotic stress resistances, which may be utilized for developing resistant varieties and rootstocks. However, for crop improvement programme, the information on possibility of crossability among different species needs to be developed for production of superior genotypes through gene transfer. Both wild and cultivated mango species having same chromosome number (2n = 40) show the possibility of inter- and intra-specific crossing. Mukherjee (1950) reported that crossability exists between the different mango species like *M.indica* L., *M. odorata, M. zeylanica, M. caloneura, M. sylvatica, M.foetida and M. caesia.* It has already been proved that *M.odorata* and *M.camptosperma* are compatible with *M. indica.* Further, in the recent past, global warming and urbanization have been a great threat to the habitats of *Mangifera* spp and several species are under threatened conditions. Survey, collection, characterization and conservation of wild mangoes and semi-cultivated forms/types have to be done. Major emphasis must be given for in situ as well as ex situ conservation of threatened and economically important species. Since the wild mangoes and their relatives are reservoirs of potential genes those may introgressed in the *M. indica* background in order to capture the market demand in the coming years.

References

Abirami K, Singh SK, Singh Room, Mohapatra T, Kumar AR (2008) Genetic diversity studies on polyembryonic and monoembryonic mango genotypes using molecular markers. Indian J Hort 65(3):258–262

Ara H, Jaiswal U, Jaiswal VS (2005) Mango (*Mangifera indica* L.). In: Jain SM, Gupta PK (eds) Protocols for somatic embryogenesis in woody plants. Springer, Dordrecht, Netherlands, pp 229–246

Barfod A (1988) Inflorescence morphology of some American Anacardiaceae. Nordic J Bot Sect Holarctic Gener Taxon 8(1):3–11

Bojappa KM, Singh RN (1974) Root activity of mango by radiotracer technique using 32P. Indian J Agric Sci 44:175–180

Bompard JM (1993) The genus *Mangifera* re-discovered: the potential contribution of wild species to mango cultivation. Acta Hortic 341:69–77

Bompard JM, Schnell RJ (1997) Taxonomy and systermatics. In: Litz R (ed) The mango, botany, production and uses. CAB International, Wallingford, pp 21–48

Brooks RM, Olmo MP (1972) Register of new fruit and nut varieties. 2nd ed. Univ. of California Press, Berkeley

Chandra D, Mukherjee SK (2000) In: Singh NP, Vohra JN, Hajra PK, Singh DK (eds) Flora of India, vol 4

Davenport TL (2007) Review: reproductive physiology of mango. Braz J Plant Physiol 19(4):363–376

Dijkman NJ, Soule MJ (1951) A tentative method of mango selection. Proc FL State Hortic Soc 64:257

Dinesh MR, Vasugi, CS (2002) Catalogue of mango germplasm. Indian Institute of Horticultural Research, Hesseraghatta Lake PO, Bangalore 560089, pp 160

Dinesh MR, Vasugi C, Ravishankar KV, Reddy YTN (2012) Mango Catalogue, ICAR-IIHR, pp 468

Dinesh MR, Ravishankar KV, Nischita P, Sandya BS, Padmakar B, Ganeshan S, Chithiraichelvan R, Sharma TVRS (2015) Exploration, characterization and phylogenetic studies in wild *Mangifera indica* relatives. Am J Plant Sci 6:2151–2160

Dinesh MR, Rajan S, Singh SK, Singh IP, Ravishankar KV, Reddy BMC, Parthasarathy VA, Sthapit B, Ramanatha Rao V, Sandya BS (2015) Heirloom/seedling mango varieties of India – potentialities and future. Indian J Pl Genet Resour 28:17–30

Eiadthong W, Yonemori K, Sugiura A, Utsunomiya N, Subhadrabandhu S (1999) Analysis of Phylogenetic Relationships in *Mangifera* by Restriction Site Analysis of an Amplified Region of cpDNA. Sci Hortic 80:145–155. https://doi.org/10.1016/S0304-4238(98)00222-2

Hooker JD (1978) Bishen S, Mahendra PS (eds) Flora of British India, vol 2. Dehra Dun, India, pp 13–20

Iyer CPA (1991) Recent advances in varietal improvement in mango. Acta Hortic 291:109–132

Iyer CPA, Subbaiah MC, Subramanyam MD, Rao GSP (1989) Screening of germplasm and correlation among certain char- acters in mango. Acta Hortic 231:83–88

Iyer CPA, Subramanian TR (1989) Genetic resources activities concerning tropical fruit plants. In: Paroda RS, Arora RK, Chandel KPS (eds) Plant genetic resources: Indian perspec- tive. NBPGR, New Delhi, India, pp 310–319

Jesper K, Groeninckx I, Steven D, Timothy JM, Brigitta B (2008) the phylogenetic utility of chloroplast and nuclear DNA Markers and the phylogeny of the *Rubiaceae* Tribe *Spermacoceae*. Mol Phylogenet Evol 49:843–866. https://doi.org/10.1016/j.ympev.2008.09.025

Juliano JB, Cuevas NL (1932) Floral morphology of the mango (*Mangifera indica* L.) with special reference to the Pico variety from the Philippines. Philippine Agriculturist 21:449–472

Kostermans AJGH, Bompard JM (1993) The mangoes: their botany, nomenclature, horticulture and utilization. Aca- demic, London, UK

Leroy JF (1947) La polyembryonic chez les Citrus son interet dans la culture et Lamilioration. Revenue Internationale de Botanique Appliquee Paris 27:483–495

Litz RE (2004) Biotechnology and mango improvement. Acta Hortic 645:85–92

Lopez-Valenzuela JA, Martinez O, Parades-Lopez O (1997) Geographic differentiation and embryo type identification in *Mangifera indica* L, cultivars using RAPD markers. Hort Science 32:1105–1108

Maheshwari P (1934) The Indian mango. Curr Sci 3:97 Maheshwari P, Rangaswamy NS (1958) Polyembryony andin vitro culture of embryos of Citrus and *Mangifera*. Indian J Hortic 15:275–282
Maheshwari P, Rangaswamy NS (1958) Polyembryony and in vitro culture of embryos of Citrus and Mango. Ind J Hort 15:275–282
Malik, PC (1951) Morphology and biology of the mango flower. Indian J Hortic 14:1–23
Mitra SK (2001) Mango. In: Parthasarathy VA, Bose TK, Deka PC, Das P, Mitra SK, Mohandas S (eds) Biotechnology of horticultural crops, vol 1. Naya Prokash. Calcutta, India, pp 238–245
Mukherjee SK (1948) The varieties of mango (*M. indica* L.) and their classification. Bull Bot Soc Bengal 2:101–133
Mukherjee SK (1949a) The mango and its wild relatives. Sci Cult 15:5–9
Mukherjee SK (1950) Mango: its allopolyploid nature. Nature 166:196–197
Mukherjee SK (1951) The origin of mango. Indian J Genet 2:49–52
Mukherjee SK (1953) The mango—its botany, cultivation, uses and future improvements, especially as observed in India. Econ Bot 7:130
Mukherjee SK (1958) The origin of mango. Indian J Hortic 15:129–134
Mukherjee SK (1985) Systematic and ecogeographic studies on crop genepools 1. Mangifera L, IBPGR, Rome
Mukherjee SK (1997) Introduction: botany and importance. In: Litz R (ed) The mango; botany, production and uses. CAB International, Wallingford, UK
Mukherjee SK, Singh RN, Majumder PK, Sharma DK (1968) Present position regarding breeding of mango (*Mangifera indica* L.) in India. Euphytica 17:462–467
Mukherjee SK (1949b) A monograph on the genus Mangifera L. Lloydia 12:73–136
Mustaffa MM (1993) Studies on genetic diversity in some mango and papaya genotypes using isozyme analysis. PhD Thesis, University of Agricultural Sciences, Bangalore, Karnataka, India
Naik KC, Gangolly SR (1950) Classification and nomenclature of South Indian mangoes. The Madras Department of Agri- culture Printing Press, Madras, India, p 311
Pandey SN (1984) International check list of mango cultivars. Division of Fruits and Horticultural Technology, Indian Agricultural Research Institute, New Delhi-110012
Pandey SN (1986) Mango cultivars: nomenclature and registration. Acta Hort 182:259–264
Pandey SN (1988) Nomenclature and registration of mango cultivars. Acta Hort 231:125–131
Pandey G (1988) Role of malformins and auxins in the floral malformation of mango (*Mangifera indica* L.). Ph.D Thesis, GBPUA and T, India
Pandey SN (1998) Mango cultivars. In: Mango Cultivation. Ram Prakash Srivastava(ed) International Book Distributing Co., Charbagh, Lucknow, pp 39–99
'Pandey, S.N. and Dinesh, M.R. SN (2010) Mango. Indian Council of Agricultural Research, New Delhi, pp 30–97
Pandey SN, Dinesh MR (2010) Mango. Published by Indian Council of Agricultural Research, New Delhi, p 153
Pandey SN (1998b) Mango cultivars. In: Srivastav RP (ed) Mango cultivation. International Book Distributing Company, Lucknow, India, pp 39–99
Popenoe W (1917) The pollination of the mango. USDA Bull, p 542
Popenoe W (1920) Manual of tropical and sub-tropical fruits.Macmillan, New York, NY, USA
Ram S, Bist LD, Lakhanpal SC, Jamwal IS (1976) Search of suitable pollinizer for mango cultivars. Acta Hortic 57:253–263
Ravishankar KV, Chandrashekara P, Sreedhara SA, Dinesh MR, Anand L, Prasad GVSS (2004) Diverse genetic bases of Indian polyembryonic and monoembryonic mango (*Mangi-fera indica* L) cultivars. Curr Sci 87(7):870–871
Rebeiro SMR, Queiroz JHD, Queiroz MELR, Santana HM (2007) Antioxidants in mango (Manigera indica) pulp. Plant Foods Hum Nutr 62(1):13–17
Roy B, Visweswariya SS (1951) Cytogenetics of mango and banana. Report of Maharashtra Association for Cultivation of Science, Pune, India
Sankaran M, Dinesh MR, Chaitra N, Ravishankar KV (2018) Morphological, cytological, palynological & molecular characterization of certain *Mangifera* species. Current Sci 115(7):1379–1386
Sen PK, Mallik PC, Ganguly BD (1946) Hybridization of mango. Indian J Hortic 4:4
Seward AC (1912) Dictyledonous leaves from Assam. Records of the Geological survey of India 42, 100
Sharma DK (1987) Mango breeding. Acta Hortic 196:61–67 Sharma DK, Singh RN (1970) Self incompatibility in mango (*Mangifera indica* L.). Hortic Res 10:108–115
Sharma DK, Majumder PK, Singh RN (1972) Inheritance pattern in mango (*Mangifera indica* L.). In: Proceedings of symposium on recent advances in horticulture, UP Institute of Agricultural Sciences, Kanpur, UP, India, pp 66–68
Shaw J, Lickey EB, Beck JT, Farmer SB, Liu W, Miller J, Siripun KC, Winder CT, Schilling EE, Small RL (2005) The Tortoise and the Hare II: relative utility of 21 noncoding chloroplast DNA sequences for phylogenetic analysis. Am J Bot 92:142–166. https://doi.org/10.3732/ajb.92.1.142
Shaw J, Lickey EB, Schilling EE, Small RL (2007) Comparison of whole chloroplast genome sequences to choose noncoding regions for phylogenetic studies in angiosperms: The Tortoise and the Hare III. Am J Bot 94:275–288. https://doi.org/10.3732/ajb.94.3.275
Shukla MA, Babu RU, Mathur VK, Srivastava DK (2004) Diverse genetic bases of Indian polyembryonic and monoembryonic mango (Mangifera indica L) cultivars. Curr Sci 10(85):870–871
Singh L (1960) The Mango. Botany, cultivation, and utilization

Singh LB (1960a) The mango—botany, cultivation and utilisa- tion. Leonard Hill, London, UK
Singh RN (1954a) Studies on floral biology and subsequent developments of fruit in the mango varieties Dashehari and Langra. Indian J Hortic 11:1
Singh RN (1960b) Studies in the differentiation and development of fruit buds in mango (*Mangifera indica* L.). IV. Periodical changes in the chemical composition of shoots and their relation with fruit-bud-differentiation. Hort Adv 4:48–59
Singh RN, Majumder PK, Sharma DK (1962) Self incompat- ability in mango var, Dashehari. Curr Sci 31:209
Singh SN, Singh SP (1952) Studies on the pollen storage and longevity of pollens of some fruits and vegetables. Agric Anim Husb Dept UP 2:3–11
Singh RN (1954b) Studies on floral biology and subsequent development of fruits in the mango varieties Dashehari and Langra. Indian Hort 11:1
Spencer JL, Kennard WC (1955) Studies on Mango fruit set in Puerto Rico. Trop Agric 32:323–330
Sturrock TT (1968) Genetics of mango polyembryony. Proc FL State Hortic Soc 81:311–314
Vavilov NI (1926) Centres of origin of cultivated plants. Bulle Appl Bot Genet Plant Breed 16:1–248
Volkov RA, Kostishin S, Ehrendorfer F, Schweizer D (1996) Molecular Organization and Evolution of the External Transcribed rDNA Spacer Region in Two Diploid Relatives of *Nicotiana tabacum* (Solanaceae). Plant Syst Evol 201:117–129. https://doi.org/10.1007/BF00989055
Weeraratne WAPG, Samarajeewa PK, Nilanthi RMR (2005) Genetic diversity of Etamba in Sri Lanka. Trop Agric Res Ext 8:107–112
Weeraratne WAPG, Samarajeewa PK, Nilanthi RMR (2005) Characterization of mango germplasm using random amplified polymorphic DNA and morphological markers. Ann Sri Dep Agric 7:289–300
Whitemore TC (1975) Tropical rain forests of the Far East. Clarendon
Yadav IS, Rajan S (1993) Genetic resources of mango. In: Chadha KL, Pareek OP (eds) Advances in horticulture, vol 1. Fruit, part 1. Malhotra Publishing House, New Delhi, pp 77–93

Mango Propagation

3

Victor Galán Saúco, Alberto Carlos de Queiroz Pinto, Sisir Kumar Mitra, Fabio Gelape Faleiro, and Francisco Ricardo Ferreira

Abstract

Mango trees can be propagated both by sexual and asexual ways, but the existence of polyembryonic and monoembryonic mango plants conditionates its way of propagation. Air layering, cuttings, and even micropropagation can be used for mangoes, however, practically all commercial mango plantings are established nowadays from mangoes propagated by grafting or budding procedures using polyembryonic mangoes as rootstocks. Despite the existence of mango nurseries where all the operations from sowing to grafting are done directly in the soil without using any close structure, even without irrigation, modern mango production is realized in nurseries with all seedling establishment, from sowing to grafting of plant, made in soil substrate in propagation beds and polyethylene bags under modern protected environment and automatic ferti-irrigation systems. The most common grafting techniques as well as other techniques for vegetative propagation are described. The procedures for careful selection of mango seedlings for rootstock use, scion preparation, packing, and transport of budwood as well as the regulations for exchange of seeds and grafting material between countries are also indicated in this chapter.

V. Galán Saúco (✉)
Consultant on Tropical Fruits, Ing. Agr, Ph.D, Research Professor (Retired), Instituto Canario de Investigaciones Agrarias, Canary Islands, Spain
e-mail: vgalan46@gmail.com

A. C. de Queiroz Pinto
Consultant on Tropical Fruits, Eng. Agr. Ph.D. Researcher from Embrapa (Retired), Visitor Professor, University of Brasilia-DF, Brasilia-DF, Brazil
e-mail: alcapi@terra.com.br

S. K. Mitra
Board Member of International Society for Horticultural Science (ISHS), Brasilia, Brazil
e-mail: sisirm55@gmail.com

F. G. Faleiro
Eng. Agr. Dr. Researcher of Embrapa Cerrados, Brasília-DF, Brazil
e-mail: fabio.faleiro@embrapa.br

F. R. Ferreira
Eng. Agr. Dr. Researcher of Embrapa Genetic Resources and Biotechnology, Brasilia-DF, Brazil
e-mail: francisco.ferreira@embrapa.br

C. Kole (ed.), *The Mango Genome*, Compendium of Plant Genomes,
https://doi.org/10.1007/978-3-030-47829-2_3

3.1 Introduction

The mango is commercially cultivated in all the continents both in the tropics and subtropics from 37° north till 33° south latitude. According to the FAO statistics, mango is actually produced in more than 100 countries (www.fao.org). Despite the number of mango producing countries has stabilized in the last decade, world mango production has increased considerably, 15.7×10^6 tons in 1990; 25.0×10^6 tons in 2000; 38.0×10^6 tons in 2010, 50.6×10^6 tons in 2017. This continuous increase in production has been possible in great part due to planting or top working of more productive cultivars for which sound methods of propagation are required.

Mango trees can be propagated both by sexual and asexual ways, but as can be seen below the existence of polyembryonic and monoembryonic mango plants conditionates its way of propagation.

3.2 Mango Nurseries: Structure and Organization

Nurseries for mango propagation have different possibilities where all the operations from sowing to grafting are done directly in the soil without using any close structure, using traditional irrigation systems, or even without irrigation still exist in rural areas in many underdeveloped countries. However, modern mango nurseries made all seedling establishment, from sowing to grafting of plant, in soil substrate in propagation beds and polyethylene bags under modern protected environment and automatic ferti-irrigation systems.

Different propagation structures for modern mango production such as polyhouses with forced ventilation and evaporative cooling system, shade nets (Fig. 3.1), and net houses as well as mist propagation units were described in a recent paper on mango propagation by de Queiroz Pinto et al. (2017). These modern propagation structures are also useful for propagation of other tropical fruit trees and it will be redundant to describe them again here.

Fig. 3.1 Shade Net structure used in modern commercial mango propagation nurseries

3.3 Substrates for Mango Propagation

Substrate quality has marked influence on mango propagation. As indicated by de Queiroz Pinto et al. (2017), a good nursery substrate should have the following characteristics:

(1) Light weight to facilitate transport but hold the plants firmly mainly around the root system.
(2) Able to retain water but also allowing drainage and aeration.
(3) It should have necessary nutrients for proper growth and development of mango and should be also easily available to the plants.
(4) It should be free from weed seeds, toxic chemicals, and soil-borne fungus, bacteria, and pests.
(5) Sterilization should not change the characteristics of the substrate.

Sanitation of substrate is one of the most important factors for the success of high-quality graft mango, and this can be done either by pasteurization, solarization, or by chemical fumigation. In the first case, a temperature of about 70 °C for 30 min is considered sufficient to kill almost all disease-producing organisms (Sahdu 2005). Fumigation is most useful for destroying harmful bacteria, fungi, and nematodes in a relatively small quantity of soil that is used for propagation of plants. Methyl bromide was the most widely used soil fumigant for mango for many years, but its use is nowadays prohibited in most countries. A chemical alternative to eliminate pathogens from the medium that can be applied even when the plant is growing in the media consists of drenching the medium with certain fungicides such as Captan or Fytolan at a dosage of 1 g/liter of water (de Queiroz Pinto et al. 2017)

Besides the use of clean and sterile medium, all other parts and tools of the nursery must also be disinfected to avoid recontamination of the medium. General recommendations to achieve this purpose include the use of 2% formaldehyde, chlorinated water, and alcohol or even boiling water.

3.4 Monoembryonic Versus Polyembrionic Cultivars

Seeds of monoembryonic mango cultivars give rise to plants from sexual embryo and due to its heterozygotic character will not develop uniform plants and are not valid for true-to-the-type propagation of these cultivars. On the other hand, polyembryonic cultivars propagated by seeds, which are originated from nucellar embryos produce true-to-type progenies identical to the mother plant and, therefore, they can be used either for propagating polyembryonic cultivars or to establish mango rootstocks for grafting on them the desired monoembryonic cultivar. It should be noticed, however, that the guaranty of homogeneity is not total, even if we eliminate the potentially viable sexual embryo produced from open-pollination which may originate the development of a seedling distinct from the mother plant. Generally the sexual embryo is less vigorous than the nucellar embryo which makes selection of the most vigorous seedling a safe practice to ensure uniformity, but it is not always the case- some polyembryonic rootstock can even produce tetraploid progenies (Galán Saúco et al. 2002)—and it may be necessary to complement selection at least by careful observation of morphological characteristics to eliminate off-types. With these precautions grafted trees on polyembryonic rootstocks guarantee homogeneity and, also, favor early or precocious bearing because of the stress effect of the bud or graft itself. It has been indicated, in some cases particularly to reduce the initial cost in high-density plantings or to reduce the time for reaching full productivity in the subtropics the convenience of direct field planting of polyembryonic rootstocks or simply polyembryonic seeds for posterior in situ grafting (Galán Saúco 2008; Oosthuyse 2017). However, the potential loss of homogeneity derived from using rootstocks not 100% polyembryonic or from not carefully managed makes this practice not always recommendable (Galán Saúco 2018). Therefore, mango propagation by budding, and mainly by grafting method, is the most common method used for commercial mango production in most of the producing countries.

3.5 Seed Selection, Preparation, and Sowing

Mango seed selection and preparation procedures have been clearly described by several authors (Castro Neto et al. 2002; Galán Saúco 2008; Ram and Litz 2009. Donovan et al. 2016; de Queiroz Pinto et al. 2017).

The seeds for mango propagation used as rootstocks should be free of pests and diseases and collected from healthy and productive polyembryonic selected trees. According to the review made by Galán Saúco (2019) salinity and dwarfism are the main characteristics desired in most counties for a rootstock, although the latest is not necessary in the subtropics due to the slower growth of trees under subtropical conditions (Galán Saúco 2017a).and, with the advent and success of modern mango pruning techniques, it is possible to keep trees at a manageable size without losing productivity which explain why in countries like Brazil the dwarf character has been put as secondary aim in the mango breeding program (A.C. de Queiroz Pinto 2020, Personal communication). Other characteristics demanded for a rootstock include good compatibility with the cultivars, good ability to absorb nutrients, particularly iron and calcium, tolerance to flooding, tolerance to dry conditions, and favoring fruit quality. It should, of course, not be detrimental to yield or, even better, it should improve yield and/or yield efficiency (yield by canopy unit area). However, in the majority of countries, rootstocks are selected by the facility of obtaining seeds choosing local and well-adapted polyembryonic trees exhibiting heavy and uniform production annually and long before introduced in the area. A few exceptions to this are the case of Israel where the main rootstock used is 13/1, also polyembryonic, but selected especially for salt tolerance (Gazit and Kadman 1980) or in India, Pakistan, Bangladesh, Oman, China, and Hawaii where monoembryonic seedlings are also used as rootstocks. Some *Mangifera* spp., namely, *M. lalijiwa* and *M.kasturi* compatible with mango are also used as rootstocks in Hawaii and Indonesia. The explanation for the use of monoembryonic rootstocks in the area of origin of mangoes in South East Asia may be that the majority of plantings are of old orchards with low-density system where uniformity of the rootstock is not as important as in modern plantings and are also only used in some parts of China and in Hawaii when polyembryonic seeds are not available (Galán Saúco 2017b).

The fruits, whose seeds will be collected for propagation, should be fully mature, but not overripe, to avoid germination inside the fruit or rotten. They should be as large as possible to get a strong rootstock, since there is a direct and positive relation between seed size and vigor of the seedlings (Giri and Chaudhri 1966). In countries, where seed weevil (*Sternochetus mangiferae*) is present, special care should be taken to eliminate any seed affected by this insect. It is also recommended to avoid the fruits for seed purposes mainly from countries where temperatures are below 15 °C during fruit set and beginning of the fruit growth because of the common occurrence of embryo abortion. Generally, these fruits are of small size (Galán Saúco 2008), although some of them can reach commercial size, but having almost no seed or seed with no viable embryo.

The preparation of the seeds includes the following successive steps:

(1) Extraction from the fruits.
(2) Washing and elimination of all rest of pulp.
(3) Drying either in the shade for 1–2 days or with the help of an adequate ventilation system.
(4) Removal of the endocarp with a pruning scissor or a sharpened knife avoiding damaging the cotyledons, which may decrease germination success.

de Queiroz Pinto et al. (2017) indicate that the removal or not removal of the thin tissue covering the cotyledons has not been noticed to have influence on the germination success. de Queiroz Pinto and de Carvalho Genú (1996) also report the existence of a tool named Endocarp Remover

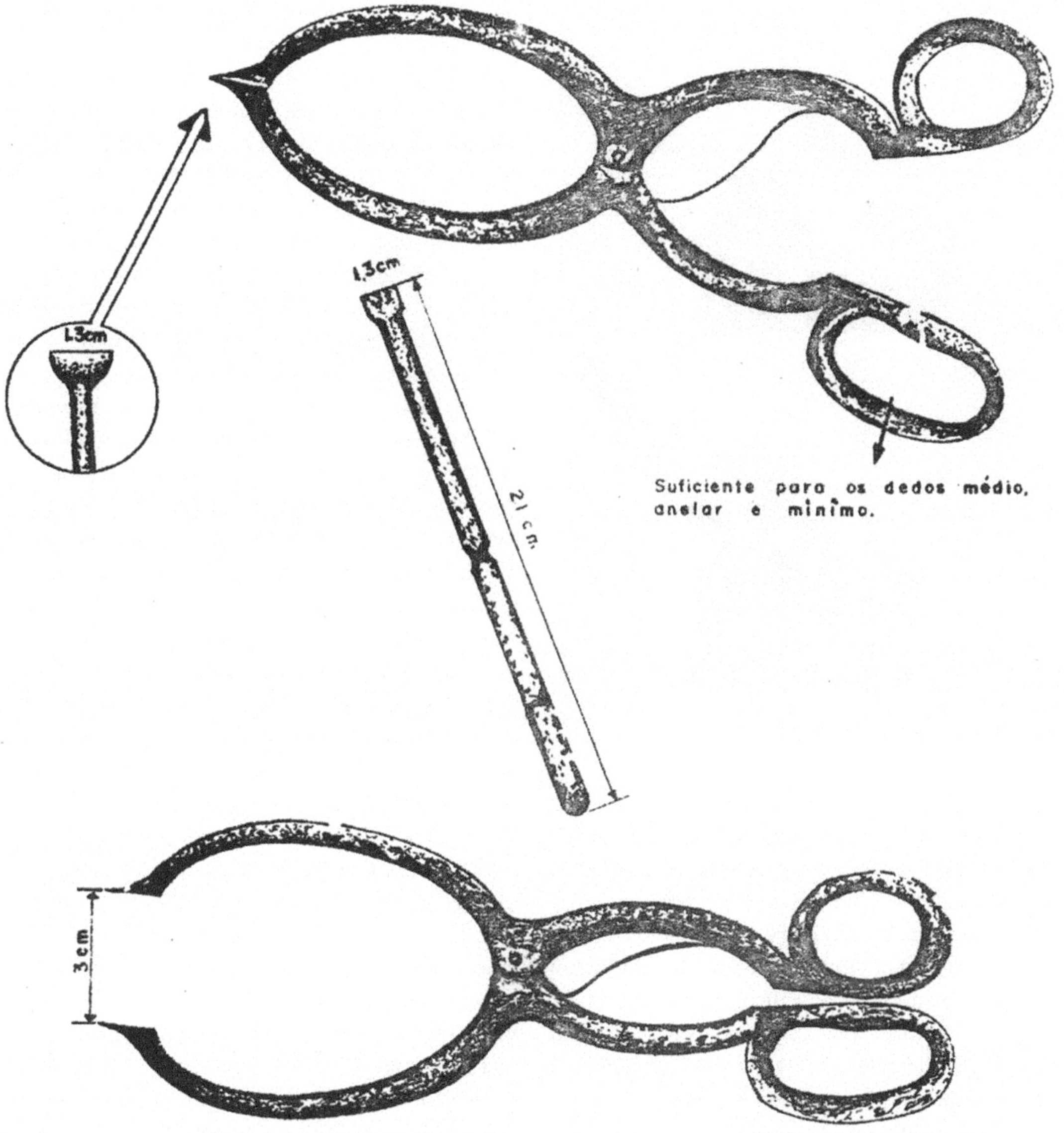

Fig. 3.2 Endocarp remover may help to pull out mango cotyledon free of injury from inside the endocarp. Figure inside the circle: Final thin part of 13 mm like a duck beak. Translation for the Spanish legend: Enough for 3–4 medium fingers

which facilitates the operation to remove the endocarp (Fig. 3.2). Although there are reports indicating that seed viability can extend till 84 days by placing them in sterile cotton impregnated with deionized water (Ram 1997) or up to 70 days by placing the mango seeds in charcoal dust in a desiccator set at 50% relative humidity (Caines 2020), the University of Florida Cooperative Extension Service (Sauls and Campbell 1994) say that mango seed retains viability for only a short time and seeds older than a couple of weeks frequently will not germinate. Because of this short viability period it is strongly recommended that seeds are sown as soon as possible after fruit harvest. Warm temperatures above 30 °C should be avoided during storing because they will provoke a quick germination (Parisot 1988). For sending seeds to

Fig. 3.3 Placing the seed in the bag

distant places it has been recommended to place them between layers of vegetal charcoal (Singh 1960), humid layers of newspaper or any other isolated material (Galán Saúco 2008).

Mango seeds should be planted by burying the ventral part (concavity portion) of the cotyledon into the soil substrate and then cover the convex portion with the soil substrate (Fig. 3.3) and immediately irrigate with water preferably added with an appropriate fungicide solution

In modern mango nurseries, seeds are sown directly into an appropriate substrate in nursery beds of around 25 cm depth of light soil substrate or be sown—this is the more appropriate method – into substrate inside of plastic bags. The seeds should be buried into the substrate and covered with a substrate layer of maximum 1.5 cm in the appropriate position discussed above. These two methods may avoid problems of damage in the root system which promotes losing of plants during transplanting from the nursery to the field found in traditional mango nurseries. Mckenzie (1994) indicates that a good soil substrate must have a full air porosity (FAP) value between 15 and 20%.

Transplanting from the seedling beds to the plastic bags where the seedlings are later grafted should be done soon after germination which occurs normally 2–3 weeks after sowing. To avoid deformities in the area of the union of root and stem due to the emergence of various seedlings from polyembryonic seeds it is advisable to select as quick as possible the best seedlings (erect and without deformity) eliminating all the rest by cutting them at the stem base.

The size of the bag depends on the time they will be maintained in the nursery before planting and/or selling, but usually they have a size between 300–350 mm length, 200–222 mm of diameter and 0.15–0.20 mm thick dimensions. These plastics bags should have lateral and bottom holes to avoid flooding substrate. However, in the subtropics where, by climate reasons, the grafted plants may stay in the nursery a longer period, polyethylene bags should better have a minimum diameter of 250 mm, a length of 400 mm, and the same thickness as above (Castro Neto et al. 2002; Galán Saúco 2008; de Queiroz Pinto et al. 2019). To avoid problems, some nurseries use open bottom bags placed in a type of railing structure allowing good substrate aeration and facilitating pruning of the taproot with the consequent developing of a better secondary and tertiary root system which improves rootstock vigor. Mckenzie (1994) also recommends chemical application of cupric products in the interior faces of the plastic bags for pruning the taproot.

Continuous seedling selection must be practiced removing all of them showing abnormalities like yellow leaves, short and compact internodes, small and twisted leaves in the apex, root deformations, or those which flowers precociously in the nursery.

Fertigation alone or combined with direct incorporation of granulate and slow releasing fertilizers into the soil substrate of seedling-beds or polyethylene bags is usually applied in modern nurseries. However, it is not recommending the exclusive use of slow releasing fertilizer when hot temperatures exist to avoid a faster release of nutrients not desirable for seedling development. Seedlings should be ready for grafting when reaching 400–500 mm height with a stem thickness of 8–12 mm at the middle of the height, which normally occurs in modern nurseries between 8 and 12 months after sowing (de Queiroz Pinto et al. 2018).

3.6 Propagation Methods

Air layering, cuttings, and even micropropagation can be used for mangoes, however, practically all commercial mango plantings are established nowadays from mangoes propagated by grafting or budding procedures using polyembryonic mangoes as rootstocks.

Several types of grafting have been used for long time from approach grafting (inarching) and different buddings types to the most common, splice veneer, cleft, softwood, and side grafting. They are well described in Ram and Litz (2008), Galán Saúco (2008), Donovan et al. (2016), and de Queiroz Pinto et al. (2018), and, essentially, they do not differ from grafting procedures for other subtropical trees, but the most common in modern mango propagation will be again described below.

As indicated by de Queiroz Pinto et al. (2018) there are five steps to do correct grafting which, slightly modified by us, can be seen below:

(1) Selection of plants and seeds for the establishment of seedlings to be used as rootstocks.
(2) Selection of plants of the desired cultivar (mother plant) from which scions should be taken.
(3) Picking, packing, and transport of scions to the nursery.
(4) Grafting in the nursery.
(5) Maintenance of the grafted plant until planting.

3.6.1 Selection of Mango Seedlings for Rootstock Use

The two main aspects considered nowadays for a good rootstock are tolerance to salinity and dwarfing or semi-dwarfing habit (Galán Saúco 2017b). Other important aspects for a mango rootstock include vigor, high yield potential, good compatibility with scion variety, environmental adaptability, and resistance to pests and diseases.

Despite the noticeable impact on the growth, yield, tolerance to salinity, nutrient absorption as well as fruit quality of mango cultivars, rootstock studies and its relationship with the selected scion cultivars are perhaps the most important, field of research lacking on mango. This fact may partially explain why polyembryonic rootstocks long time introduced in the countries (named Criollo type varieties in the Latin-American countries) are generally used for mango propagation in the tropics since they are more adapted to the different areas where they are found. These seedlings are widespread and found easily in tropical regions and, for mango growers, they are a source of selected homogeneous seedlings with a more regular bearing condition and they will probably continue to be the only option as rootstock until new specific studies be developed on each region (Galán Saúco 2019).

3.6.2 Scion Preparation at Mother-Plant, Packing and Transport

As well explained by de Queiroz Pinto et al. (2018), the scions usually used for grafting should have a size between 8 and 12 cm which roughly correspond to terminal shoots of about 4–6 months old under tropical conditions. According to Donovan et al. (2016), the scions should have at least 3 buds and be about 10 cm long and the same diameter as the rootstock and those with swollen buds just before bursting are the most suitable for grafting.

Before grafting all the leaves should be removed and the petioles cut. It has been strongly recommended to do this operation 1 week before cutting the scions from the branch in order to promote the swelling of the scion buds which will improve the grafting success (Castro Neto et al., 2002). It is also possible to obtain a good success in grafting by cutting the petioles immediately after severing the scion from the branch as it is usually done in the Canary Islands (V. Galán Saúco 2020, Personal communication).

In case of transport of the grafts to other places then prolonging time between cutting and grafting, scions should be protected from desiccation, which can be done by different ways (Galán Saúco 2008; de Queiroz Pinto et al. 2017) like:

(1) Immersion of the bottom of the scions (around 2.5 cm) into a parafilm liquid solution at about 51 °C.
(2) Packing and tying several scions together in toilet paper and then spray with tapwater.
(3) Wrapping them individually with a non-sticky adherent (cellofim or similar) and place them into a Styrofoam box.

Scions prepared by any of the mentioned methods can be kept in the lower portion of a house refrigerator (at 5 °C) in a sealed and labeled plastic bag where they can be conserved without losing its freshness by around 1 month (Donovan et al. 2016).

3.6.3 Regulations for Exchange of Seeds and Grafting Material Between Countries

Until the early 1970s, germplasm was considered a human good with no restriction existing for its transfer and exchange of plant material from one country to another and even between different parts of the same country including grafting materials either for commercial purpose or for scientific research. However, germplasm exchange it is subjected nowadays to a whole set of legal and bureaucratic regulations. After advent of the Biodiversity Convention, developed in Rio de Janeiro in 1972, and especially after its implementation in 1992, germplasm became the property of the country where it occurs, and as such, based on this premise, the exchange of germplasm started to be practiced with restrictions. Several international and national laws have been established to regulate the exchange of genetic resources or material for the propagation of plant species, mainly the laws and rules that consider bio-security to prevent entry of new pests in the country. One of the decisions taken after this Convention was the requirement for a document called Material Transfer Agreement (MTA), which allows the exchange of genetic material. Therefore, in order to import or export mango propagating material, seeds, cuttings, and scions (grafts), the donor and the recipient of the mango propagulum must sign the MTA document, which sets out the rules for using this exchange material. In addition, the propagation material of mango must be attached to a Phytosanitary Certificate issued by the donor country according to the requirements of the recipient country and should be subjected to quarantine after arrival to avoid undesirable introduction of pests, diseases, or weeds. To complicate more the process there are, in most countries, a whole set of general laws, rules regulations and even taxes related to importation of goods which delay the whole process of transfer of plant material between countries which the consequence in many cases of losing of the plant material (Ferreira et al. 2019).

3.6.4 Common Grafting Techniques

3.6.4.1 Splice or Whip and Tongue Grafting

This widely used grafting technique is perhaps the easiest method of grafting. This is usually done on 1–2 years-old rootstocks with a diameter of 0.6–1.0 cm. They are pruned at 10–20 cm above the soil making a slant cut of 3–4 cm long at the top part of the rootstock. A similar diagonal cut is made in the scion, which should be of the same diameter than the rootstock. Both exposed surfaces are then tied together with grafting tape (Fig. 3.4a–c). About 90% of success can be obtained with this technique, but the method has also proven useful with seedlings 3–9 months old.

Fig. 3.4 a Cutting the scion in two. b Insertion of the scion into a. c Wrapping the scion and rotstock with a plastic band

3.6.4.2 Veneer Grafting

Several authors, cited by Ram and Litz (1997), indicated that this is the most successful type of grafting. It has been done using rootstocks of different age from 3 months to 2 years which are prepared by making first a slanting cut of around 5 cm long. A second oblique cut is later made at the base of the first cut which permits the removal of a piece of wood with bark. 3–6-month-old scions shoots of 1.0–1.5 cm of diameter which are defoliated 3–15 days before grafting leaving the petioles attached are used. For doing grafting the base of the scion wood is fitted into the rootstock in such a way that the cambium layers of both scion and rootstock are in close contact tying finally both securely with grafting tape.

3.6.4.3 Cleft Grafting

This technique can be used for rootstocks of greater diameter and even for top working. It is normally done with 5–7 cm or more rootstocks which are cut back to around 45 cm above the ground and then split to a depth of about 5 cm long in the center of the stem with the help of a knife and a mallet keeping the cut open, if necessary, by inserting a hard wooden wedge. A 15–20 cm long scion coming from terminal shoots older than 3 months is wedged securely to a depth of 6–7 cm with the cleft of the scion slipped into the split part of the stock providing that the thicker border of the cleft is placed towards the external part of the rootstock and securing that the back of it meet firmly with the scion. Another scion is inserted diametrically opposed to the first scion. Once this is done, the wooden wedge is removed and the cut surfaces sealed with grafting wax.

3.6.4.4 Softwood Grafting

It is done by direct insertion of a scion into the bronze colored stem of a rootstock in such a way that the cut surfaces should adjust perfectly by practicing an appropriate cut on both scion and rootstock. About 100% of success is obtained in many cases especially if the leaves are not removed below the grafting union. Softwood grafting has been recommended in dry hot weather and places with low precipitation.

3.6.4.5 Side Grafting

It is normally practiced with 1.0–1.5 cm diameter rootstocks using around 10 cm long scions defoliated about one week prior to grafting which are inserted into the wedge of the rootstock and then tied with grafting tape. The top part of the rootstock should be removed after the graft union between stock and scion has occurred. It has been reported to be useful for humid coastal areas of India and also for mild weather in absence of strong winds, heat, and heavy rain.

It should be however mentioned that all these grafting techniques can be more successful in modern nurseries with environmental control conditions than in open-air nurseries with low or without a good environmental control.

3.6.5 Care and Maintenance of the Grafted Plant Until Planting

For improving success of mango grafting it is advisable to cover, immediately after grafting, the graft and the scion with a plastic bag about 200 mm long, 40 mm wide, and 0.01 mm thick or similar tied just below the graft union to prevent desiccation of the scion (Fig. 3.5). In case that the grafted plants were exposed directly to the sun the plastic bag should be covered with a white paper bag to reduce heat stress. All the shoots and branches that eventually grow out from rootstock must be removed. Both plastic and paper bags should be eliminated after the scion start to grow 1–2 cm which usually takes between 15 and 30 days (Fig. 3.6). After this grafted plant should remain in the nursery for at least two periods of growth for proper hardening, after which will be ready for planting in the field (Fig. 3.7).

Fig. 3.5 Covering the grafting union to protect it

Fig. 3.6 Mango graft after removal of protection

In order to keep identified the grafting point for longer time some nurseries paint this area. Normal nursery practices of watering, fertilizing, root-pruning, etc. should be carried out during the time the plants remain in the nursery. Hardening, which involves exposing the plant to regular sunlight, and reducing the frequency of watering, must be practised before leaving the nursery for field planting. For doing it the grafted plants should be gradually exposed during a few weeks to full sunlight to ensure a good response to the changing environmental conditions, but

Fig. 3.7 A high-quality commercial graft

care should be taken during hardening to avoid putting plants directly on the ground outside, where they could, particularly under subtropical conditions, experience damaging low or high temperatures. Hardening is, obviously, more important when the plants have been grown under modern environmental conditions than when the nursery is established in open-air nurseries.

Labeling the graft plant is a very important final step and this is done with a hard paper or hard plastic (about 80 mm long × 60 mm width) in which the names of the orchard and the rootstock and scion variety besides the date of grafting must be written on the label.

3.6.6 Other Techniques for Vegetative Propagation

Mango can be also propagated through cuttings and air layers. Although there are some reports indicating 100% of success for cuttings (Ram 1997; Reuveni and Castoriano 1997) and up to 70% for air layers using growth regulators (Chhonkar and Singh 1972), these techniques are not generally used for commercial mango propagation. It may however be useful in case of the need of using a monoembryonic rootstock because of its resistance or tolerance to biotic or abiotic stresses. Although this circumstance has not yet be occurred in mango since no such tolerance or resistance has been detected in any monoembryonic rootstock it may be useful in the future because of the possibility of using other *Mangifera* species showing tolerance to flooding, cold conditions, or even resistance to anthracnose (Bompard and Schnell 1997), some of which are graft and sexually compatible with mango (Campbell 2004; Ledesma et al. 2017).

Information about different methods of rooting can be also found in Ram and Litz (2008) but the main recommendations for rooting cuttings as summarized by Galán Saúco (2008) from literature review are as follows:

(1) Use preferably hardwood etiolated cuttings of around 15 cm length and 1.2–1.8 cm of diameter with 3–5 buds keeping 1–2 leaves in the apex,
(2) Use substrates of low pH (4.5–7) and incorporate rooting hormones like indole butyric acid or naphthalene acetic acid at dosages between 5,000 and 15,000 ppm and slow-release fertilizers.
(3) Use nebulization and bottom heat (best temperature between 25 and 30 °C).

3.6.7 Micropropagation

Although a high homogeneity of planting material is achieved from conventional propagation on a polyembryonic pattern, micropropagation may have a special interest either to clonally propagate monoembryonic patterns (Thomas 1999) and even to propagate cultivars (Hussain et al. 2000; Kidwai et al. 2009; Mishra et al. 2010; Bimal and Singh 2017).

Micropropagation is a technique that can be used to obtain plants with phytosanitary quality, positively related with fruit quality and productivity. Additionally, micropropagation and tissue culture can allow the rapid multiplication of genotypes of interest from a small number of explants obtained at any time of the year (Andrade et al. 2005). These micropropagation possibilities are also important to support genetic improvement programs based on conventional and biotechnological methods, which requires the development of an efficient plant regeneration system (Krishna and Singh 2007).

The potential of tissue culture of mango to produce a large number of somatic embryos of both mono and polyembryonic mango cultivars has been documented (Bimal and Singh 2017). There have been several references concerning the micropropagation of many mango cultivars using different types of explants and culture medium (Litz 1986; Litz and Lavi 1997; Thomas 1999; Hussain et al. 2000; Pateña et al. 2002; Andrade et al. 2005; Al-Busaidi et al. 2016; Bimal and Singh 2017). In many of these studies, some problems of micropropagation protocols are still reported, such as the difficulty in explants asepsis, the callus oxidation due to the exudation of phenolic compounds, high rates of somaclonal variation, and the low reproducibility of the methodologies for different mango cultivars (Andrade et al. 2005).

Despite these problems, there are successful experiences in the micropropagation of mango (Ribeiro et al. 2010). Ara et al. (1999) successfully promoted the germination of somatic embryos in cotyledon stage encapsulated with alginate. Xiao et al. (2004) obtained regenerated plants from immature zygotic embryos of mango through direct somatic embryogenesis. The conversion of somatic embryos to plants occurred 4 weeks after inoculation of somatic embryos in culture medium.

There is an advance in technological knowledge about mango tissue cultures, however, there is still no optimized protocol for micropropagation and its practical use is still incipient. Therefore, further basic studies are still necessary to establish the proper culture medium for tissue culture, besides the methodologies to reduce costs and somaclonal variations that might occur. The scientific advances obtained in the areas of micropropagation, organogenesis, and somatic embryogenesis open important perspectives for biotechnological breeding and for the economic process to micropropagation of different mango cultivars in a high scale and with high genetic and phytosanitary quality.

References

Al-Busaidi KT, Shukl M, Al-Burashdi AH, Al-Blushi GS, Al-Jabri MH, Al-Kalbani BS, Al-Hasani HD (2016) *In vitro* regeneration of mango (*Mangifera indica* L.) cv. Baramasi through nucellar embryogenesis. J Hort Forest 8:37–43

Andrade SEM, de Queiroz Pinto, AC, Faleiro FG, Cordeiro, MCR, Ramos VHV, Teixeira JB (2005) Desenvolvimento e avaliação de protocolos para descontaminação de explantes de MangueiraVisando à Micropropagação. PlanaltinaEmbrapa Cerrados, 24 p. (Boletim de Pesquisa e Desenvolvimento/ Embrapa Cerrados No 153). Disponívelem: https://ainfo.cnptia.embrapa.br/digital/bitstream/CPAC-2009/27991/1/bolpd_153.pdf

Ara H, Jaiswal U, Jaiswal VS (1999) Germination and plantlet regeneration from encapsulated somatic embryos of mango (*Mangifera indica* L.). Plant Cell Rep 19:166–70

Bompard JM, Schnell R (1997) Chapter 2. Taxonomy and Sistematics. In: Litz RE (ed) The mango, botany, production and uses. CAB International, Wallingford, Oxon, UK, pp 21–48

Bimal R, Singh K (2017) Biology and biotechnology of mango (*Mangifera indica* L.) with reference to *in vitro* cell and tissue culture in endangered and endemic cultivars: a review. J Cell Tiss Res 17:6285–6292

Caines K (undated) Is a 2-year-old mango seed OK for planting? Home GuidesSF Gate,http://homeguides.sfgate.com/2yearold-mango-seed-ok-planting-98415.html. Accessed 22 Jan 2020

Campbell RJ (2004) Graft compatibility between *Mangifera* species and *Mangifera indica*'Turpentine' rootstocks and their subsequent horticultural traits. Acta Hort 645:311–313

Castro Neto M, Teixeira de Fonseca N, Santos Filho HP, Cavalcante Junior AT (2002) Chapter 6. Propagaçao e padrao da Muda. In: de Carvalho Genú PJ, de Queiroz Pinto AC (eds) A Cultura da Mangueira. Brazil, Embrapa Informaçao, Tecnológica Brasilia, pp 117–136

Chhonkar VS,Singh RK (1972) Propagation of *Mangifera Indica* L. by air-layering. Acta Hort 24:89-92, DOI:10.17660/ActaHortic.1972.24.14. https://doi.org/10.17660/ActaHortic.1972.24.14

Donovan N, Bally I, Cooke T (2016) Nursery manual for citrus and mango. In: House S (ed) Australian Centre for International Agricultural Research. www.aciar.gov.au. Accessed 21 Jan 2020

Ferreira FR, Benito NP, Silva ML da, Albuquerque M do SM, Marques AS dos A (2019) Intercâmbio e quarentena de recursos genéticos In: Paiva SR, Albuquerque M.do SM, Salomao AN, Jose SCBR, Moreira JR de A (eds) Recursos genéticos: o produtor pergunta, a Embrapa responde. Embrapa Recursos Genéticos e Biotecnologia, Brasília, DF, pp 141–156

Galán Saúco V (2008) El Cultivo del Mango, 2nd edn. Mundi Prensa, Madrid, Spain, p 340

Galán Saúco V (2017a) Chapter 6. Mango cultivation practices in the subtropics. In: Galán Saúco V, Lu P (eds) Achieving sustainable cultivation of mangoes. Burleigh Dodds Science Publishing, Cambridge, UK, pp 165–183 (ISBN: 978-1-78676-132. www.bdspublishing.com

Galán Saúco V (2017b) Mango rootstocks. Literature review and interviews. (www.mango.org/research-resources)

Galán Saúco V (2018) Establishing a mango orchard: Procedures and costs of establishment. (www.mango.org/research-resources)

Galán Saúco V (2019) Mango rootstock. A review. Acta Hort 1 244: 1–15

Galán Saúco V, Coello Torres A, Grajal Martín MJ, Juárez J, Navarro L, Fernández Galván D (2001) Occurrence of spontaneous tetraploid nucellar mango plants. HortScience 36(4):755–757

Gazit S, Kadman A (1980) 13-1 Mango rootstock selection. HorstScience 57:81–87

Giri A, Chaudhri MY (1966) Relation of mango stone weight to its germination and seedlings vigour. Pak J Sci 18:148–150

Hussain A, Jaiswal U, Jaiswal VS (2000) Plant regeneration from protoplasts of mango (*Mangifera indica* L.) through somatic embryogenesis. Plant Cell Rep 19:622–627

Kidwai NR, Jain MB, Chaturvedi HC (2009) Role of thidiazuron in *in vitro* induction of embryogenesis in nucellar tissue of *Mangifera indica* L. var. Dashehari, leading to plantlets. Curr Sci 96:1119–1124

Krishna H, Singh SK (2007) Biotechnological advances in mango (*Mangifera indica* L.) and their future implication in crop improvement—A review. Biotechnol Adv 25(3):223–243

Ledesma N, Campbell RJ, Hass M, Campbell TB (2017) Interspecific hybrids between *Mangifera indica* and related species. Acta Hort 1183:83–88

Litz RE (1986) Mango. In: Evans DA, Sharp WR, Ammirato PV (eds) Handbook of Plant Cell Culture, vol 4. Techniques and Applications. Macmillan Publishing Company, New York, USA, pp 612–625

Litz RE, Lavi U (1997) Biotechnology. In: Litz RE (ed) The mango, botany, production and uses. CAB International, Wallingford, Oxon, UK, pp 401–423

Litz RE, Gómez-Lim MA, Lavi U (2009) Chapter 18. Biotechnology. In: Litz RE (ed) The mango, botany, production and uses. CAB International, 2nd edn. Wallingford, Oxfordshire, UK, pp 641–669

McKenzie CB (1994) Plant Growing media and nutrition in mango nurseries. S.A. Mango Growers' Assoc Yearb 14:11–13

Mishra M, Shree Y, Pati R, Seal S, Shukla N, Kamle M, Chandra R, Srivastava A (2010) Micropropagation of *Mangifera indica* L. cv. Kurakkan through somatic embryogenesis. Indian J Genet 70(1):85–90

Oosthuyse SA (2017) Chapter 8. Management of an ultra-high-density mango orchard. In: Galán Saúco V, Lu P (eds) Achieving Sustainable Cultivation of Mangoes. Burleigh Dodds Science Publishing, Cambridge, UK, pp 205–227

Parisot E (1988) Study of the growth rhythm in young mango plants. Part 1. Description, germination and storage of polyembryonic mango seeds. Fruits 43 (2):97–10

Pateña LF, Carlos-Refuerzo LR, Barba RC (2002) Somatic embryogenesis and plantlet regeneration in mango (*Mangifera indica* L.). *In Vitro* Cell DevBiol – Plant 38(2):173–177

de Queiroz Pinto AC, Galán Saúco V, Mitra SK, Ferreira FR (2018) Mango propagation. Rev Bras Frutic 40(1): e-586. https://doi.org/10.1590/0100-29452018586

de Queiroz Pinto AC, de Carvalho Genú PJ (1996) Idéiassimples e práticas para uso na exploração frutífera. IV. Eliminador de Endocarpo. Comunicado Técnico, n. 21, 2p. Embrapa Cerrados, Planaltina-DF, Brazil

Ram S (1997) Propagation. In: Litz RE (ed) The mango, botany, production and uses. CAB International, Wallingford, Oxon, UK, pp 363–400

Ram S, Litz R (2009) Chapter 11. Crop production: Propagation. In: Litz RE (ed) The mango, botany, production and uses. CAB International. 2nd edn. Wallingford. Oxfordshire, UK, pp 367- 403

Reuveni O, Castoriano M (1997) Beneficial effects of slow release fertilizers incorporated into the rooting medium of mango cuttings. Acta Hort 455(1):512–517

Ribeiro JM, Bastos DC, Melo NF, Oliveira EAG, Teixeira Pinto MS (2010) Produção de mudas micropropagadas de videira, mangueira e goiabeira. Petrolina: EMBRAPA Semiárido, 21p. (Documentos/Embrapa Semiárido Nº 232). Disponívelem. https://ainfo.cnptia.embrapa.br/digital/bitstream/item/29164/1/Juliana1.pdf

Sadhu MK (2005) Plant Propagation. New Age International (P) Limited, India, p 277

Singh LB (1960) The mango, botany, cultivation and utilization. World Crops Books, Inter Science Publishers, New York, p 438

Sauls JW, Campbell CW (1994) Mango propagation fact sheet HS-58 Horticultural Sciences Department, Florida Cooperative Extension Service. Institute of Food and Agricultural Sciences, University of Florida, USA

Thomas P (1999) Somatic embryogenesis and plantlet regeneration from nucellar tissue of monoembryonic mango. J Hort Sci Biotechnol 74:135–139

Xiao JN, Huang XL, Wu YJ, Li XJ, Zhou MD, Engelmann F (2004) Direct somatic embryogenesis induced from cotyledons of mango immature zygotic embryos. Vitro Cell Dev Biol Plant 40:196–199

Genetic Resources in Mango

4

Shailendra Rajan, Manish Srivastav, and Heiplanmi Rymbai

Abstract

Mango (*Mangifera indica* L.) is an important fruit crop of the world and predominantly grown in tropics and subtropics of Asia, Africa, and America. Apart from thousands of varieties in Asia, there are 69 species of *Mangifera* in wild gene pools. In India, the primary centre of diversity, mango cultivation commenced a few thousand years ago and seedling propagation for several centuries and out-crossing emanated the valuable gene pool. Characterization and evaluation of mango germplasm is an imperative step for its utilization in breeding programs being undertaken in different parts of the world. Genetic resources have been effectively used for conventional breeding mainly seedling selection and introgressing genes through hybridization for evolving new varieties. Mango is subject to various production constraints, both biotic and abiotic. Some of these are global in nature, while others are of national or local importance. Genetic resources need to be screened as donor for resistance genes against stresses. Probability of getting new varieties through seedling selections is becoming meager as the newer orchards are being planted using grafted plants.

4.1 Introduction

Genetic resources can be defined as all materials that are available for improvement of a cultivated plant species (Haussmann et al. 2004). Mango genetic resources constitute all the diverse genetic material having value as a resource for present and future generations of human being. It includes total pool of genetic variability that exists in the species or within species with which the crop plant is sexually compatible (Holden et al. 1993). There have been many factors which have led to an increased realization of the importance and conservation of indigenous as well as exotic genetic resources. It is more pertinent with respect to the 'King of fruit' mango since out of the 70 plus *Mangifera* species hardly one dozen is under cultivation globally, while many are under the constant threat of extinction in their regions of origin due to wrath of natural

S. Rajan (✉)
ICAR- Central Institute for Subtropical Horticulture, Rehmankhera, Kakori, Lucknow 226101, Uttar Pradesh, India
e-mail: srajanlko@gmail.com

M. Srivastav
Division of Fruits and Horticultural Technology, ICAR- Indian Agricultural Research Institute, New Delhi 110012, India
e-mail: msrivastav@iari.res.in

H. Rymbai
Scientist, Division of Horticulture, ICAR Research Complex for NEH Region, Umiam 793103, Meghalaya, India
e-mail: rymbaihort@gmail.com

C. Kole (ed.), *The Mango Genome*, Compendium of Plant Genomes,
https://doi.org/10.1007/978-3-030-47829-2_4

calamities, climate changes, etc. before they are actually collected, studied, documented, and conserved for immediate or long-term use by the mankind. Wild relatives are genetically much more diverse than related cultivated types. Wild gene pools and thousands of mango varieties extend ambit for characterization, evaluation, conservation, and utilization. India as the primary centre of diversity and secondary gene pools in other parts of the world exists. The characterization and assessment of mango germplasm has been carried out in different areas of the world. For traditional breeding, primary and secondary gene pool has been used for the development of new varieties. Mangoes are subject to different biotic and abiotic production constraints. For many production problems, genetic resources can be exploited to search for genes.

4.2 Diversity and Distribution

India, together with several other Southeast Asian countries is growing mangoes for thousands of years thus have access to huge genetic resources in Asia. Mango genetic resources play an important role in developing newer varieties as per the farmers need and it will become more vital in near future. In mango, so far, natural genetic variability has been exploited for improving productivity as well as introducing new varieties with specific traits. Maintenance of mango gene pool of cultivated and uncultivated types is important in present day context and more for future breeding programs because with the advent of new biotechnological tools, genetic manipulations techniques available, these resources might be used for desirable genes.

Enormous genetic diversity exists in primary and secondary centres of domestication of *M. indica* in terms of tree performance, tree vigour, flowering behavior, fruit quality, resistance to disease and pests, etc. However, limited efforts have been made to quantify the genetic variation existing in nature. Moreover, most of the molecular studies are confined to ex situ germplasm collections. Majority of the workers described variability in terms of morphological characters, mostly related with fruits, and less emphasis has been given to flower and vegetative characters. Large variation exists in tree vigour, fruit size (50-2000 g), fruit shape (more than 15 types of shapes have been recorded) (Ram and Rajan 2003).

Based on polyembryony, cultivated diversity in mango can be classified into two major groups: (i) subtropical with monoembryonic varieties (Indian types) (ii) tropical with polyembryonic varieties (Southeast Asian types). Polyembryonic cultivars spread along the western coast of India may have been introduced from Southeast Asia into Goa, either from their Malacca colonies in the Malay Peninsula or from Timor in the Indonesian Archipelago (Mukherjee 1997).

The Malay Peninsula, the Indonesian Archipelago, Thailand, Indo-China, and the Philippines are the places of greatest diversity in wild relatives of mango (Mukherjee 1985; Bompard 1989; Kostermans and Bompard 1993). Many mango wild relatives have small fruits with hard, acidic pulp, large stones, plentiful fiber, and resinous astringent substances that are found near the fruit peel. Along with *M. indica* at least 26 other species of the genus present in Southeast Asia are grown to produce edible fruits (Gruezo 1992).

The Indian subcontinent is considered as the most important centre of diversity for *M. indica* and has a large array of cultivated varieties. *M. indica* is distributed in most of the tropical and subtropical parts of India. *M. indica* occur as wild in forests as well as in cultivated areas. In India, wild mangoes are found almost all over the tropical and subtropical hilly forests and ravines of up to 1000 m and the Andaman Islands. Yadav and Rajan (1993) reported significant variability in wild mangoes in different parts of the country (Fig. 4.1).

Seedling races originated from monoembryonic mango stones are the most important component of diversity present in India. Almost all commercial mango cultivars have arisen as a result of the selection from these seedlings. Seedling trees are generally hardy and can withstand water stress (Yadav and Rajan 1993).

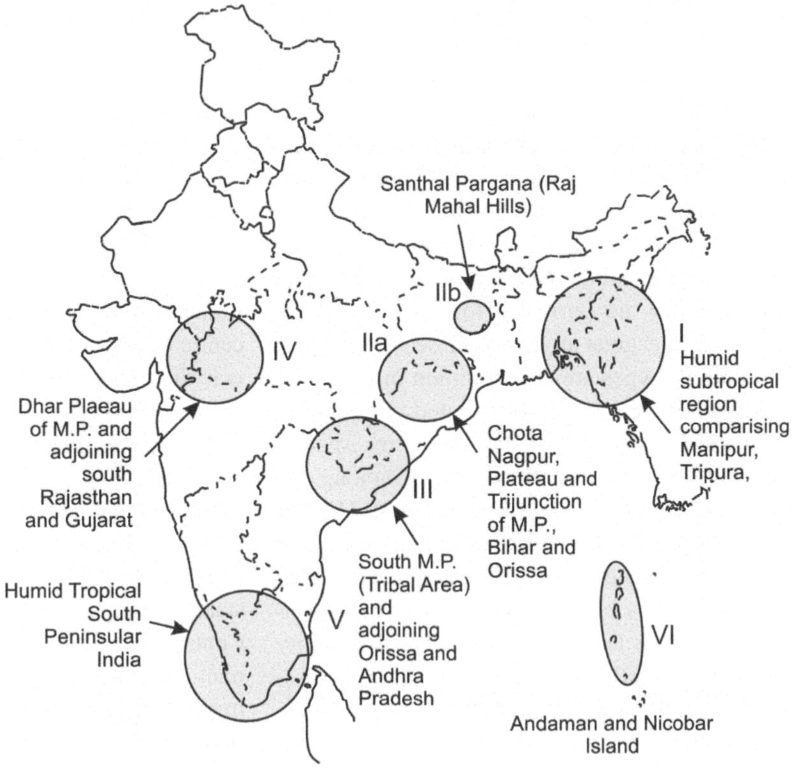

Fig. 4.1 Important zones of variability in wild *Mangifera indica* in India (Yadav and Rajan 1993; Ram and Rajan 2003)

With the increasing awareness, vegetatively propagated types (grafts) are replacing seedling use for new plantation.

M. indica grown in India has many best quality fruit-bearing varieties. There are three major diversity centres where large number of varieties evolved, viz., Lucknow-Saharanpur belt of Uttar Pradesh, the Murshidabad area of western Bengal, and the Hyderabad in Andhra Pradesh. Large variations in cultivated forms are still available in old orchards of these areas. Planting of seedling trees was continued till mid of last century (Rajan et al. 2016). Diversity in the form of over a thousand different cultivated varieties of mango are existing in India (Rajan and Hudedamani 2019). In Indian market, about 30 cultivated varieties are playing major role. Eco-geographical adaptation of commercial varieties is one of the key factors for many cultivars and regional consumer preference.

In India, Uttar Pradesh has the greatest diversity in cultivated types. In western region, apart from important commercial varieties, which are particularly suited to humid districts from Ratnagiri (Maharashtra) to Valsad and Surat (Gujarat) in the West Coast, wild types are found in Western Ghats particularly appimidi types (Chadha 1996). In Goa, in addition to cultivars like Mankurad, Hilari, and Fernandin, diversity exists in the form of local cultivated types. In Central India, particularly in Madhya Pradesh, commercial varieties of northern, southern, and western India are grown. Moreover, apart from the commercial varieties of other states, diversity in the form of local varieties like Dahiyar, Karela, Sunderja, and Madhu Kuppi is available. In some cultivars like Sunderja, variability within cultivar also exists.

The southern region of India has rich diversity of mango in Andhra Pradesh, Karnataka, Tamil Nadu, and Kerala. In the coastal districts, diversity in the form of juicy varieties (Rasalu), viz, Peddarasam, Chinnarasam, Cherukurasam, Panchadara Kalasa, Firangiludwa, and Kothapalli Kobbari are available. In Tamil Nadu, along with the popular varieties like Neelum, Mulgoa,

Kalepad, Bangalora, and Peter, substantial area under local varieties exists. Kerala has several polyembryonic races of mango and some varieties obtained from other states have adopted well. In eastern states, Bihar, West Bengal, and Odisha have rich diversity. Besides cultivated types, wild types are also found in Odisha and Bihar. In Odisha, varieties of Andhra Pradesh, *e.g.,* Banganapalli, Neelum, Totapuri, and Suvaranrekha are grown. However, the state represents the great diversity in seedling types. Several wild types are also common in this state.

Mangoes are well distributed throughout Bangladesh. It is one of the major tree species indicating its spread and popularity among the rural people (Abedin and Quddus 1990). Most of the commercial cultivars are of Indian origin and adapted in Bangladesh. In Bangladesh, seed propagation has been continuing for many centuries resulting large variations. The variability has been mostly created by natural cross-pollination (Ram and Rajan 2003).

In Indonesia, considerable diversity is available in *M. indica* and other edible *Mangifera* species because most growers in the villages are still maintaining even inferior cultivars around their houses (Kusumo 1996; Vasudeva et al. 2015). Sizeable variability in *M. indica* is found in Thailand. Variability is available in the form of three main types based on the stage of consumption, viz, green mango, ripe mango, and dual-purpose mango. In Philippines, diversity in *M. indica* is distributed throughout the country (Coronel 1996). Significant diversity in polyembryonic types is available in this country (Ram and Rajan 2003). In China, *M. indica* is distributed in Guangdong, Hainan, Yunnan, Guangxi, Fujian, Taiwan, and Southwest Sichuan. Sporadic cultivation is also found in the south of Zhejiang (Mukherjee 1985; Shupei and Yanqing 1996).

Nowadays, mango is commercially grown throughout the tropics and in many subtropical areas of the world (Rajan and Hudedamani 2019). It is cultivated in wide range of agro-ecological conditions, at the equator and at latitude of 35-37 ^{o}N in southern Spain. Widespread distribution of seedling selections made at Florida from the superior germplasm imported from different mango areas has made it an important fruit of subtropics and tropics (Mukherjee 1997).

4.2.1 Primary and Secondary Gene Pools

Indian subcontinent is considered as the primary centre of origin for *M. indica* and later the mango was distributed in other continents creating secondary centres of variability. In India, mango cultivation commenced a few thousand years ago and seedling propagation for several centuries and out-crossing emanated the valuable primary gene pool. Creation of diversity coincided chronologically with the dispersal of cultivated mango, starting with Southeast Asia, later to Africa, Central and South America. In recent decades, mango has become a common crop in most of the tropics and subtropics, nevertheless, primary gene pool confines to Asia and particularly to Indian Subcontinent. Hundreds of varieties contribute toward mango production in Asia which accounts about 60% of the world production.

During late nineteenth and early twentieth century, dozens of mango varieties were evolved in Florida establishing the western hemisphere mango industry (Schnell et al. 2006). These varieties are as per the preferences of the Western consumers' taste, aroma, and red peel color. Breeding system of mango prefers out-crossing. Availability of several genotypes from diverse geographical locations resulted into the development of secondary gene pool in Florida. This gene pool immensely contributed for the establishment of Florida mango industry, furthermore, in western hemisphere.

Several molecular characterization studies results based on a considerable number of cultivars have also conceptualized various groups of mango cultivars (Shamili et al. 2012; Dillon et al. 2013; Tsai et al. 2013; Sherman et al. 2015; Singh et al. 2016; Ajayi et al. 2019; Warschefsky and von Wettberg 2020). This grouping is not only based on the chronology of their introduction in different countries but also on the molecular basis and genetic relatedness. In most of the countries, mango cultivar collections are secondary collections. Cultivars were introduced from the country of origin indirectly to other area. Mostly, this process has taken place for evaluating suitability of cultivar for commercial production in mango-growing countries. In most of the mango-growing areas of western hemisphere, seedlings are not used for planting new orchards, and therefore further evolution of the diversity could not take place.

A good number of cultivars moved from one country to another thus at most of the places diversity was not naturally created because grafting was used for multiplication. In general, grouping of the diversity by molecular studies was based on the grouping of country of origin. Phenotypic and genotypic analysis of the diversity has clearly indicated that the place of their origin is related with the group to which they belong. Analysis of the primary and secondary diversity based on molecular tools has also revealed almost similar grouping results.

Collections of mango cultivars have been named as Indian, Floridian, Thai, Pakistani, Filipino, Spanish, Colombian, etc. by researchers working on diversity analysis by molecular markers. High variability has been reported in the Thai accessions (Eiadthong et al. 2000). Most of the Thai cultivars have polyembryony, thus it seems many of these varieties have given rise to zygotic seedling variants, different from mother genotype, widening genetic variability. Genetic diversity in cultivated mango cultivars in Pakistan is similar to North Indian gene pool as most of the varieties were from India and being commercially cultivated.

India, being primary centre of diversity, a lot of variation exists among regionally established varieties. Rajan et al. (2013) reported similarity in climatic suitability of north and east Indian varieties. Most of the Pakistan cultivars are from north and eastern mango gene pool. Being a part of India and free exchange of germplasm, Pakistan mango industry is based on north and east Indian gene pool. At molecular level also the similar results have been reported (Ahmad et al. 2008).

4.2.2 Wild Relatives of *M. indica*

Changing consumer preference and climate change has confounded the need of developing new mango varieties through the introduction of novel existing genes from diverse available sources. Germplasm collections in the field genebank and wild relatives are the major gene sources for developing the disease-resistant and locally adapted varieties with several desirable characteristics. During beginning of the twentieth century, it has been realized that the wild relatives can serve as an important genetic resources in building breeding programs (Vavilov 1926; Tanksleyj and McCouch 1997; Meilleur and Hodgkin 2004). Compared to domesticated germplasm, crop wild relatives often exhibit better resistance to biotic and abiotic stress (Hajjar and Hodgkin 2007). Although advantageous traits from crop wild relatives have been used to improve major crops for more than eight decades, the genetic diversity of mango wild relatives remains unutilized.

In addition to the abundance in Malaysia's peninsular area (Kostermans and Bompard 1993), *Mangifera* genus comprising approximately 69 species are mainly found on the islands of Borneo, Java, and Sumatra. *M. khasiana* Pierre, *M. sylvatica* Roxb, *M. camptosperma* Pierre, *M. andamanica* King were recorded in various parts of northeastern India sub-Himalayan tracts. *M. sylvatica* is known to occur in East Sikkim, in West Bengal and in the islands of Meghalaya, Assam, and the Andaman (Mukherjee 1949b; Agharkar and Roy 1951; Yadav and Rajan 1993). Wild *Mangifera* gene pool existing in natural habitat has been well documented by Mukherjee (1985) and Kostermans and Bompard (1993). Information on their occurrence, distribution, and utilization in Asia has been accounted by Iyer (1991), Ram and Rajan (2003); Rajan and Hudedamani (2019). Most of the *Mangifera* species are useful as sources of gene(s) for one or another desirable trait. These species can be utilized in crop improvement program or as rootstocks through proper exploitation of their special attributes.

Mangifera species useful for crop improvement program fall into two categories; 1) species having edible fruits with other desirable characters, and 2) species that may not have edible fruits but serve as gene source for improvement of specific trait like resistance to anthracnose, tree decline, floral and vegetative malformation, powdery mildew, hoppers, gall midge, etc. (Iyer 1991). The role of wild relatives of mango is more important with respect to trait-specific breeding compared to varietal diversity present in *Mangifera indica*. Kosterman and Bompard (1993) reported 69 *Mangifera* species with maximum species diversity available in peninsular Malasyia. In India, six *Mangifera* species are available, of which few are under threat of extinction, viz., *M. griffithii, M. camptosperma,* and *M. zeylanica.* Similarly, the diversity within *Mangifera indica* in terms of seedling origin varieties has also been eroded due to extensive infrastructural development across the India territory. *Mangifera* species and wild types have tremendous value in trait-specific mango breeding (Table 4.1).

4.2.3 Use of Wild Relatives

Under natural habitat, wild relatives of *M. indica* are confined to the countries where extensive breeding program is presently not operating. The introductions of these wild relatives to the field gene banks in Florida, Australia, and India have been made with limited utilization for breeding. Iyer (1991) detailed several *Mangifera* wild species producing edible fruits, viz., *M. altissima, M. longipes, M. caesia, M. macrocarpa, M. cochinchinensis, M. odorata, M. foedita, M. pajang, M. griffithii, M. pentandra, M. indica, M. sylvatica, M. lagenifera,* and *M. zeylanica.* A good number of wild relatives have been identified with useful traits. *M. caesia* can be utilized for its fragrance; *M. gedebe, M. inocarpoides,* and *M. decandra* can be exploited as rootstock for waterlogged conditions; *M. pajang* can be looked for varieties with peelable fruits, whereas, *M. similis* has free stone mangoes (Iyer 1991, Kostermans and Bompard 1999, Ram and Rajan 2003).

Several of the wild relatives in southeast Asian countries are valued for their aromatic fruits utilized for making various dishes. These wild relatives are also rich in bioactive compounds like mangiferin and lupeol which are considered important for human health. Clinical trials have demonstrated various health benefits of this native compound of *Mangifera* (Tangah et al. 2017). Wild species are also used for their medicinal value. Bark of *M. zeylanica,* a native species to Sri Lanka, is being used in traditional medicine for the treatment of cancer. Cytotoxicity to the cancer cell lines has been reported (Ediriweera et al. 2016). In some parts of the southeast Asia, due to high rainfall, *M. indica* cannot grow successfully; however, *M. odorata* is being grown successfully in these areas. *M. casturi* Kosterm, a common species in South Kalimantan and East and Central Kalimantan, with typical dark-colored fruits, deep orange pulp, unique flavor, aroma and very tasty is also suited to ever wet climates (Ram and Rajan 2003).

Not many attempts have been made to know that wild species of the *Mangifera* genus can be easily crossed with cultivars of *M. indica* for the transfer of desirable alleles. So far, only limited numbers of possible interspecific crosses were attempted. Initiatives at Florida to produce new hybrids with reduced susceptibility to disease using wild relative's, viz., *M. lalijiwa, M. quadrifida, M. rubropetala,* with *M. indica* were successful. New hybrids between *Mangifera* species and *M. indica* can provide the first step toward these goals (Ledesma et al. 2017). *M. laurina* was used as the pollen parent in a breeding program with a *M. indica* hybrid in Australia resulting 60 successful hybrids (Bally et al. 2013).

4.2.4 *Mangifera* Species as Rootstock

Mangifera species have also been used as rootstock. There are reports of grafting experiments between *M. indica* and *M. foetida, M. odorata,*

Table 4.1 *Mangifera* species having direct bearing on mango improvement

S. No.	Species	Special Characteristics	Source
1	*M. caesia*	Milky white soft pulp having sweet taste and fragrance	East Borneo, Karangan Reserves
2	*M. decandra*	Tolerant to water-logged conditions and can be used as rootstock	Sumatra, Borneo Pakambaru (Indonesia)
3	*M. gedebe*	Tolerant to water-logged conditions and can be used as rootstock	Borneo, Pegu, Myanmar, Thailand, Indo-China, Thailand
4	*M. indica var. mekongensis*	Fruits twice a year	Saigon, Vietnam
5	*M. inocarpoides*	Tolerant to water-logged conditions and can be used as rootstock	Papua, Morehead
6	*M. paiang*	Fruits can be peeled like banana	Borneo and Kapit Sarawak (Malaysia)
7	*M. similis*	Free stone fruits	Kalimanthan (Borneo)
8	*M. mangifica*	Fibreless	Western Malaysia
9	*M. rufocostat* and *M. swintonioides*	Off season bearing habit	Sumatra, Borneo, and Malay Peninsula
10	*M. foetida*	Sweeter, less fibrous pulp	Sumatra, Kalimantan (Borneo) and Java
11	*M. odorata*	Fragrance, firmness, ability to grow under high rainfall, excessive humidity and sour-sweet taste	Philippines, Peninsular Thailand, Western Malaysia, Vietnam
12	*M. casturi*	Prolific bearer with small, black, and sweet fruits	South Kalimanta (Borneo)
13	*M. altissima*	Tolerant to insects (hoppers, tip and seed borers)	Solomon Islands, New Britain, New Guinea, Moluccas, Sulawesi, Lesser Sunda Islands and the Philippines
14	*M. pentandra*	Prolific bearer due to high proportion of hermaphrodite flowers	Northern Malay Peninsula
15	*M. laurina*	Immune to anthracnose	West Java and tropical Malaysia
16	*M. orophila*	Restricted to mountain forests and known as mountain species	Malay Peninsula and Sumatra
17	*M. dongnaiensis*	Restricted to mountain forests and also known as mango of mountain	South Vietnam

Source Mukherjee (1985), Angeles (1991), Iyer (1991), Bompard (2009), Usman et al. (2001)

M. kemanga, *M. caesia* in Java (Ochse and Bakhizen 1931), *M. odorata* on *M. indica* in Philippines (Wester 1920), *M. indica* on *M. zeylanica* in Sri Lanka (Gunaratman 1946). Similarly, *M. indica* 'Madu' in Java and *M. laurina* in Sabah, Borneo have been used as rootstocks for *M. casturi*. Grafting of *M. caesia* on *M. indica* (Wester 1920) and *M. indica* on *M. kemanga* or *M. caesia* (Ochse and Bakhuizen 1931) were unsuccessful. This may be due to the fact that these two species have distinct bark growth and have remote affinity with common mango. From different experiments, it has been postulated that closely related species of *M. indica* within a subgenus *Mangifera* have better scion and roostock compatibility. In the inundated riverbanks of West Kalimantan (Borneo), *M. laurina* is occasionally used as rootstock for *M. indica,* and this has been tried by the Department of Agriculture in Sabah (Lamb

1987). Grafts of *M. casturi, M. griffithii, M. laurina, M. odorata, M. pentandra,* and *M. Zeylanica* on *M. indica* showed high success percentage (Campbell 2004). There are several species of *Mangifera* growing well permanently inundated areas, viz., *M. decandra, M. gedebe, M. quadrifida, M. griffithii, M.gracilipes, M. microphylla, M. paludosa, M. parvifolia* can serve as potential rootstocks of mango on poorly drained soils or areas suffering from prolonged flood (Bompard 2009). *Mangifera* species like *M. orophilla* and *M. dongnainsis* may be used in future as mango rootstocks for sub-mountainous areas.

4.2.5 *Mangifera* Species for Pest and Diseases Resistance

The most devastating disease which causes huge economic loss to mango production is anthracnose. *M. laurina* a species having subglabrous and laxly flowered panicles is well adapted to area with perpetual wet climates and found to be immune to anthracnose disease and can be used for imparting resistance in future bred varieties (Bompard 1993). Sharma and Chaudhry (1976) reported that wild type trees from Tripura state of India were free from mango malformation. *M. altissima* is another species which has resistance/tolerance to major insects and pests of mango like mango hopper, tip, seed borers, etc. There is an urgent need to screen the *Mangifera* species for their reaction against major diseases and insect pests.

4.2.6 Extension of Mango Cultivation in Non-traditional Areas

Wild mango species reported from Malaysia and Vietnam like *M.orophila* and *M. dongnaiensis* are restricted to mountain forests at 1,000 to 1,700 m above mean sea level. These species could be explored to extend the possibility of mango cultivation in temperate regions (Iyer and Schnell 2009). Similarly, there are several species of *Mangifera* that can be grown in permanently inundated areas, viz., *M. gedebe, M. quadrifida, M. griffithii*. Other species of the section *Rawa*, viz., *M.gracilipes, M. microphylla, M. paludosa, M. parvifolia* can serve as potential rootstocks of mango on poorly drained soils or in areas liable to prolonged flood (Bompard 2009).

4.3 Varieties and Cultivars

Mango is very rich in terms of varietal diversity. There are more than a thousand varieties of mango growing across the world propagated either through seedling as well as vegetative (Ram and Rajan 2003). However, most of them have originated as a chance seedling and maintained asexually (Singh 1960; Ram and Rajan 2003). Due to high level of heterozygosity in mango, each and every seedling is unique in its fruit characteristics. Use of mango seedlings for planting new orchards was continued till the middle of twenty-first century. Although later on, at many places new orchards where planted using grafts, even then the seedling orchards were available in India. Multiplication through seedling resulted in a large number of varieties in India, Florida, and Southeast Asia. Cultivation for a long time over an extended area, reflecting different selection criteria and genetic responses to varied environmental influences are responsible for high degree of diversity represented by mango varieties. Based on the mode of propagation, *M. indica* germplasm can be classified into two groups.

4.3.1 Seedling Races (Wild and Cultivated)

This group includes plants derived from both monoembryonic and polyembryonic mango stones. Almost all the mango cultivars are the result of selection from these seedlings. As compared to cultivars this group exhibits wide variability with regard to fruit and tree characters. In India and other parts of the world, the seedlings of monoembryonic cultivars are not much cared, but in places where the polyembryonic varieties are commercially grown, seedlings are important. A number of polyembryonic varieties have been reported in mango (Singh 1960) mostly from coastal areas. No commercial variety in North India is polyembryonic (Ram and Rajan 2003). In about 20 polyembryonic varieties (indigenous and exotic) available, except Kensington, most of them have inferior fruit quality (Yadav and Rajan 1993) with some exceptions of southeast Asian varieties.

4.3.2 Horticultural Varieties (Vegetatively Propagated) and Cultivars

The varieties of *M. indica*, which produce the best quality fruits, are mostly of Indian origin. About a thousand monoembryonic varieties occur in India and those from other regions of the world are not more than 50 (Mukherjee 1949a). Significant clonal variations have been observed in the commercial varieties being grown in different tracts. In the same variety, there are clones with very big sized fruits whereas small sized fruits are also not unusual. Attempts have been made to collect the variability in these clones (Rao and Mukherjee 1982; Singh et al. 1986; Chadha 1989).

Large-scale seedling population of mango gave rise to several outstanding types but due to lack of grafting method till the early twentieth century most of the promising seedlings could not be asexually multiplied. Promising seedling varieties were limited to the single tree and thus could not attain the commercial status. Grafting enabled Indians to multiply outstanding seedlings for planting orchards. The process gave rise to multiplication of hundreds of varieties through grafting. Vegetative propagation helped in establishing seedling varieties as a cultivar.

In addition to various seedling varieties, over one thousand vegetatively multiplied mango cultivars have been registered. Most of these were selected as chance seedlings and maintained asexually afterward. Most of such cultivars come from India, and numbers are limited to other parts of the world. There are approximately 30 commercial cultivars in India. Most of them are of limited adaptability and have a regional preference for production, performance, and consumers (Yadav and Rajan 1993). In different regions of India, however, the situation is slowly replaced by the standard cultivars ensuring higher returns. Mango varieties are categorized into broad groups based on their place of origin; the Indian mango varieties and Indo-Chinese group which originated in south East Asia (Evans et al. 2017). However, another group of mango varieties having their origin from Florida, USA are commercially cultivated in central and south America, Europe, and Mexico. The present day Floridian mango varieties have been originated as open pollinated seedling of south Indian variety Mulgoa and Sandersha (*syn.* Totapuri). Important commercial mango varieties from different countries are summerized in Table 4.2.

The commercial cultivars in Bangladesh are Gopal Bhog, Khirsapat, Himsagar, Langra, Fazli, Mohon Bhog, Kohinoor, and Kua Pahari. Most of these were selected from superior chance seedlings in India and subsequently adapted in Bangladesh. In China, about 130 cultivars are available and originated from the following three sources: (i) Traditional local cultivars (ii) introductions from southeast Asia, south Asia, Africa, and Australia, and (iii) Novel germplasm selected by various mango producing provinces/regions in China (Ram and Rajan 2003). Almost all local mango cultivars in Philippines have polyembryonic seeds which enable them to preserve their phenotypic identities for generations. About eight distinct varieties have been recognized: 'Carabao', 'Pico', 'Pahutan', Mampalang, 'Digos', 'Binoboy', 'Dudul', and 'Senora'. Numerous mango varieties have been introduced from foreign countries. However, except for two varieties which are better consumed at the green stage, none has surpassed the eating qualities of the 'Carabao' and 'Pico' mangoes (Coronel 1996).

In Sri Lanka, both monoembryonic and polyembryonic cultivars are grown. Cultivars Willard, Pandithasekera, Neelam, Chembatan, Walamba, and Ambalavi are among

Table 4.2 Most important mango cultivars in major producing countries

Country	Cultivars
Australia	'Kensington Pride', 'Banana', 'Earlygold', 'Glenn', 'Haden', 'Irwin', 'Keitt', 'Kent', 'Zill'
Bangladesh	'Aswina', 'Fazli', 'Gopal Bhog', 'Himsagar','Khirsapati', 'Langra', 'Kishan Bhog', 'Kohinoor', 'Kua Pahari', 'Mohan Bhog'
Brazil	'Adams', 'Embrapa 141', 'Ametista', 'Apple', 'Ataulfo', 'Augustus', 'Bourbon', 'Brooks', 'Davis Haden', 'Delicioso', 'Espada', 'Extrema', 'Fernandeza', 'Florigon', 'Glenn', 'Haden', 'Iturbae', 'Kent', 'Palmer', 'Rosa', 'Santa Alexandrina', 'Santa Cruz', 'Sensation', 'Smith', 'Tommy Atkins', 'Vitoria', 'Zill'
China	'Baiyu', 'Guixiang', 'Huangpi', 'Huangyu', 'Macheco', 'Sannian', 'Yuexi No. 1'
Costa Rica	'Haden', 'Irwin', 'Keitt', 'Mora', 'Tommy Atkins'
Columbia	'Albania', 'Bocado de Reina', 'Cauca', 'Chancleto', 'Haden', 'Irwin', 'Keitt', 'Kent', 'Lorito', 'Mango de Azúcar', 'Mango Manzanita', 'Magdalena River', 'Mariquita', 'Palmer', 'Tommy Atkins', 'Van Dyke', 'Vellenato', 'Zill'
Cuba	'Bizcochuelo', 'Cecil', 'Delicioso', 'Filipino', 'Haden', 'Huevo de', 'La Paz', 'Kent', 'Macho', 'Manga Amarilla', 'Manga Blanca', 'San Diego', 'Smith', 'Toro'
Haiti	'Francine', 'Madame Francis', 'Francis', 'Julie'
India	'Alphonso', 'Banganapalli', 'Bombay', 'Bombay Green', 'Chausa', 'Dashehari', 'Rataul', 'Fazli', 'Fernandian', 'Himsagar', 'Kesar', 'Kishen Bhog', 'Langra', 'Mallika', 'Mankurad, 'Mulgoa', 'Neelum', 'Pairi', 'Suvarnarekha', 'Totapuri', 'Vanraj', 'Zardalu', 'Amrapali', 'Bangalora', 'Gulabkhas'
Indonesia	'Arumanis', 'Dodol', 'Gedong', 'Golek', 'Madu', 'Manalagi', 'Cengkir', 'Wangi'
Israel	'Haden','Tommy Atkins','Keitt', 'Maya', 'Nimrod', 'Kent', 'Palmer'
Kenya	'Apate', 'Apple', 'Batwawi', 'Boribo', 'Boubo', 'Dodo', 'Gleen', 'Haden' 'Irwin', 'Kent', 'Kimaji', 'Kisikio Punda', 'Ngowe', 'Zill'
Malaysia	'Arumanis', 'Kuala Selangor 2', 'Golek', 'Apple Rumani', 'Malgoa', 'Apple Mango', 'Maha-65', 'Tok Boon'
Mexico	'Agua', 'Ajo', 'Alcanfor', 'Alcanforado', 'Amate', 'Amatillo', 'Ataulfo', 'Canela', 'Chico', 'Coche', 'Cuero', 'Haden', 'Irwin', 'Jocote o Cachetico', 'Keitt', 'Kent', 'Manilila', 'Manilon', 'Manzana', 'Manzanillo', 'Melon', 'Ora', 'Palmer', 'Papaya', 'Pepino', 'Peter', 'Pija', 'Pina', 'Platano', 'Pomarrosa', 'Sensation', 'Sin nombre 1', 'Sin nombre 2', 'Sin nombre 3', 'Tapanero', 'Tecolote', 'Tommy Atkins', 'Trinadad', 'Van Dyke' and 'Viejita'
Peurto Rico	'Amini', 'Colombo Kidney', 'Divine', 'Blanco', 'Haden', 'Mangotino', 'Mayaguezano' and 'Tete Nene'
Myanmar	'Aug Din', 'Ma Chit Su', 'Sein Ta Lone', 'Shwe Hin Tha'
Pakistan	'Anwar Ratol', 'Chausa', 'Dashehari', 'Gulab Khas', 'Langra', 'Siroli', 'Sindhri', 'Suvarnarekha', 'Zafran', 'Baganapalli'
Philippines	'Carabao', 'Manila Super', 'Pico', 'Binoboy', 'Carabao', 'Dudul', 'Pahutan', 'Senora'
Singapore	'Apple Mango', 'Arumanis', 'Golek', 'Kaem Yao', 'Mangga Dadol'
South Africa	'Brooks', 'Chené', 'Crimson Pride', 'Fascell', 'Glenco', 'Heidi', 'Joa', 'Keitt', 'Kent', 'Manzanillo', 'Neldica', 'Osteen', 'Peach', 'President', 'Saber' 'Sensation', 'Shelly', 'Smith Seedling', 'Tommy Atkins', 'Zill'
Spain	'Osteen', 'Tommy Atkins', 'Sensation', 'Glenn', 'Gomera-1', 'Gomera-3', 'Palmer', 'Lippens', 'Irwin', and 'Valencia', 'Pride' and 'Kensington Pride'
Sri Lanka	Karutha Colomban, Willard, Vellai Colomban, Petti amba, Malwana amba, Parrot Mango and Peter Pasand, Dapara, Hingurakgoda
Taiwan	'Jin-hwung', 'Keitt', 'Tommy Atkins', 'Tainong No. 1' and 'Tsar Swain'

(continued)

Table 4.2 (continued)

Country	Cultivars
Thailand	'Nam Doc Mai', 'Ngar Charn', 'Okrong', 'Rad', 'Choke Anand', 'Kao Keaw', 'Keow Savoey', 'Pimsenmum'
United States of America	'Adams', 'Anderson', 'Apple', 'Davis Haden', 'Dixon', 'Earlygold', 'Edward', 'Eldon', 'Fairchild', 'Florigon', 'Ford', 'Gibbons', 'Glenn', 'Haden', 'Irwin', 'Julie', 'Keitt', 'Kent', 'Lily', 'Lippens', 'Osteen', 'Palmer', 'Simmonds', 'Smith', 'Springfels', 'Tommy Atkins', 'Zill'
Vietnam	'Combodiana'

Source Malo (1970); Pandey (1986); Ram and Rajan (2003); Human and Rheeder (2004); Durán-Zuazo et al. (2004); Mukherjee and Litz (2009), Njuguna et al. (2009); Human et al. (2009); Galvez-Lopez et al. (2010); Pleguezuelo et al. (2012); Rajan and Hudedamani (2019); http://*Mangifera*.res.in/varieties.php

monoembryonic ones, whereas Peterpasand, Kohuamba, Karutha Colomban, Dampara, Parrot and Vellaicolomban are polyembryonic (Ram and Rajan 2003). Medagoda and Herath (1984) reported that Petti Amba also produces polyembryonic seed and the highest incidence of polyembryony is shown in Parrot mango. Similar to India, different cultivars do not perform in the same way in different parts of the island. Varieties for different agro-ecological regions are identified and the recommendation of varieties for the dry zone, intermediate zone, and wet zone are different.

In China, at present the germplasm which have become the principal cultivated varieties in the provinces/regions in South China include Huangpi, Wupi, Neelum, Okrong, Nam Doc Mai, Dashehari, Carabao, Baiyu, Dabaiyu, Liuxiang, Macheso, Sannian, Zihua, Guixiang, Chuanmang, Yexi No.1, Guire No.10, Irwin, Keitt, Haden, Zill, Sensation, Caixuan, Tainong Nos.1 and 2, Chun Hwang, Honghua, etc. Varieties Huangpi, Taoye, Honghua, and Sannian have been widely used as rootstocks. Some germplasms have been used as breeding material. For example, crossing of Neelum with Golek at Guangxi University of Agriculture has resulted in novel germplasm, such as Guixiang, Lupi, and Xiaotian. The SCCTC has also screened from Neelum seedling tree a new selection Longjingda, which yields very large fruit.

Introductions of polyembryonic varieties from Southeast Asia (Philippines, Cambodia), India, and other sources created an important secondary centre of diversity of *M. indica* in Florida. 'Haden' a seedling of Indian variety 'Mulgoba' gave rise to varieties like 'Glenn', 'Lippens', 'Osteen', 'Parvin', 'Smith', 'Springfels', 'Tommy Atkins', and 'Zill'. 'Saigon' seedling selections made from 'Cambodiana', and introductions from Indo–China resulted into the cultivars like 'Alice', 'Herman', and 'Florigon'. Compared to outstanding Indian cultivars, the Florida mango cultivars are highly adaptable to many agro-ecological regions and are regular bearer. These cultivars are highly attractive and develop red blush at maturity, firm flesh, a high flesh to seed ratio and regular bearing habit are reasons for their increasing popularity. Florida cultivars, *e.g.*, 'Tommy Atkins', 'Keitt', etc. are also moderately resistant to anthracnose, an added advantage during prolonged storage and export (Knight 1993). In many countries, now cultivation of Florida cultivars is well adopted, providing the base for international mango trading.

Therefore, only about six dozens cultivars are commercially grown in the world, of which the export quality commercial cultivars are Alphonso, Dashehari, Banganpalli, Chausa, and Kesar of India; Bourbon, Carlota, and Haden of Brazil; Mabrouka of Egypt, Madame Frances of Haiti; Haden and Maya of Israel; Apple, Borino, and Ngowe of Kenya; Haden, Keittoand Manila of Mexico; Anwar Rataul, Samerbehist Chowsa and Fazli of Pakistan; Carabao of the Philippines; Haden, Keitt and Zill of South Africa; Hindi Be Sennar and Kitchener of Sudan; nam Doc Mai and Okrong of Thailand; Julie of Trinidad and

Venezuela (Pandey and Dinesh 2010; Rajan et al. 2020).

In India, the same variety is grown in different regions and fruits produced in geographical regions are recognized for quality due to specific climate of the location. Alphonso of Konkan region; Kesar from Gujarat; Banganpalli of Vishakhapatnam region in Andhra Pradesh; Dashehari of Malihabad, Mall, and Kakori; Appemidi of Shimoga and Chikmanglur distirict; Kesar in Marathwada; Zardalu in Bhagalpur; Khirsapati (Himsagar), Laxman Bhog, Fazil of Malda have been issued Geographical Indication (Sharma and Rajan 2018).

4.3.2.1 Classification and Characterization of Varieties

Maries (1902) made attempts to collect and describe Indian mango varieties scientifically on the basis of fruit characters in the beginning of the twentieth century. Other early attempt in Puerto Rico and Hawaii were made to classify the varieties by Collins (1903) and Higgins (1906), but no definite system of classification could be evolved. Later, varieties of Bhagalpur (Bihar) were described by Woodhouse (1909) following system of classification mainly based on fruit characters. The varieties were classified in three broad groups: (i) round fruit, (ii) Bombay type, and (iii) long oval or rectangular fruits, longer than broad. In Florida, Rolfs (1915) classified the mango varieties into nine main groups based on the fruit and stone characters and growth habit of the tree. The varieties were grouped into groups: (i) Turpentine, (ii) Elcanor group, (iii) Pineapple, (iv) No.-11 group, (v) Bombay group, (vi) Cambodiana group, (vii) Sandersha group, (viii) Mulgoba group, (ix) Gola group. Wester (1920) suggested making the system of classification should be comprehensive and evolving the mango varieties not only from India but also from the other mango-growing regions. Agreeing with this view, Wester (1920) described important mango varieties of Philippines and a descriptive list of Indian mango varieties was also given. Based on the system of Woodhouse, 89 varieties of Bombay state were classified by Burns and Prayag (1921) into three major classes on their fruit shape basis as round-fruited, long-fruited, and indefinite.

Sturrock (1951) developed an artificial key based on fruit characters for the identification of mango varieties grown in Florida. Hartless (1913) for the first time used floral characters as criteria for classification. Popenoe (1920) classified 300 varieties, based on the descriptor list of the varieties from all parts of the world on the basis of fruit characters, coloration of panicle axis, laterals and pubescence of the panicle branches, and the number of embryo in the seed.

Naik and Gangolly (1950) described 335 south Indian varieties, based on system which seems to be natural (Singh 1960). Besides fruit characters, other characters were also given importance by Naik and Gangolly (1950). On the basis of primary characters mentioned above, south Indian varieties are grouped in main groups. Secondary characters were used to distinguish sub-group of classes. Tertiary characters are used to distinguish varieties within each classes or sub-groups (Naik and Gangolly 1950). On the basis of fruit shape, varieties are categorized into six main groups or cohorts. Considering leaf tip and folding of leaf, each main group is further divided into four classes giving rise to 24 classes.

Majority of the attempts made to classify mango variety show that the fruit shape is the most important and stable character for distinguishing varieties from each other. Various fruit shapes of mango have been described by Naik and Gangolly (1950) and Gangolly et al. (1957). The size of fruit is also of relative importance, fruit dimensions and weight are helpful in distinguishing varieties. Among the fruit characters, the presence and form of beak is of prime importance for the classification of mango varieties. Wide variation in the form of beak is present. It may be absent in some varieties and prominent with various shapes of the fruit. This character is genotype specific and shows least variation under various environments. Apart from the shape of the fruit, characters like fruit base, shoulder, sinus, apex, and cavity of stalk insertion are important characters. Variation in these characters have been depicted in monographs of Naik and Gangolly (1950) and Gangolly et al. (1957).

Illoh and Olorode (1990) used multivariate methods for the grouping of varieties grown in Nigeria. On the basis of flower and fruit characters, they grouped the varieties in four groups. Rajan et al. (1998 and 1999b) used hierarchical approach for analyzing diversity in canopy characteristics of mango and classified important varieties based on leaf component and diffuse non-interceptance values of the tree canopy. Rajan et al. (1999c) applied multivariate analysis of growth and yield performance of Dashehari scion on 24 varieties as rootstock.

During last three decades, with the advancement in molecular techniques, a good number of studies have been undertaken for characterizing varieties. Initial studies on isozyme banding pattern (Degani et al. 1990), random amplified polymorphic DNA (RAPD) analysis (Schnell et al. 1995), and DNA finger print information (Adato et al. 1995) have been used for grouping the cultivars and estimating gene diversity in different groups of the cultivars. Use of molecular markers as tools for analyzing the diversity of mango has been well documented.

Using group constellation, cultivars were grouped into three distinct clusters for selecting promising parents (Rajan et al. 2009). Mango cultivars can be categorized into several groups on the basis of their suitability for different usage. The group may be as table and sucking types. Majority of the cultivar contain none to scanty fiber and firm flesh make them suitable for table purpose, whereas varieties with high juice content and fiber are used for sucking purpose. Some of the varieties are excellent sucking types with high juice content but sucking types are mostly limited to India and Bangladesh (Ram and Rajan 2003).

4.4 Germplasm Collection and Introduction

Exploration and collection of mango germplasm in India started in the sixteenth century (Singh 1960). The search for better fruit quality mangoes was the prime objective of these collections. Scientific approach for the collection of mango germplasm was made after the establishment of Fruit Research Stations at Sabour, Kodur, Saharanpur, Sangareddy, and Vengurla. With the establishment of Indian Institute of Horticultural Research, Bangalore, efforts were made to collect mango varieties of south and west India (Chadha 1996). At the Central Institute Subtropical Horticulture, Lucknow, a concerted effort was made to have a National collection of mango germplasm representing germplasm from different parts of the country.

Majority of the attempts made for the survey and collection of mango germplasm are limited to in situ evaluation for various characters with a special emphasis on fruit quality. However, several reports on the survey and collection of germplasm attempted in different parts of the country are available (Patwardhan 1918; Burns and Prayag 1921; Mukherjee 1948, 1949a, b; Gangolly and Singh 1950; Mukherjee 1950a, b; Naik and Gangolly 1950; Singh and Singh 1957; Teotia and Srivastava 1961; Singh and Jawanda 1962; Teaotia and Singh 1963; Kashyap and Jyotshi 1969; Teaotia and Singh 1971; Dhaki 1972; Saha 1972; Bose et al. 1974; Sadhu and Bose 1976; Sharma and Choudhary 1976; Das and Hota 1977; Hoda and Mandal 1979; Kulkarni 1979; Mukherjee et al. 1983; Tyagi 1986; Ramaswamy 1988; Yadav 1989; Gupta et al. 1993; Lal and Gupta 1993; Choudhari et al. 1993; Singh et al. 1993; Thimmaraju 1993. Mukherjee (1948) collected and described 77 varieties of mango from Murshidabad. Saha (1972) and Mukherjee et al. (1983) conducted surveys for the collection of superior types in West Bengal.

Very few systematic attempts have been made to collect the large variability available in Bihar. Das and Hota (1977) and Parida and Rao (1989) studied diversity in mango gene pool in Odisha. From the collection of 500 varieties from different districts of the state, 35 varieties were selected for desirable characters. Expeditions were also made by Sharma and Chaudhary (1976), Yadav (1989), Singh et al. (1993) to study the variability in mango in North eastern region. Singh and

Jawanda (1962) studied the seedling populations of mango in Punjab for selection of superior types. One-hundred-fifty mango varieties of Bihar and Uttar Pradesh were collected and described by Singh and Singh (1956). Teotia and Singh (1963, 1971) and Teaotia and Srivastava (1961) collected and studied the variability in mangoes of eastern Uttar Pradesh. Mangoes of south and central India have also attracted the attention for collection. Burn and Prayag (1921) collected the varieties of mango and described them. Vanraj and Kesar were selected in Gujarat and the later Kesar has become an important commercial variety in many mango-growing areas (Chadha 1996).

In Bangladesh, mango germplasm collection started in 1946 and maintained at the Central Horticulture Station, Dhaka. At Regional Horticulture Research Station, Nawabgonj, collection of known cultivars started in 1985. Many of the superior clones/seedlings identified in the National Mango Shows were collected and planted (Hossain 1996).

In Hainan (China), during 1959 to 1967 surveys for mango genetic resources, collection, and introduction of exotic germplasm were made. The South China Academy of Tropical Crops (now CATAS) and other institutions or departments also introduced germplasm subsequently from other countries through various channels. Varieties introduced in between 1963 and 1980 from Cuba, Indonesia, Sri Lanka, and Thailand included Haden, Maden, Mecico, Emperado, Arumanis, Golek, Okrong, Nam Doc Mai, Keaw Saweuy, Chaokun thip, Pemsenmun, No.811, No. 814, etc. In Taiwan, five varieties including Haden, Irwin, Zill, Keitt, and Kent were introduced from the USA. Over 30 varieties from the mango producing countries, 12 varieties from Costa Rica were collected. Nearly 100 genotypes have been collected in Taiwan (Zhen 1989; Shu et al. 2000).

In Indonesia, efforts were made for the collection of genetic resources of mango at Cukurgondang and 224 accessions have been collected. In 1987, Purnomo collected 19 *Mangifera* species from South Kalimantan and East Kalimantan (Kusumo 1996).

Mango varieties from foreign countries were perhaps first introduced into the Philippines during prehistoric times. Recorded introductions were made in early 1900s when numerous varieties were introduced from India. One of these Indian varieties, the 'Katchamitha', gained acceptance by the Filipinos. During Second World War, several collections were destroyed. However, in the mid 1970s, the Institute of Plant Breeding in U.P. Los Banos was able to retrieve a few of these mango varieties, not from Lamao but from other nearby sources. The college of Agriculture of U.P. Los Banos collected mango varieties from Hawaii and planted these around the university campus (Coronel 1996).

In Thailand, Ministry of Agriculture and Crop Experiment Station, Sisaket Horticultural Research Centre, Bangkok; Noi, Horticultural Research Station, Chachoengsao; Rubber Research Centre, Surat; Thani Horticultural Research Centre, and Tachai Horticultural Research Station have made collections (Vangnai 1996). In Sri Lanka, Plant Genetic Resources Centre (PGRC) specially established for this purpose in 1989 leads the collection program (Medagoda and Jayawardena 1997).

4.5 Characterization

Morphological characterization is the easiest and simple characterization process that allows study of plant variation using visual attributes. The morphological characters mainly help in identifying elite mango varieties as well as superior donor parents for different horticultural traits. The fruit characters combined with several other traits were used in the classification of varieties (Burns and Prayag 1921). Even primary, secondary, and tertiary characters were utilized for characterization (Singh and Singh 1956; Pandey 1984; Rajan et al. 1999a, b). The fruit characters were used for grouping the cultivars based on their performance and their region of adaption (Rajan et al. 2013a, b).

Before the development of the normative DUS evaluation guide by a Task Force appointed by the PPV&FR authority (PPV&FRA 2008),

IPGRI descriptors were used for visual characterization of mango (IPGRI 2006; Pinto et al. 2006). The biochemical markers (Subedi et al. 2004), morphological and molecular techniques combined (Mussane et al. 2011; Ramessur and Ranghoo-Sanmukhiya 2011) were used for confirming the phylogenetic relationship and geographic distribution of different *Mangifera* sp. (Eiadthong et al. 1999).

DNA markers have been used for characterization of diversity in *M. indica* and wild *Mangifera* species (Eiadthong et al. 1999; Gonzalez et al. 2002; Weerarathne et al. 2005; Bajpai et al. 2008; Shamili et al. 2012; Dillion et al. 2013; Dinesh et al. 2015). Adato et al. (1995) used variable number tandem repeats (VNTRs) markers for DNA finger printing and genetic analysis of mango genotypes and hybrids. Various PCR approaches were used for the study of genetic variation, parentage, and grouping (Srivastava et al. 2007). Studies on morphology and molecular characterization indicate almost similar pattern in grouping (Dinesh et al. 2015).

The use of a good number of sequence-based molecular tools started with the advent of the sequencing techniques and are also used to classify mango germplasm for different objectives. The isolation of novel ripening cDNA clones was based on gene cloning (Lycett et al. 1997), while mRNA isolation and characterization were used during mixing of the mango to classify differently expressed genes (Saiprasad et al. 2004). Latest studies on molecular characterization focus mainly on identification of the candidate fruit color and ripening genes, genes active in biosyntheses of phenylpropanoid/anthocyanin, and specific isoforms (CISH 2015). The Mango Genome already published would be a valuable source for mango genomics workers (Singh et al. 2016; Li et al. 2020; Wang et al. 2020).

Evaluation of mango cultivars in India for different characters has been reported by several workers: Juicy cultivars of south India by Kulkarni (1979); sucking varieties of Uttar Pradesh and Bihar (Teaotia and Singh 1963; Singh 1972; Gupta et al. 1993); raw sweet and pickling varieties (Kulkarni 1979); dwarfing (Yadav and Rajan 1993); leaf area index (Rajan et al. 1999b); fruit color (Iyer and Subramanayam 1991); pulp content (Chadha 1996); TSS (Yadav et al.1987; Rameshwar et al. 1979); time of ripening (Iyer and Subramaniam 1986), regularity in bearing (Anon 1979); malformation (Prasad et al. 1965; Mishra 2004) have been studied.

Germplasm has been evaluated on the basis of anatomical and physiological parameters to correlate them with the tree vigour (Mukherjee and Das 1976; Srivastava et al. 1980; Singh et al. 1986; Kurian and Iyer 1992; Srivastav et al. 2009). Mango germplasm in India has been screened for varietal resistance against different diseases. Germplasm resistant/tolerant to leaf blight, die back and bacterial canker has been reported (Verma and Singh 1970; Shekhawat and Patel 1975; Kishun and Chand 1986).

Other reports on the evaluation aspects from Dhaka (Ahmad 1973), Binodpur (Hossain 1974), Chittagong (Ahmad et al. 1988), Hathazari (Golder et al. 1992) are available. In Thailand, critical evaluation of the germplasm is undertaken using morphological and agronomical characters, and certain biochemical patterns are also being characterized. The National Research Council of Thailand had many projects on the development of molecular techniques for identification of mango varieties through application of DNA finger printing (Vangnai 1996).

Characterization of mango germplasm in Indonesia at Cukurgondang comprises a total of 224 accessions. Some cultivars have more than one accession, such as Arumanis has five, Golek has eight, and Manalagi has three. so that there are in all 181 cultivars. The cultivars were planted and evaluated on cv. Madu as rootstock. The characterization based on the morphological description of 181 cultivars was reported by Kusumo et al. (1975) and Kusumo and Suminto (1987).

Purnomo et al. (1990) evaluated the accessions maintained at Cukurgondang (Indonesia). Twenty mango cultivars were evaluated for suitability as table fruit or processing based on the taste, fiber, and water content (Suhardjo and Suhardi 1986). The results showed that cultivars, namely, Manalagi, Gedong, Alphonso, Danas

Madu, Kidang, Sengir, and Bangalora are good for table purpose while the other 13 cultivars are good for processing.

4.5.1 Variability in *M. indica*

The list of 1595 cultivars of mango in world (Pandey 1998) was revised with the names of 1663 cultivars by Pandey and Dinesh (2010). Mango has rich diversity in Gangetic Plains, Eastern Peninsular region, Eastern and Western Ghats and Deccan plateau of India. A list of the names of cultivars available in the world as probable gene sources for dwarfness, fruit size, red peel color, high pulp content, high content of total soluble solids, long shelf-life of fruit, regularity in fruit bearing, earliness and lateness in fruit maturity and good processing quality have been mentioned by Pandey and Dinesh (2010). The sources of resistance to biotic and abiotic stresses are available at the cultivar level to the limited extent. Commercial and export quality mango cultivars of the world have been enlisted by Pandey (1998) and Pandey and Dinesh (2010). These varieties are obviously excellent gene sources for high yield and fruit quality parameters. Dinesh and Vasugi (2002) cataloged 151 cultivars of mango. They have evaluated cultivars for 54 characters as per the International Plant Genetic Resources Institute (IPGRI) Descriptors.

Dinesh et al. (2012) using Biodiversity International Descriptors had developed barcodes for the above-mentioned additional varieties through molecular characterization. These varieties include some polyembrynic varieties and 34 pickle-making varieties known as Appemidi from Western Ghats.

4.6 Conservation

For crops, various germplasm conservation methods are followed such as field gene bank, seed storage, in vitro storage and cryopreservation of shoot tips and pollen (Rajan and Hudedamani 2019). In mango, with current available knowledge, ex situ field genebank and on-farm conservation have contributed the most.

4.6.1 In Situ Conservation

Kostermans and Bompard (1993) recommended for both ex situ and in situ conservation of *Mangifera* spp. Although, in situ conservation has advantages, its big drawback is that the nature reserves in Asia may not be secure. There is an urgent need to conserve the diversity in the zone of natural variability. Wild seedling races are reported to occur almost throughout India in the natural forests of Tripura, Orissa, Chota Nagpur, and Western Ghat (Yadav and Rajan 1993). In the forest of the northeastern India, many of the seedlings still have a lot of primitive features such as polyembryony and dwarf habit. Work with ecotypes in Manipur for wild mango trees reveals that primitive characteristics persist and out-crossing with commercial varieties has not taken place (Yadav and Rajan 1993).

In India, in situ conservation of wild forms of *M. indica* was suggested in the hill region of Orissa, Chhotanagpur, and the plateau adjoining Madhya Pradesh, forests bordering Myanmar in Manipur valley and Western Ghat (Mukherjee 1949b; Yadav and Rajan 1993). In situ conservation efforts in Indonesia have been confined to some protected forests and National Forest Parks for the conservation of wild plants and animals but mango species were not given particular attention (Kusumo 1996). In situ conservation of *M. sylvatica* Roxb. is possible in one or several areas of greater Chittagong Hill Tracts (Bangladesh). It deserves urgent attention since forest is declining fast.

4.6.1.1 On-Farm Conservation

'The objective of on-farm conservation is to allow natural genetic introgression between domesticated crop and its wild relatives' (Harlan 1992) and to maintain crop evolu-tion in farmer's fields, farms, home gardens, and landscapes (Bellon and Etten 2014). Considerable efforts under the Asian Development Bank (ADB) sponsored initiative were made to conserve the fruit genetic resources of ten countries, including India, China, Bangladesh, Indonesia, Malaysia, Nepal, Philippines, Sri Lanka, Thailand and Vietnam, under the 2010 initiative 'Conservation and Use of Native Tropical Fruit Species and Biodiversity in Asia' (Mal et al. 2011).

In India, a significant number of accessions were collected and conserved in ICAR institutes under UNEP-Global Environment Agency (GEF) funded by project in five States Amravati (Maharashtra) and Chittoor (Andhra Pradesh), Malihabad (UP), Pusa (Bihar) and Sirsi (Karnataka) on the Mango (Gajanana et al. 2015a; Sthapit et al. 2016).

Custodian farmers played an important role in in situ on-farm diversity conservation in mango. Custodian farmers are the farmers actively supporting, adopting and promoting agricultural biodiversity and related knowledge over time, in the farms and communities (Dinesh et al. 2015; Gajanana et al. 2015a; Sthapit et al. 2015; Vasudeva et al. 2015; Rajan et al. 2016).

In 36 rural communities of India, Indonesia, Malaysia, and Thailand, the identification and documentation of tropical fruit tree traditional knowledge was undertaken. There are several motivations to the farmers for on-farm diversity conservation. The analysis in India, Indonesia, Malaysia, and Thailand showed that many farmers conserve rich variation with few rare, special fruit tree species or varieties, motivated mostly by conservation philosophy (Sthapit et al. 2015). Personal, social, cultural/religious, natural and biological traits also played important role in motivating the farmers for conserving the mango diversity apart from economic factor (Gajanana et al. 2015a). 'National Fruit Catalogue of Tropical Fruit Diversity' was developed based on the identification of unique varieties along with details of custodian farmers (Dinesh et al. 2014).

For on-farm conservation of cultivated and wild germplasm of fruit trees, awareness is required on diversity conservation. Fruit Tree Diversity Fairs in India are being arranged for general awareness about the diversity of fruit trees as a participatory method to recognize indigenous varieties with different traits and encourage conservation. The diversity fairs also helped in recognizing the custodian farmers (Gajanana et al. 2015b).

The heirloom varieties have a significant diversity and the study was undertaken in India under UNEP-GEF TFT project for their identification and conservation. Such varieties may be directly used for commercial cultivation or as parents to add characteristics through breeding (Rajan et al. 2014; Dinesh et al. 2015). Although several edible *Mangifera* species, obsolete and old varieties are being maintained in home gardens in mango-growing countries of Asia but scientific backup for on-farm conservation is lacking. However, this program needs strengthening for the conservation of valuable diversity available in genus *Mangifera*. Since the conservation by this means require incentives to the farmers, it should be supported through international organizations. It is important to be aware of the complementarity of the different techniques in the conservation of gene pool. Conservation in field gene banks should be used as a backup in case material is lost in situ (Husk 1998). Nevertheless, farmers are still keeping inferior cultivars in their home gardens, especially in the mango production centres because of their use value (Vasudeva et al. 2015).

4.6.2 Ex Situ Conservation

4.6.2.1 Field Gene Bank

Presently, establishment of field gene bank is the most important way for the conservation of mango germplasm. The plants are conserved in the field as living inventory. It is easier to conserve in the field, characterize and use in breeding programs. However, field genebank may be subjected to pressure from land use, prevalent diseases and pests, which may result into germplasm depletion.

In India, few decades back, germplasm of *M. indica* was collected and assembled as working collection in the field genebank at different locations with major emphasis to the accessions with desirable traits be exploited as variety or to be used in breeding program (Table 4.3).

Table 4.3 Field genebank sites of mango in India

Northern India	ICAR- Central Institute for Subtropical Horticulture, Lucknow ICAR- Indian Agricultural Research Institute, New Delhi Horticulture Experiment & Training Centre, Saharanpur (Uttar Pradesh) Horticultural Experiment & Training, Centre, Basti (Uttar Pradesh) G.B. Pant University of Agricultural Technology, Pantnagar (Uttarakhand) Fruit Research Station, Rewa (Madhya Pradesh)
Western India	Regional Fruit Research Station, Vengurla (Maharashtra) Agricultural Experiment Station, Paria (Gujarat) Gujrat Agricultural University, Navsari (Gujarat) Mahatma Phule Agricultural University, Rahuri (Maharashtra)
Southern India	ICAR- Indian Institute of Horticultural Research, Bengaluru (Karnataka) University of Agriucultural Sciences, Dharwad (Karnataka) Fruit Research Station, Sangareddy (Andhra Pradesh) Fruit Research Station, Anatharajupet, Kodur (Andhra Pradesh)
Eastern India	Central Horticultural Experiment Station, Ranchi, (Bihar) Agricultural Research Institute, Sabour Bhagalpur (Bihar) Regional Research Center, ICAR-CISH, Malda (West Bengal) Fruit Research Station, Krishanagar (West Bengal)

A closer look to the accessions available in field gene banks established at various places indicates that substantial diversity of cultivated *M. indica* is conserved, but these collections do not contain any representative variability of other *Mangifera* species like *M. sylvatica, M. andamanica, M. gedebe* and *M. khasiana* which need to be collected from Manipur, Indo-Burma region and Andaman and Nicobar islands (Ram and Rajan 2003). Mango germplasm in India is safely duplicated at few centres because germplasm of one region behaves differently in other region.

In Thailand, the Department of Agriculture (DOA) have established Tropical Fruit Germplasm Resources Centre at Khao Chong, Muang District of Trang Province. Besides, other collections of mango are located at U-thong Field Crop Experiment Station, Sisaket Horticultural Research Centre, Bangkok; Horticultural Research Station, Chachoengsao; Rubber Research Centre, Surat; Thani Horticultural Research Centre, and Tachai Horticultural Research Station. More than 80 varieties are being maintained and evaluated at Pakchong Experiment Station of Kasetasrt University, Nakhon Ratchasima province (Vangnai 1996)

In Philippines, Department of Horticulture, at U.P. Los Banos, had established a fruit gene bank. It had 52 mango varieties and four other *Mangifera* species (*M. altissima, M. caesia, M. foetida,* and *M. odorata*). A duplicate collection of some 58 foreign mango varieties was established in 1981 at the campus of the Western Luzon Agricultural College in San Marcelino, Zambales. An independent collection of foreign varieties was established at Davo National Crop Research and Development Centre of the Bureau of Plant Industry in Bago Oshiro, Davao City. Varieties not previously maintained at the Institute of Plant Breeding were propagated and planted at the National Collection Site in U. P. Los Banos (Coronel 1996).

Live collections of the crops have long been maintained by the Sri Lankan Department of Agriculture. However, attempts have been made to systematically organize these collections into field gene banks to conserve the variety of fruit crops that are multiplied vegetatively. The effort to collect and conserve local germplasm is not given a high priority (Medagoda and Jayawardena 1997).

The Bangladesh Society for Horticultural Science established a Fruit Plant Repository on a 200 hectare of land at Madhupur, Tangail about 120 km away from the capital city. However, ex situ collections at the Research Stations at BARI, Joydebpur; Fruit Research Station (FRS),

Binodpur; Regional Horticulture Research Station (RHRS), Hathazari, Chittagong; Horticulture Centre, Baluria, Natore and Horticulture Centre, Kallyanpur, Nawabgonj are maintained.

4.6.2.2 In Vitro Conservation

Germplasm maintenance of mango in field gene banks has its own limitations because of large land and other resource requirements. The other possible way for the conservation of the gene pool through in vitro conservation is still in fancy because till date tissue culture technique in mango is not available. Pollen storage of important mango varieties suggested by Iyer and Subramaniam (1989) and Dutta et al. (2013) may be used in hybridization program for gene transfer. Pollen storage can reduce the lag period which is required for grafted plants to attain flowering stage and pollen of desired parents could be exchanged for breeding.

4.7 Documentation

In India, initial characterization consisted of recording data on desirable characters by the users. First written evidence of evaluation and documentation of mango germplasm is available from *Ain-i-Akbari* written in the sixteenth century (Singh 1960). Burns and Prayag (1921) documented the germplasm of mango from central India. Description of 335 varieties was documented and published by Naik and Gangolly (1950) in 'A Monograph on classification and nomenclature of South Indian Mangoes'. Singh and Singh (1956) described 150 mango varieties of Uttar Pradesh and Bihar in a publication titled 'A Monograph of the mangoes of Uttar Pradesh' published in two volumes. Gangolly et al. (1957) also described 210 important varieties grown in the country from states of Punjab, Haryana, and Uttar Pradesh in North India; Andhra Pradesh and Tamil Nadu from South; Maharashtra and Gujarat from West; and Bihar, Odisha and West Bengal from East India and documented this in the form of a monograph, 'The Mango'. On the basis of comprehensive studies, Mukherjee (1948) described 72 varieties of Bengal.

In addition, local varieties have also been documented from different regions (Teaotia and Srivastava 1961; Teaotia and Singh 1963; Kashyap and Jyotishi 1969; Teaotia and Singh 1971; Saha 1972; Bose et al. 1974; Sadhu and Bose 1976; Parida and Rao 1989).

Sri Lanka Database management for PGR activities for inventory as well as for utilization is now operational at PGRC (Medagoda and Jayawardena 1997). At present registration work on mango is being continued. Hossain and Ahmed (1994) published a 'Monograph on Mango Varieties of Bangladesh'. It contains detailed studies of 72 mango cultivars of Govt. farms and private orchards of greater Rajshahi district. The description of cultivars in Indonesia was documented (Kusumo and Suminto 1987; Purnomo et al. 1990).

Documents prepared under the aegis of the International Plant Genetic Resources Institute (IPGRI) on eco-geographical study of known *Mangifera* genetic resources (Mukherjee 1985) are available. Another document by Kostermans and Bompard (1993), in the latest revision of the taxonomy of *Mangifera* was the consequence of joint IBPGR-IUCN (International Union for the Conservation of Nature)-WWF (World Wildlife Fund) project initiated in 1985 on wild mangoes on the island of Borneo and in the Malay peninsula (Bompard 1989), the regions that held the highest concentrations of *Mangifera* species. The detailed documentation of the recognized 69 species, many of which were collected during the course of this project has formed the basis of the future line of work on the genus' status reports, compiled by IPGRI, on the genetic resources of mango in Asia: South Asia (Bangladesh and India); Southeast Asia (Indonesia, Philippines, Thailand) and East Asia (China) and Sri Lanka deal with the mango gene pool diversity in these important mango-growing countries (Ram and Rajan 2003).

Based on characterization of mango germplasm under subtropical conditions, Rajan et al. (1999a) catalogued 252 Indian mango varieties using 56 descriptors. E-catalogue on 184 varieties of India was developed as open and searchable offline database (Rajan et al. 2012). Documentation of status of genetic resources is important for their management, protection and utilization. Documentation is also crucial for germplasm exchange and sharing of benefits also. Besides, the information on germplasm published as reports, informal publication, information supplements, books, journals and research papers is important. The electronic medium is making information more readily available enabling countries to know about germplasm wealth (Rajan et al. 2020).

The data on field gene banks can tell us what else needs to be conserved and help to determine appropriate germplasm management needs. The exchange data availability is a measure of the success of the documentation system of the gene bank. IPGRI has advocated standard documentation of genetic resource information with its mandate to improve the conservation and use of plant genetic resources. IPGRI has published descriptor lists in respect of many crops (Ram and Rajan 2003).

Through catalogues, the genetic variation of mango is recorded and shared. A catalog of 252 mango cultivars (Rajan et al. 1999a, b) and two mango catalogs with information for 404 accessions (Dinesh et al. 2014) were published, while the Philippines released catalog on mangoes (Rajan et al. 2014).

The Mango Resources Information System (www.mangifera.org) includes information about the 682 accessions with details of fruit, leaf and other characteristics, as well as the 26 expressed sequence tags (ESTs) and 285 nucleotide sequences present in mangoes. This allows rapid knowledge access and was used as a quick method to retrieve data from a large collection of information (Rajan et al. 2013a, b). National Mango Database developed at ICAR-CISH has several modules on genetic resources of mango including wild relatives, distribution, molecular data, Field Gene Banks and IP issues and visited by millions of visitors (Rajan et al. 2020).

In most of the countries, there is still no definite and effective strategy/policy of ex situ and in situ conservation for mango genetic resources. Human habitat is expanding fast due to population pressure leading to subsequent household fuel crisis and environmental hazards. Therefore, there is an urgent need to manage the genetic erosion of the mango genetic resources. Despite several advances in biotechnological approaches for the conservation of germplasm, presently, field genebank is the only alternative for the conservation of *Mangifera* genetic resources. The genetic improvement of mango hitherto has depended on the utilization of the genetic variability found within a single species, *i.e.*, *M. indica* thus Mukherjee (1985) stressed that 'a concerted sampling strategy should be devised for ex situ samples to meet urgent needs for use in research for improvement of the crop through breeding or as rootstocks. Sources of resistance to mango malformation, anthracnose, powdery mildew, gall midge are urgently needed'.

4.8 Genetic Erosion

Consumer's preferences are associated with changing market needs for fresh fruit and value added products thus making mango cultivation more or less commercial venture. Large-scale use of clonally multiplied mango cultivars has diminished the genetic variability and chances of seedling selection, a method which resulted development of countless varieties in western and eastern Hemisphere. Climate change and urbanization triggers the depletion of genetic diversity in the wild gene pool. Many workers have understood the value of biodiversity protection and its genetic potential. Several mango populations are endangered by habitat loss (Rajan and Hudedamani 2019). Several reports have indicated that four species of the common mango, viz., *M. pajang*, *M. zeylanica*, *M. lalijiwa* and *M. odorata* have been listed as endangered, while, the Kalimantan mango (*Mangifera casturi*) has been listed as extinct even in its wild habitat (IUCN 1998a, b, c, d, e; Rhodes and Maxted 2016; IUCN 2017). Since, variability available in mango is still unexplored and many of the indigenous land races are threatened with extinction, collection of this diversity was suggested (Yadav and Rajan 1993).

Although mango is most important fruit in Asia, there is a potential for competition between the cultivated and non-commercial varieties for arable land. The largest mango producing countries including India have the most potential for its cultivation to be displaced by other more profitable crops. India being a major centre of diversity in mango, there is also significant erosion of the gene pool owing to urbanization, industrialization and deforestation (Yadav and Rajan 1993). Other *Mangifera* spp. have their centre of diversity and origin in Southeast Asia, a region that has accomplished great economic development in recent years at the cost of genetic erosion in wild relatives (Ram and Rajan 2003).

In Thailand, *M. indica* has narrow base of genetic variation which is the main cause of genetic erosion in mango cultivars. Few principal commercial varieties are grown for commercial purpose. The introduction of selected clonal varieties to replace seedling varieties is also causing genetic erosion in the cultivated mango (Vangnai 1996). In Indonesia, genetic diversity is still maintained by the villagers in the form of inferior cultivars around houses which is under the threat of extinction due to the need to build houses or clearing of land for growing other crops. In situ conservation efforts in Indonesia have been confined to some protected forests and National Forest Parks for the conservation of wild plants and animals but mango species are not given particular attention as yet. Nevertheless, inferior cultivars in the home gardens of farmers, especially in the mango production centres such as Pasuruan, East Java and Indramayu, West Java may be under threat (Kusumo 1996).

In Bangladesh, many of the elite varieties (which are not commercial varieties) are in potential threat of erosion. The predominance of one major variety, the 'Carabao', is threatening the existence of the minor varieties in Philippines. 'Carabao', 'Pico', 'Pahutan', 'Binoboy', 'Dudol', 'Senora', 'Digos' and 'Mampalang' mangoes are the eight recognized local mango varieties. *Mangifera* species are also greatly threatened. *M. laurina* and *M. monandra* are getting more rare (Coronel 1996).

4.9 Utilization of Genetic Resources

Characterization and evaluation of mango germplasm is an imperative step for its utilization in breeding programs. Available germplasm has been considerably characterized to identify desired traits mostly based on morphological and molecular markers. Moreover, germplasm conservation for future utilization also requires dependable information. Comprehensive details of mango cultivars, characterization of desired traits or genes and evaluation information on germplasm based on morphological as well as molecular markers will lead to better utilization.

Mango germplasm is being extensively used in breeding being conducted at different locations spread all over India. The main emphasis has been to develop regular bearing, colored varieties and to avoid the formation of spongy tissue in cultivar Alphonso (an export variety). Cultivar Neelum has been used to impart regular bearing habit, cv. Vanraj for colored skin character and cv. Banganapalli to avoid spongy tissue formation. As a result of massive hybridization programs, more than 45 mango varieties have been developed, including Amrapali (dwarf, regular bearing), Mallika (regular bearing, processing) and Ratna (regular bearing, free from spongy tissue). Variety Sindhu is a recently developed hybrid with papery and soft stone. Ambika (regular bearer and red colored) and Arunika (dwarfer than Amrapali, regular bearer and red colored) developed from ICAR-CISH. Elaichi has been used for incorporation of resistance against mango malformation (Rajan et al. 2005). Few studies based on fruit traits have been conducted for selecting promising parents (Rajan et al. 2009).

Rootstocks response to soil salinity, a significant abiotic stress, restricts mango production worldwide. Salt resistant polyembryonic rootstock makes mango cultivation possible in salt and problematic soils. Collection and screening of polyembryonic mango types from tsunami affected areas of the South Andaman district where trees are under natural selection pressure for salt tolerance and screening of collections against high sodium in sodic soils ex situ (Damodaran et al. 2013). The six accessions, viz., GPL-1, GPL-3, ML-3, ML-4, ML-2, and GPL-4 exhibited the high-soil sodium and pH tolerances. Damodaran et al. (2018) studied the field tolerance of sodium, potassium, antioxidants, and biochemicals for polyembryonic accessions in salt-impaired soil in relation to their dynamics. Study resulted in the identification of two rootstocks (ML-2 and ML-6) suitable for mango production in salt-affected soils.

Wild species are valuable source of resistance genes for many of the common pathogens of cultivated mango. Wild relatives are reported source of genes. Initiatives at Florida to produce new hybrids with reduced susceptibility to disease using wild relatives, viz., *M. lalijiwa, M. quadrifida, M. rubropetala* with *M. indica* were developed. New hybrids between *Mangifera* species and *M. indica* can provide the first step toward these goals (Ledesma et al. 2017). *M. laurina* was used as the pollen parent in a breeding program with a *M. indica* hybrid in Australia resulting 60 successful hybrids (Bally et al. 2013).

4.10 Genetic Resources Management Approaches

In the development of new varieties, genetic resources play a significant role. In many of the field gene banks, the majority of mango germplasm plantations are old and traditional, planted at wider distances with dense canopy, which have led to a significant reduction in yields. Such collections require assessment to determine future gene donors. The germplasm, which can act as the basic material for the development of insect/disease-resistant varieties must be collected for biotic stress management.

Concerted studies are needed to identify climate-resistant germplasm that can be used directly for cultivation or in breeding program. In the context of climate change, priorities must be reviewed and strategy must be prepared both for in situ and ex situ conservation. Duplication of the accessions in field gene banks increases the institution's burden and therefore it is necessary to analyze and reduce duplication to simplify conservation efforts in addition to helping to identify low-cost techniques. Extensive surveys may be carried out in diversity-rich regions to identify and collect trait-specific mango germplasm.

Studies can also strengthen attempts to minimize threats to wild populations by collection and maintaining germplasm from rich areas of genetic diversity. Extending opportunities to help local custodian farmers for on-farm conservation efforts would provide safe livelihood choices for them. Climate change has also made it difficult for crops in areas to explore the idea of creating conservation sites outside the current germplasm distribution. In order to manage diverse genetic trait-specific germplasm to achieve broad-based performance, core and mini-core collections need to be produced and evaluated.

References

Abedin MZ, Quddus MA (1990) Household fuel situation, home garden and agroforestry practices at six agro-ecologically different locations of Bangladesh. In: Abedin MZ, Chun KL, Omar MA (eds) Homestead Plantation and Agroforestry in Bangladesh. BARI/RWEDP/FAO/Winrock International, Joydebpur, Bangakok/Dhaka, pp 19–53

Adato A, Sharon D, Lavi U, Hillel I, Gazit S (1995) Application of DNA fingerprints for identification and genetic analysis of mango (*Mangifera indica*) genotypes. J Am Soc Hort Sci 120:259–264

Agharkar SP, Roy B (1951) On the origin and distribution of cultivated mangoes. Indian J Genet 11:48

Ahmad KU (1973) Results of past researches of the Division of Horticulture, BARI. Bangladesh Hort 1:17–25

Ahmad KU, Majumder AA, Islam QAKM (1988) Physico-organoleptic attributes of some commercial mango grown in Chittagong. Bangladesh Hort 16 (1):56–60

Ahmad RI, Tabbasam N, Malik AU, Malik SA, Mehboob-ur-Rahman Zafar Y (2008) Assessment of genetic diversity among mango (*Mangifera indica* L.) genotypes using RAPD markers. Sci Hortic 117:297–301. https://doi.org/10.1016/j.scienta.2008.04.009

Ajayi II, Olawuyi OJ, Ayodele AE, Faneye AO (2019) Molecular relationship among *Mangifera indica* L. (Mango) varieties using simple sequence repeat (SSR) marker. J Adv Biol Biotechnol 22(4):1–16. https://doi.org/10.9734/jabb/2019/v22i430120

Angeles DE (1991) *Mangifera altissima.* In: Coronel RE, Weshjehi EWM (eds) Edible fruits and nuts. Plant Res. SE Asia 2 (PROSEA), Pudoc, Wageningen, Netherlands, pp 206–207

CISH-Annual Report (2015) Central Institute for Subtropical Horticulture (CISH). Annual Report 2015-16. Lucknow, India

Anon (1979) Research report on mango workers meeting, Panaji, Goa, 2-5th May, 1979 Tech Doc No 13 Project coordinated fruit improvement project. Lucknow, India, pp 216–218

Bajpai A, Srivastava N, Rajan S, Chandra R (2008) Genetic diversity and discrimination of mango accessions using RAPD and ISSR markers. Indian J Hort 65 (4):377–382

Bally ISE, Grice C, Dillon NL, Akem CN, Lakhesar D, Stockdale K (2013) Screening and breeding for genetic resistance to anthracnose in mango. Acta Hortic 992:239–244. https://doi.org/10.17660/ActaHortic.2013.992.31

Bellon MR, Etten JV (2014) Climate change and on-farm conservation of crop landraces in centres of diversity Mauricio. In: Jackson M, Ford-Lloyd B, Parry M (eds) Plant genetic resources and climate change. Rome/Recta Cali: Bioversity International/ Bioversity International, Regional Office for the Americas, pp 137–150

Bompard JM (1989) Wild *Mangifera* species in Kalimantan (Indonesia) and in Malaysia. Final report on the 1986–1988 collecting missions. WWF Project No. 3305. IBPGR, IUCN-WWF, Rome

Bompard JM (1993) The genus *Mangifera* rediscovered: the potential contribution of wild species to mango cultivation. Acta Hortic 341:69–77

Bompard JM (2009) Taxonomy and systematics. In: Litz RE (ed) The mango: botany, production and uses. CAB International, New York, USA, pp 21–48

Bose SK, Pathak RK, Seth JN, Verma VK (1974) Mango varieties in valley areas of the central Himalyan region of Uttar Pradesh and their quality. Progress Hort 6: 31–40

Burns W, Prayag SH (1921) The book of mango. Bombay Department of Agriculture, India

Campbell RJ (2004) Graft compatibility between *Mangifera* species and *Mangifera indica* L. 'Turpentine' rootstocks and their subsequent horticultural traits. Acta Hortic 645:311–313

Chadha KL (1989) Mango research in India - New development. Indian J Hort 46:279–294

Chadha KL (1996) Status report on genetic resources of mango in India. IPGRI Project No. B06, IPGRI Regional Office for Asia, the Pacific and Oceania, Singapore

Collins GN (1903) The mangoes in Puerto Rico. United States Department of Agriculture Bulletin No. 28

Coronel RE (1996) Status report on genetic resource of mango in the Philippines. IPGRI project No. B06. Singapore, IPGRI Regional Office for Asia. The Pacific and Oceania

Damodaran T, Rajan S, Kumar R, Sharma DK, Misra VK, Jha SK, Rai RB (2013) Post-tsunami collection of polyembryonic mango diversity from Andaman islands and their *ex situ* reaction to high sodium in sodic soil. J Appl Hort 15:21–25. 10.37855/ jah. 2013. v15i01.04

Damodaran T, Rajan S, Jha SK, Misra VK, Sahu A (2018) Screening polyembryonic mango accessions for salt tolerance and assessing dynamics of sodium (Na^+), potassium (K^+), and antioxidants. Fruits 73 (4):228–235

Das RC, Hota BN (1977) Some indigenous mango varieties of Orissa. Orissa J Hort 5(2):35–52

Degani C, El-Batsri R, Gazit S (1990) Enzyme polymorphism in mango. J Am Soc Hort Sci 115:844–847

Dhaki AJ (1972) Saurashtra's five leading mangoes. Indian Hort 17:9–10

Dillon NL, Bally ISE, Wright CL, Hucks L, Innes DJ, Dietzgen RG (2013) Genetic diversity of the Australian National Mango Genebank. Sci Hortic 150:213–226. https://doi.org/10.1016/j.scienta.2012.11.003

Dinesh MR, Rajan S, Singh IP, Singh A, Singh SK, Vasudeva R, Vinoth S, Patil P, Parthasarathy VA, Reddy BMC, Sthapit B (2014) National fruit catalogue of tropical fruit tree diversity (*Mangifera, Citrus* and *Garcinia*). GEF, UNEP, Bioversity and ICAR, India

Dinesh MR, Rajan S, Singh SK, Singh LP, Ravishankar KV, Reddy BMC, Parthasarathy VA, Sthapit B, Rao VR, Sandya BS (2015) Heirloom/seedling mango varieties of India - Potentialities and future. Indian J Plant Genet Resour 28 (1):139–152

Dinesh MR, Vasugi C, Ravishankar KV, Reddy YTN (2012) Mango Catalogue, ICAR-IIHR. p 468

Dinesh MR, Vasugi CS (2002) Catalogue of mango germplasm. Indian Institute of Horticultural Research, Hesseraghatta Lake PO, Bangalore 560089, India, p 160

Durán-Zuazo VH, Raya AM, Aguilar-Ruiz J, Tarifa DF (2004) Impact of salinity on macro and micronutrients in mango (*Mangifera indica* L cv. Osteen) with different rootstocks. Spanish Agri Res 2:121–133

Dutta SK, Srivastav M, Chaudhary R, Lal K, Patil P, Singh SK, Singh AK (2013) Low temperature storage of mango (*Mangifera indica* L.) pollen. Sci Hortic 161:193–197. https://doi.org/10.1016/j.scienta.2013.06.022

Ediriweera MK, Tennekoon KH, Samarakoon SR, Thabrew I, de Silva ED (2016) A study of the potential anticancer activity of *Mangifera zeylanica* bark: Evaluation of cytotoxic and apoptotic effects of the hexane extract and bioassay-guided fractionation to identify phytochemical constituents. Oncol Lett 11:1335–1344. https://doi.org/10.3892/ol.2016.4087

Eiadthong W, Nakatsubo F, Yonemori K, Sugiura A, Utsunomiya N, Subhadrabandhu S (2000) Chemotaxonomic studies on some *Mangifera* species by chemical compositions in the bark. Acta Hortic 509:143–152

Eiadthong W, Yonemori K, Sugiura A, Utsunomiya N, Subhadrabandhu S (1999) Identification of mango cultivars of Thailand and evaluation of their genetic variation using the amplified fragments by simple sequence repeat-(SSR-) anchored primers. Sci Hortic 82:57–66

Evans EA, Ballen FH, Siddiq M (2017) Mango production, global trade, consumption trend and postharvest processing and nutrition. In: Siddiq M, Brecht JK, Sidhu JS (eds) Handbook of mango fruit—Porduction, postharvest science, processing technology and nutrition. Wiley Blackwell, UK, pp 1–16

Gajanana T, Dinesh M, Rajan S, Vasudeva R, Singh SK, Lamers HA, Parthasarathy V, Sthapit B, Rao VR (2015a) Motivation for on-farm conservation of mango (*Mangifera indica*) diversity in India–A case study. Indian J Plant Genet Resour 28:1–6. https://doi.org/10.5958/0976-1926.2015.00001.7

Gajanana TM, Rajan S, Singh LP, Dinesh MR, Vasudeva R, Singh SK, Lamers H, Parthasarathy VA, Sthapit B, Rao VR (2015b) Fruit diversity fair and on-farm conservation—An Indian experience. Indian J Plant Gene Resour 28(1):80–86

Gálvez-López D, Salvador-Figueroa M, Adriano-Anaya ML, Mayek-Pérez N (2010) Morphological characterization of native mangos from Chiapas, Mexico. Subtrop Plant Sci 62:18–26

Gangolly SP, Singh R, Katyal SL, Singh D (1957) The Mango. Indian Council of Agricultural Research, New Delhi, India, p 320

Gangolly SR, Singh D (1950) Distribution of the mango (*Mangifera indica* L.) and its varieties. Indian J Hort 7:7–16

Golder PC, Ahmad KU, Saha SR (1992) Qualitative expression of some mango varieties grown in the eastern region of Bangladesh. Bangladesh J Agric Res 17(1):96–98

Gonzalez A, Coulson M, Brettell R (2002) Development of DNA markers (ISSRs) in mango. Acta Hortic 575:139–143

Gruezo WS (1992) *Mangifera* L. In: Verheij EWM, Coronel RE (eds) Plant resources of south-east Asia No.2 Edible Fruits and Nuts. Pudoc-DLO, Wegeningen, Netherlands, pp 203–206

Gunaratman SC (1946) The cultivation of mango in the dry zone of Ceylon (Part II). Trop Agri 102:23–30

Gupta PN, Rai M, Rana RS, Lal B (1993) Genetic diversity of mango (*Mangifera indica*) in western Uttar Pradesh. Paper presented in Golden Jubilee Symposium on Horticulture Research: Changing Scenario, Bangalore, India, p 2

Hajjar R, Hodgkin T (2007) The use of wild relatives in crop improvement: a survey of developments over the last 20 years. Euphytica 156(1–2):1–13. https://doi.org/10.1007/s10681-007-9363-0

Harlan JR (1992) Crops and man (2nd edn). American Society of Agronomy, Madison, WI, 284 p. https://doi.org/10.1017/s0889189300004938

Hartless AC (1913) The flowering of mango. Agri J India 8:9–12

Higgins JE (1906) The mango in Hawaii. Hawaii Agricultural Experiment Station, Honolulu, Bull No. 12:31

Hoda MN, Mandal G (1979) Collection, maintenance and evaluation of mango germplasm. Mango Workers Meeting, Panaji, Goa, 2–5th May, p 25

Holden J, Peacock J, Williams T (1993) Genes, crops, and the environment. Cambridge Press, NY, USA

Hossain A (1974) Characteristics of Bangladesh mango grown at Rajshahi. An unpublished M.Sc. Ag (Horticulture) Thesis, Horticulture Department, Bangladesh Agricultural University, Mymensingh

Hossain AKMA (1996) Status report on genetic resources of mango in Bangladesh. IPGRI project no. B06. Singapore: IPGRI Regional Office for Asia, The Pacific and Oceania

Hossain AKMA, Ahmed A (1994) A monograph of mango varieties in Bangladesh. HRC-BARI and FAO/UNDP Mango Improvement Project. p 3

Human CF, Rheeder S (2004) Mango breeding: Results and successes. Acta Hortic 645:331–335

Human CF, Rheeder S, Sippel AD (2009) New cultivars and hybrid selections from the mango breeding program of the agricultural research council in South Africa. Acta Hortic 820:119–126

Husk AVD (1998) *In* situ conservation of tropical fruit germplasm. In: Arora RK, Rao V Ramanatha (eds) Tropical fruits in Asia: diversity, maintenance, conservation and uses. Proceedings of IPGRI-IIHR-UTFANET Regional Training Course held, At at IIHR, Bangalore, India, 18–31 May, 1997. IPGRI South Asia Office, New Delhi, pp 127–140

Illoh HC, Olorode O (1990) Numerical taxonomic studies of mango (*Mangifera indica* L.) varieties in Nigeria. Euphytica 51:197–205. https://doi.org/10.1007/BF00039719

IPGRI (2006) Descriptors for mango (*Mangifera indica* L.). International Plant Genetic Resources Institute, Rome, Italy

IUCN (1998a) World Conservation Monitoring Centre. Mangifera casturi, The IUCN Red List of Threatened Species

IUCN (1998b) World Conservation Monitoring Centre. Mangifera lalijiwa, The IUCN Red List of Threatened Species

IUCN (1998c) World Conservation Monitoring Centre. Mangifera odorata, The IUCN Red List of Threatened Species

IUCN (1998d) World Conservation Monitoring Centre. Mangifera pajang, The IUCN Red List of Threatened Species

IUCN (2017) Red List of Threatened Species. Version 2017-1

IUCN (1998e) World Conservation Monitoring Centre. Mangifera zeylanica, The IUCN Red List of Threatened Species

Iyer CPA (1991) Recent advances in varietal improvement in mango. Acta Hortic 291:109–132

Iyer CPA, Schnell RJ (2009) Breeding and genetics. In: Litz RE (ed) The mango: botany, production and uses. CAB International Publisher, New York, USA, pp 67–96

Iyer CPA, Subramanian TR (1989) Genetic resources activities concerning tropical fruit plants. In: Paroda RS, Arora RK, Chandel KPS (eds) Plant genetic resources: Indian prospective. ICAR- NBPGR, New Delhi, India, pp 310–319

Iyer CPA, Subramanyam MD (1986) Varietal collection and evaluation of mango germplasm. Research Report Fruit Research Workshop, Dapoli, Maharashtra, India, 18–20 Dec, pp 1–11

Iyer CPA, Subramanyam MD (1991) Breeding mango for developing new varieties. Acta Hortic 291:151–153

Kashyap R, Joytshi RP (1969) Some local quality mangoes of Madhya Pradesh. Acta Hortic 24:29–30

Kishun R, Chand R (1986) Studies on bacterial diseases of fruit crops. Annual Report, ICAR-IIHR, Bangalore, India

Knight RJ Jr (1993) Evaluating important fruit characters in mango germplasm. Fruit Var J 47:25–31

Kostermans AJGH, Bompard JM (1993) The mango: botany, nomenclature, horticulture and utilization. Academic Press, London, UK

Kulkarni VJ (1979) Collection of new mango cultivars for survey and introduction. Mango Workers Meeting, Panaji. Goa, India, 2–5th May, 1979, pp 3–4

Kurian RM, Iyer CPA (1992) Stem anatomical characters in relation to tree vigour in mango (*Mangifera indica* L.). Sci Hortic 50:245–253

Kusumo S (1996) Status report on genetic resources of mango in Indonesia. IPGRI Project No. B06, IPGRI Regional Office for Asia, the Pacific and Oceania, Singapore

Kusumo S, T Suminto (1987) Description of mango production in Indonesia. CRIH Jakarta, Indonesia V +122

Kusumo SR, Poernomo, Soehendro and Suminto T (1975) Mangga (*Mangifera indica* L.). Hort Res Inst, Jakarta, Indonesia, VIII +144

Lal B, Gupta PN (1993) Genetic diversity of mango in Western Uttar Pradesh. Changing Scenario. Hoticultural Society of India, Golden Jubilee Symposium on Horticulture Research, p 16

Lamb A (1987) The potential of some wild and semi wild fruit trees in Sabah and the progress made by the department of agriculture. Sabah in establishing a germplasm pool, Paper presentated at Forest Research Institute, Kepong, Malaysia

Ledesma N, Campbell RJ, Hass M, Campbell TB (2017) Interspecific hybrids between *Mangifera indica* and related species. Acta Hortic 1183:83–87. https://doi.org/10.17660/ActaHortic.2017.1183.12

Li W, Zhu X, Zhang Q, Li K, Zhang D, Shi C, Gao L (2020) SMRT sequencing generates the chromosome-scale reference genome of tropical fruit mango. Mangifera indica, BioRxiv. https://doi.org/10.1101/202002(22)pp960880

Lycett GW, Zainal ZB, Los M, Findlay CL, Tucker GA (1997) Novel ripening- specific cDNA clones from mango fruit. Acta Hortic 455:277–286

Mal B, Rao VR, Arora RK, Sajise PE, Sthapit BR (2011) Conservation and sustainable use of tropical fruit species diversity: Biodiversity's efforts in Asia, the Pacific and Oceania. Indian J Plant Gene Resour 24:1–22

Malo SE (1970) Mango and avocado cultivars present status and future developments. Proc Fla State Hort Soc 83:357–362

Maries C (1902) Indian mangoes. J Royal Hort Soc 26:755

Medagoda I, Herath HE (1984) Selection of mango rootstocks in the nursery stage. Krushi 6:3

Medagoda I, Jayawardena SDG (1997) Status report on genetic resources of mango in Sri Lanka. IPGRI Project No. B06. Singapore: IPGRI Regional Office for Asia, The Pacific and Oceania

Meilleur BA, Hodgkin T (2004) *In situ* conservation of crop wild relatives: status and trends. Biodiversity Conserv 13:663–684

Mishra SP (2004) Reaction of different mango varieties/hybrids to floral malformation. Indian J Mycol Plant Pathol 34(1):113–116

Mukherjee SK (1949a) The mango and its wild relatives. Sci Cult 15:5–9

Mukherjee SK (1950a) Cytological investigations on the mango (*Mangifera indica* L.) and the allied Indian species. Proc Natl Inst Sci India 16:287–303

Mukherjee SK (1985) Systematic and eco-geographic studies on crop genepool: I. Mangifera L. International Board for Plant Genetic Resources, Rome, Italy, p 86

Mukherjee SK (1997) Introduction: Botany and Importance. In: Litz RE (ed) The mango: botany, production and uses. CAB International, Wallingford, UK, pp 1–19

Mukherjee SK, Chakrabarty S, Sandhyakhan SK, Saha P (1983) Survey of mangoes of West Bengal. Indian J Hort 40:7–13

Mukherjee SK, Das D (1976) Screening of mango seedlings for use as dwarfing rootstock. Prog Hort 8:5–11

Mukherjee SK, Litz RE (2009) Introduction: Botany and Importance. In: Litz RE (ed) The mango: botany, production and uses, 2nd edn. CABI International, USA, pp 1–18

Mukherjee SK (1948) The varieties of mango (*Mangifera indica* L.) and their classification. Bull Bot Soc Bengal 2:101–133

Mukherjee SK (1949b) A monograph on the genus *Mangifera* L. Lloydia 12:73–136

Mukherjee SK (1950b) Wild mangoes of India. Sci Cult 15:469

Mussane CRB, Biljon AV, Herselman L (2011) Morphological and genetic characterization of mango varieties in Mozambique. African Crop Sci Conf Proceed 10:387–391

Naik KC, Gangolly SR (1950) A monograph on classification and nomenclature of South Indian mangoes. Press, Madras, India, Superintendent, Govt, p 311

Njuguna JK, Wepukhulu SB, Wanjala S (2009) Mango cultivar evaluation programme in Kenya. Acta Hort 820:113–136

Ochse JJ, Bakhuizen RC (1931) Fruits and fruiticulture in the Dutch East Indies. GKolff, Batavia (Jakarta), Indonesia

Pandey SN (1984) International checklist of mango cultivars. Division of Fruits and Horticultural Technology. ICAR-Indian Agricultural Research Institute, New Delhi, India, p 284

Pandey SN (1986) Mango cultivars: nomenclature and registration. Acta Hortic 182:259–264

Pandey SN (1998) Mango cultivars. In: Srivastava RP (ed) Mango cultivation. International Book Distributing, Charbagh, Lucknow, pp 39–99

Pandey SN, Dinesh MR (2010) Mango. Indian Council of Agricultural Research, New Delhi, pp 30–97

Parida GN, Rao DP (1989) Classification and selection of some elite mangoes in Orissa. Acta Hortic 231:89–92

Patwardhan GB (1918) Climbing mango in the Deccan. Poona Agriculture College Magazine 10:68

Pinto ACQ, Campbell R, Rachel CS, Rajan S, Dinesh MR, Mal B (2006) Descriptors of mango (*Mangifera indica* L.). International Plant Genetic Resources Institute (IPGRI), Rome, Italy, p 60

Pleguezuelo CRR, Zuazo VHD, Fernández JLM, Tarifa DF (2012) Physico-chemical quality parameters of mango (*Mangifera indica* L.) fruits grown in a mediterranean subtropical climate (SE Spain). J Agric Sci Technol 14:365–374

Popenoe W (1920) Manual of tropical and subtropical fruits. Macmillan, New York, pp 100–126

PPV&FRA (2008) Guidelines for the conduct of tests for distinctiveness, uniformity and stability. Mango (*Mangifera indica* L.) Protection of Plant Variety and Farmers Right Authority. Department of Agriculture Cooperation, Ministry of Agriculture and Farmers Welfare, New Delhi, GOI, p 17

Prasad A, Singh H, Shukla TN (1965) Present status of malformation. Indian J Hort 22:254–265

Purnomo S, Prahardini PER, Wijadi RD (1990) Classification of production and quality performance of Cukurgondang mango varieties. Penel Hort 5(1):107–129

Rajan S, Dinesh MR, Ravishankar KV, Bajpai A, Ahmad I, Singh A, Singh SK, Singh UP, Vasudeva R, Reddy BMC, Parthasarthy VA, Sthapit B (2014) Heirloom varieties of important tropical fruits: A community initiative to conservation Lucknow: ICAR-CISH. Lucknow, India, p 34

Rajan S, Dinesh MR, Ravishankar KV, Bajpai A, Vasugi C, Bhagwan A, Chandra R (2012) Catalogue of Indian mango varieties: Vol I. Central Institute for Subtropical Horticulture, Lucknow

Rajan S, Hudedamani U (2019) Genetic resources of mango: Status, threats, and future prospects. Conserv Util Hort Genet Resour p 217–249

Rajan S, Kumar R, Yadava LP, Sharan R, Bhal C, Verma JP (2013a) Variability pattern in mango (*Mangifera indica* l.) accessions of diverse geographical origins. Acta Hortic 992:341–351

Rajan S, Lamers HAH, Lal B (2016) A set of interconnected practices which enhance and conserve mango diversity in Malihabad. Trop Fruit Tree Divers Good Pract situ On-farm Conserv p, India, p 396

Rajan S, Mishra PK, Srivastav V, Aditya K, Sagar P and Tripathi PK (2020) A study on the visitor preference for different modules of the National Mango Database. J Appl Hort 22(2). https://doi.org/10.37855/jah.2020.v22i02.06

Rajan S, Negi SS, Kumar R (1998) Clustering of mango genotypes based on canopy characteristics National

Symposium on Mango Production and Export. Lucknow, India, p 7

Rajan S, Negi SS, Kumar R (1999a) Catelouge on mango (*Mangifera indica*) germplasm. Central Institute for Subtropical Horticulture, Lucknow, India

Rajan S, Negi SS, Kumar R (1999b) Hierarchical approach for analysing diversity in canopy characteristics of mango. Sixth International Mango Symposium, Pattaya, Thailand, p 106

Rajan S, Negi SS, Kumar R (1999c) Multivariate analysis of growth and yield performance of Dashehari mango on 24 rootstocks. Sixth International Mango Symposium, Pattaya, Thailand, p 147

Rajan S, Ram K, Negi SS, Mishra AK, Sinha GC, Yadav IS (2005) Mango (*Mangifera indica* L.) germplasm with tolerance to floral malformation. Indian J Plant Genet Resour 18:300

Rajan S, Sahu TK, Yadava LP (2013b) Mango resources information system: an open access portrayal of phenotypical, genetic and chemical information on mango. Acta Hortic 992:99–104

Rajan S, Yadava LP, Kumar R, Saxena SK (2009) Genetic divergence in mango varieties and possible use in breeding. Indian J Hort 66:7–12

Ram S, Rajan S (2003) Status report on genetic resources of mango in Asia-Pacific region. IPGRI office for South Asia, National Agricultural Science Centre (NASC), New Delhi, India

Ramaswamy N (1988) Survey and isolation of 'Plus trees' of mango. Acta Hortic 231:93–96

Rameshwar A, Kulkarni V, Masihuddin, Ahmed S, Sultan MA, Ramulu SA (1979) Physico-chemical analysis and keeping quality of mango cultivars at Sangareddy. Paper presented at mango workers meeting at Panaji, Goa, 2–5th May, 1979

Ramessur AD, Ranghoo-Sanmukhiya VM (2011) RAPD marker-assisted identification of genetic diversity among mango *(Mangifera indica)* varieties in Mauritius. Intl J Agric Biol 13:167–173

Rao DP, Mukherjee SK (1982) Orchard efficiency analysis of mango (*Mangifera indica*). Indian J Hort 39:158–166

Rhodes L, Maxted N (2016) *Mangifera casturi*. The IUCN Red List of Threatened Species 2016

Rolfs PH (1915) Mangoes in Florida. University of Florida Agriculture Experimental Station Bulletin 127:105

Sadhu MK, Bose TK (1976) Studies on mango (*Mangifera indica* L.) cultivars. I. Morphological and physico-chemical studies of some potential mango cultivars of the district Murshidabad, West Bengal. Paper presented in seminar on mango and its utilization, March 6–7, 1976

Saha AK (1972) Studies on some commercial characters of important early mango varieties of Murshidabad district of West Bengal, India. Acta Hortic 24:39–44

Saiprasad GVS, Lalitha A, Ravishankar KV, Mythili JB, Nagesh M, Joshi R (2004) Isolation and characterization of mRNAs differentially expressed during ripening of mango fruits. Indian J Biotechnol 3(4):533–537

Schnell RI, Brown JS, Olano CT, Meerow AW, Campbell RJ, Kuhn DM (2006) Mango genetic diversity analysis and pedigree inferences for Florida cultivars using microsateilite markers. J Am Soc Hort Sci 13:214–224

Shamili M, Fatahi R, Hormaza JI (2012) Characterization and evaluation of genetic diversity of Iranian mango (*Mangifera indica* L, Anacardiaceae) genotypes using microsatellites. Sci Hortic 148:230–234. https://doi.org/10.1016/j.scienta.2012.09.031

Sharma DK, Choudhury SS (1976) Occurrence of an unknown wild race of *Mangifera* in Tripura. Cur Sci 45:305–306

Sharma M, Rajan S (2018) Improvising GIs significance in fruit crops for doubling farmers income. J Pharmacogn Phytochem 7:1903–1905

Shekhawat GS, Patel PN (1975) Studies on bacterial canker of mango. Zeitschrift fur Pflanzenkrankheiten und pflanzenschutz 82:129–138

Sherman A, Rubinstein M, Eshed R, Benita M, Ish-Shalom M, Sharabi-Schwager M, Rozen A, Saada D, Cohen Y, Ophir R (2015) Mango (*Mangifera indica* L.) germplasm diversity based on single nucleotide polymorphisms derived from the transcriptome. BMC Plant Biol 15:277. https://doi.org/10.1186/s12870-015-0663-6

Shu ZH, Yen CR, Ke LS, Wang DN, Lin TS, Liu MF, Shiesh CC (2000) Mango production in Taiwan. Acta Hortic 509:59–68

Shupei X, Yanqing (1996) Status report on genetic resources of mango in China. IPGRI. IPGRI Project No. B06. IPGRI Regional Office for Asia. The Pacific and Oceania, Singapore

Singh KK, Jawanda JS (1962) Mango cultivation in the Punjab. Indian Hort 6:7–10

Singh LB (1960) the mango: botany, cultivation and utilization. World Crop Book Series, Leonard Hill, London, p 38

Singh LB (1972) Biennial Bearing in mango—Retrospect and prospect. Acta Hortic 24:145–148

Singh LB, Singh RN (1956) A Monograph on the Mangoes of Uttar Pradesh. Vols I and II, Superintendent Printing and Stationery Lucknow, UP, vol I, p 150

Singh LB, Singh RN (1957) Some promising mango selections of Uttar Pradesh which need commercialization. Annual Report Fruit Research Station, Saharanpur, India 150–53:37–43

Singh NK, Mahato AK, Jayaswal PK, Singh A, Singh S, Singh N, Rai V, SV AM, Gaikwad K, Sharma N, Lal S, Srivastav M, Prakash J, Kalindindi U, Singh SK, Singh AK, Khan K, Mishra RK, Rajan S,

Bajpai A, Sandhya BS, Nischita P, Ravishankar KV, Dinesh MR, Kumar N, Jaiswak S, Iquebal MA, Kumar D, Rai A, Sharma TR (2016) Origin, diversity and genome sequence of mango (*Mangifera indica* L.). Indian J Hist Sci 51:355–368

Singh NP, Srivastava RP, Chadha KL (1986) Screening of dwarfing mango rootstocks at nursery stage on the basis of anatomical characters. Indian J Hort 43(1–2):56–60

Singh U, Singh M, Singh AP (1993) Evaluation of mango seedlings (*Mangifera indica*) in eastern Uttar Pradesh. Paper presented in Golden Jubilee Symposium Horticulture Research: Changing Scenario, Bangalore, Horticultural Society of India, p 18

Srivastav M, Kumar M, Dubey AK, Singh AK, Sairam RK (2009) Relationship between physiological parameters and vigour indices in polyembryonic genotypes of mango (*Mangifera indica*). Indian J Agri Sci 79(6):469–471

Srivastava AP, Chandra R, Saxena S, Rajan S, Ranade SA, Prasad V (2007) A PCR- based assessment of genetic diversity, and parentage analysis among commercial mango cultivars and hybrids. J Hort Sci Biotechnol 82(6):951–959

Srivastava RP, Chadha KL, Singh NP (1980) Stomatal count as an index for prediction and classification of vigour in mango rootstocks. Indian J Hort 37:10–15

Sthapit B, Vasudeva R, Parthasarathy V, Rajan S, Arsanti IW, Idris S, Songpol S, Hugo L, RamanathaVR, VR (2016) On-farm/ situ conservation of tropical fruit tree diversity: emerging concepts and practices. Indian J Plant Genet Resour 29 (3):285–288

Sthapit B, Vasudeva R, Rajan S, Sripinta P, Reddy BMC, Arsanti IW, Idris S, Lamers H, Ramanatha RV (2015) On-farm conservation of tropical fruit tree diversity: Roles and motivations of custodian farmers and emerging threats and challenges. Acta Hortic 1101:69–74. https://doi.org/10.17660/ActaHortic.2015.1101.11

Sturrock TT (1951) A study of the identification of mango varieties in Florida. Mango Studies, Florida Mango Forum, p 71

Subedi A, Bajracharya J, Regmi HN, Gupta SR, Joshi BK (2004) Ecogeographic and genetic diversity analysis of mango in Nepal. Proceeding of the fourth national workshop on Hort Kathmandu, Nepal. Horticulture Research Division, Nepal Agricultural Research Council, Kathmandu. Nepal, pp 66–70

Tangah J, Bajau FE, Jilimin W, Chan HT, Wong SK, Chan EWC (2017) Phytochemistry and pharmacology of *Mangifera pajang*: An iconic fruit of Sabah, Malaysia. Syst Rev Pharm 8:86–91. https://doi.org/10.5530/srp.2017.1.15

Tanksleyj SD, McCouch SR (1997) Seed banks and molecular maps: Unlocking genetic potential from the wild. Science 277:1063–1066

Teaotia SS, Singh RD (1963) Some important sucking mangoes of Uttrar Pradesh. Punjab Hort J 3:99–106

Teaotia SS, Singh RD (1971) Studies on mango (*Mangifera indica* L.) varieties II. Morphological and physico-chemical studies of some important table varieties. Punjab Hort J 11:240–245

Teotia SS, Srivastava RP (1961) Study on some important commercial varieties of mango of eastern U. P. Indian J Hort 18:65–69

Thimmaraju KR 1993. Preservation of biodiversity on tropical fruit crops. Paper presented in Golden jubilee Symposium Horticulture Research: Changing Scenario, Bangalore, Horticultural Society of India, p 15

Tsai CC, Chen YKH, Chen CH, Weng IS, Tsai CM, Lee SR, Lin YS, Chiang YC (2013) Cultivar identification and genetic relationship of mango (*Mangifera indica*) in Taiwan using 37 SSR markers. Sci Hortic 164:196–201. https://doi.org/10.1016/j.scienta.2013.09.037

Tyagi DN (1986) `Ideotype' for high yield potential in mango (*Mangifera indica* L.) some architectural considerations. Indian J Plant Physiol 29:267–270

Usman M, Fatima B, Jaskani MJ (2001) Breeding in mango. Int Agri Biol 3(4):522–526

Vangnai V (1996) Status report on genetic resources of mango in Thailand. IPGRI Project No. B06, IPGRI Regional Office for Asia, The Pacific and Oceania, Singapore

Vasudeva R, Sthapit B, Salma I, Changtragoon S, Arsanti IW, Gerten D, Dum-ampai N, Rajan S, Dinesh M, Singh I, Singh SK, Reddy B, Parthasarathy V, Rao VR (2015) Use, values and cultural importance of major tropical fruit trees: An analysis from 24 village sites across south and south-east Asia. Indian J Plant Genet Resour 28:17–30

Vavilov NI (1926) The origin, variation, immunity and breeding of cultivated plants. Chron Bot 13:1–6

Verma OP, Singh RD (1970) Epidemiology of mango dieback caused by *Botryodiloidia theobromae* Pat. Indian J Agric Sci 40:313–318

Wang P, Luo Y, Huang J, Gao S, Zhu G, Dang Z, Gai J, Yang M, Zhu M, Zhang H, Ye X, Gao A, Tan X, Wang S, Wu S, Cahoon EB, Bai B, Zhao Z, Li Q, Wei J, Chen H, Luo R, Gong D, Tang K, Zhang B, Ni Z, Huang G, Hu S, Chen Y (2020) The genome evolution and domestication of tropical fruit mango. Genome Biol 21:60. https://doi.org/10.1186/s13059-020-01959-8

Warschefsky EJ, von Wettberg EJB (2020) Population genomic analysis of mango (*Mangifera indica*) suggests a complex history of domestication. New Phytol 222:2023–2037. https://doi.org/10.1111/nph.15731

Weerarathne PK, Samarajeewa, Nilanthi RMR (2005) Genetic diversity of *Etamba* in Sri Lanka. Tropical Agri Res Ext, pp 107–112

Wester PJ (1920) The Mango, Bureau of Agriculture Bulletin No. 18. Bureau of Agriculture, Manila, The Philippines

Woodhouse EJ (1909) Mangoes in Bhagalpur- A preliminary note. Quart. J Department of Agriculture Bengal 2:168–187

Yadav IS (1989) Survey of *Mangifera* species in Manipur. Report Submitted to CIHNP, Lucknow, India, p 3

Yadav IS, Rajan S (1993) Genetic resources of mango. In: Chadha KL, Pareek OP (eds) Advances in horticulture, vol 1. Malhotra Publishing. New Delhi, India, pp 77–93

Yadav IS, Sinha GC, Rajan S, Kalra SK (1987) Saheb Pasand a potential mango cultivar. Prog Hort 19 (3&4):316–318

Zhen C (1989) Principal mango varieties in Taiwan and their cultural practices. Agriculture situation in Taiwan

5 Genetic Diversity Analysis of Mango

Xin Hua He, Shahril Ab Razak,
and Cong Luo

Abstract

Genetic diversity is the foundation of plant breeding, and the importance of plant genetic diversity is increasingly being recognized as a specific field. The plant genetic diversity can be collected and preserved in the form of plant germplasm resources such as gene bank, DNA library, and so forth. Mango is one of the most excellent and popular tropical and subtropical fruit crops around the world. The study of mango genetic diversity using multiple molecular markers is an important strategy for mango germplasm collection, conservation, utilization, and new variety breeding. This chapter comprehensively reviews two important areas: (i) enumeration of existing plant genetic diversity by analytical methods; and (ii) modern tools available for plant genetic diversity analysis. The current review provides essential details about genetic diversity analysis of mango by morphological, biochemical, and molecular markers. The statistical tools used for assessing genetic diversity are briefly discussed, as it can promote the understanding of analysis tools and their practicality to the researchers. This review could benefit mango researchers and producers to employ the new methods and technologies for better and rapid evaluation, for efficient utilization of germplasm from genetic resources to their applied breeding programs.

X. H. He (✉) · C. Luo
State Key Laboratory for Conservation and Utilization of Subtropical Agro-Bioresources, College of Agriculture, Guangxi University, 100 Daxue Road, Nanning 530004, People's Republic of China
e-mail: honest66222@163.com

C. Luo
e-mail: 22003luocong@163.com

S. A. Razak
Biotechnology and Nanotechnology Research Centre, MARDI Headquarters, 43400 Selangor, Malaysia
e-mail: shahrilf@mardi.gov.my

C. Kole (ed.), *The Mango Genome*, Compendium of Plant Genomes,
https://doi.org/10.1007/978-3-030-47829-2_5

5.1 Introduction

With the growing population pressure, rapid development of urbanization and modernization, biodiversity is getting threatened in both direct and indirect ways. For example, land degradation, deforestation, urbanization, coastal development, and environmental stress are collectively causing mass extinction of plant species. Mango (*Mangifera indica* L.) is one of the economically, nutritionally, and admired woody fruit crops which are widely grown in tropical and subtropical areas of the world. It has been known that *Mangifera* species originated in Southeast Asia. Due to rapid economic development in Southeast Asia, the agricultural expansion to feed the people or for harvesting valuable hardwood, a number of natural and wild mango germplasm resources or plantations have been destroyed. Therefore, this has resulted in irreversible genetic loss of mango genetic resources (Kaur et al. 1980; Mukherjee and Litz 2009; Sennhenn et al. 2014). The plant genetic diversity provides opportunity for plant breeders to develop new and improved cultivars with desirable characteristics (Govindaraj et al. 2015). Genetic diversity is the vital pillar of biodiversity and diversity within species, between species, and of ecosystems. However, the problem is that modern mango cultivars, especially, have been selected and bred mainly for high yield potential under superior production conditions, which led to narrowing the diversity of mango.

The purpose of conservation genetics is to maintain genetic diversity at multiple levels. Conservation efforts and related research rarely focus on individuals, but the genetic variation is always existing in individuals. Every individual is genetically different in nature. There are differences in genetic traits among different mango varieties, such as high or low yield, early or later maturity, high or poor fruit quality, resistant or not resistant to the pests and disease. It is extremely important to trace and utilize the mango genetic diversity and ultimately conserve the important germplasm of promising and threatened mango varieties/cultivars to broaden the genetic resource repository (Litz 2004). The correct evaluation of genetic diversity within and between plant populations is routinely performed by several techniques, such as (i) morphological observation, (ii) biochemical characterization evaluation, and (iii) molecular (or DNA) marker analysis. Khan et al. (2015) has reviewed on morphological, biochemical, and molecular characterization of mango germplasm. This paper will indicate analytical methods, tools, and techniques in genetic diversity of mango which are made widely available for utilization.

5.2 Morphological Diversity Analysis

Morphological markers are those key features that can be identified and measured visually by naked eye and based on the botanic characters, such as plant height, growth pattern, flowering habit, inflorescence shape, young leaves color, fruit-bearing behavior, fruit shape, fruit color, seeds type, and disease resistant (Bhat et al. 2010). It does not require expensive technology but large tracts of land are often demanded for these field experiments, making it possibly more expensive than biochemical or molecular assessment. These marker traits are often susceptible to phenotypic plasticity; conversely, this allows diversity evaluation in the presence of environmental variation which cannot be neglected from the genotypic variation (Govindaraj et al. 2015). These kinds of markers are still having their advantages and they are necessary for distinguishing the adult plants from their genetic contamination in the field. The most important benefits of morphological assessment are that descriptor lists are easily available for most of the major fruit crops including mango. Description of mango germplasm using morphological indices is the important step which must be carried out prior to biochemical or molecular characterization. IPGRI (2006) published the descriptors of mango, which consist of valuable passport data for the identification of various accessions/cultivars, characterization of recorded traits and initial assessment at morphological level. The mango descriptor list has further quickened the cultivar/varietal description and evaluation process. Morphological data are obligatory in primary germplasm assessment as they comprise tree, leaf, inflorescence, fruit, and seed characteristics (Knight et al. 2009; Mukherjee and Litz 2009). Characterization is the essential factor which plays an important role before exploring the genetic diversity through biochemical or molecular markers (Litz 2004). This fact cannot be denied that historically when molecular markers were not available, people often used morphological characteristics to distinguish mango cultivars or species. Because some experts have reviewed mango morphological diversity in detail (Khan et al. 2015), we only accumulate some important literatures in Table 5.1, which can be used successfully for identification, evaluation, and diversity analysis of mango genetic resources.

Table 5.1 Morphological diversities of mango

Character	Work done	References
Tree	The mango botany, cultivation, and uses	Singh (1968)
	Introduction, botany, and importance	Mukherjee (1997)
	Origin, history, and distribution	Ahmed (2004)
	Production areas	Human (2008)
Leaf	A study of the distinguishing vegetative, floral, and fruit characteristics of Punjab mangoes	Ibrahim (1952)
	Relation of growth to fruit bearing in mangoes	Khan (1960)
	A numerical taxonomic studies of the mango (*Mangifera indica* L.)	Rhodes et al. (1970)
	The Mango: botany, production and uses	Litz (2003)
	Morphological and genetic characterization of mango (*Mangifera indica* L.) varieties in Mozambique	Mussane (2010)
	Morphological and bio-chemical markers for varietal characterization and quality assessment of potential indigenous mango (*Mangifera indica* L.) germplasm	Rajwana et al. (2011)
	Morphometric analysis of mango varieties in Sri Lanka	Krishnapillai and Wijeratnam (2016)
	Morphological and anatomical characters related to dwarfness in mango (*Mangifera indica* L.)	Shenoy and Rajagopala (2016)
	Morphological and physico-chemical diversity in some indigenous mango (*Mangifera indica* L.) germplasm of Pakistan	Raza et al. (2017)
	Diversity of a large collection of natural populations of mango (*Mangifera indica* Linn.) revealed by agro-morphological and quality traits	Zhang et al. (2020)
	Diversity in leaf morphology and physiological characteristics among mango (*Mangifera indica*) cultivars popular in different agro-climatic regions of India	Rymbai et al. (2014)
Inflorescence	The flowering of mango	Hartless (1913)
	Studies on blossom biology and pollination in mangoes (*Mangifera indica* L.)	Mukherjee and Rao (1943)
	Mango (*Mangifera indica* L.) flowering physiology	Ramírez and Davenport (2010)
	Morphological, physico-chemical characterization and evaluation of mango (*Mangifera indica* L.) germplasm in Multan (Pakistan)	Ali (2013)
	Physiology of flowering – the case of mango	Sandip et al. (2015)
	Mango (*Mangifera indica* L.) pollination	Ramírez and Davenport (2016)
	Genetic differences and cluster analysis of pollen quantity of 53 mango cultivars	Xu et al. (2018)

(continued)

Table 5.1 (continued)

Character	Work done	References
Fruit	The mango	Lakshminarayana (1980)
	Variation in physico-chemical and morphogenetic characters of some mango varieties of Goa	Desai and Dhandar (2000)
	Descriptors for mango (*Mangifera indica* L.)	IPGRI (2006)
	Analysis of the correlations between fruit traits and diversity of Quality traits within Mango (*Mangifera indica* L.) germplasm resource in Nujiang hot-dry valley	Li et al. (2010)
	Morphological characteristics and distribution of wild germplasm resources of *Mangifera indica* in South and North Pan River Valley	Fan et al. (2012)
	Divergence for fruit characters in mango (*Mangifera indica* L.)	Barholia and Yadav (2014)
	Fruit trait diversity of mango (*Mangifera indica* L.) germplasm resources in Lancang River basin	Xie et al. (2018)
Seed	A study of the endosperm and embryo in *Mangifera indica* L	Sachar and Chopra (1957)
	Genetics of mango polyembryony	Sturrock (1968)
	Performance studies on polyembryonic varieties of mango (*Mangifera indica* L.)	Prasad and Prasad (1972)
	Somatic embryogenesis in angiosperms	Tisserat et al. (1979)
	Classical breeding and genetics	Iyer and Degani (1997)

5.3 Biochemical Diversity Analysis

Biochemical markers include a series of measurement indicators, such as total soluble solids, total sugars, antioxidants, flavonoids, vitamins, phenolics, non-reducing sugars, reducing sugars, and isozymes. These are widely used to identify variations in mango cultivars/varieties (Khan et al. 2015). The isoenzymes are often used in biochemical diversity analysis. Isozyme markers are codominant in nature, and the allelic variants of enzymes are detected by electrophoresis and specific staining. It demands only small amounts of plant material for its detection. However, only a limited number of enzymes as biochemical markers are available, and these enzymes are not alone but it has complex structural and special problems; therefore, the resolution of genetic diversity is limited to exploration (Govindaraj et al. 2015).

Degani et al. (1993) reported that isozymes as biochemical markers were used in mango germplasm evaluation. The different kinds of proteins have been separated through electrophoresis and various alleles are detected between them for genetic variability. Selfed and outcrossed offspring were identified using the triosephosphate isomerase (TPI) isozyme system (Daga et al. 1999). Peroxidase isozymes can be used as a biochemical genetic index for the genetic relationship among mango species (Chen and He 2000). The 11 enzyme systems were used in the identification of the six commercial Thai mango cultivars, whereas 6-phosphogluconate dehydrogenase, isocitrate dehydrogenase, glucose-6 phosphate dehydrogenase, glutamate dehydrogenase, phosphoglucose mutase, malate dehydrogenase, diaphorase, shikimate dehydrogenase, and transaminase exhibited maximum resolution (Jintanawongse and Changtragoon 2000). The peroxidase variation existed among polyembryonic mango seedlings from the same seed (Mo et al. 2005). A lot of enzymes have been found to play an important role in various fruit crops, whereas three enzymes (peroxidase, catalase, and aspartate) have most commonly been used in mango diversity evaluation (Karibasappa 1995; Khan et al. 2015). For example, peroxidase, catalase, and aspartate amino acids were used in identification and characterization evaluation of zygotic mango seedlings. Many quantitative characters clearly described the diversity in mango accessions (Khan et al. 2015).

Fruit quality and its production are mainly affected by diseases, insects, and some physiological disorders which may result to narrow genetic diversity (Jintanawongse and Changtragoon 2000). The biochemical attributes of fruit quality about 17 cultivars were discussed, among which only five mango cultivars ('Kala Chaunsa', 'Faiz Kareem', 'Camal Wala', 'Sufaid Chaunsa', and 'Late Ratole 12') displayed the distinguished physiognomies (Rajwana et al. 2011).

5.4 Genetic Diversity Analysis in Genomics Era

Markers play a crucial role in assessing the variability and diversity of the studied organism. The drawbacks of the morphological-based approach became quickly apparent. Morphological-based approach in assessing genetic diversity shows limited numbers of discriminating traits and can be influenced by environmental conditions, making it difficult to accurately evaluate the varietal variability. Genomics-based marker systems were introduced to overcome these problems. Genomics-based marker systems comprised various DNA marker types, which can be used to evaluate the genetic variation. These DNA markers can detect the variation that comes from duplication, deletion, insertion, and/or inversion, of the DNA in the chromosomes. They are inherited both in dominant and codominant patterns. Different markers have different genetic traits (Govindaraj et al. 2015). Some authors reviewed different markers techniques in detail (Semagn et al. 2006; Govindaraj et al. 2015; Khan et al. 2015). In this section, we are presenting the most widely used molecular markers in genomics era. Different molecular systems such as random amplified polymorphic DNA (RAPD), amplified fragment length polymorphism (AFLP), inter-simple sequence repeat (ISSR), simple sequence repeat (SSR), expressed sequence tag-simple sequence repeat (EST-SSR), start codon targeted (SCoT), sequence-related amplified polymorphism (SRAP), and single nucleotide polymorphism (SNP) marker will be further discussed.

5.4.1 Random Amplified Polymorphic DNA (RAPD) Marker

One of the most straightforward PCR-based techniques for genetic variation evaluation involves the use of RAPD marker. This particular PCR-based technique has been widely used for different purposes, such as to identification and classification of cultivars, breeds, or accessions (Fukuoka et al. 1992), and diversity studies (Yu and Paul 1992; Cao and Oard 1997). Technically, the process of using RAPD marker depends on the utilization of a single arbitrary primer that amplifies numerous DNA segments, resulting in an efficient and effective screening of polymorphism or a wide variation of loci in the DNA.

However, this particular technique has several drawbacks that should be considered, especially its reproducibility and nature of inheritance in a dominant mode. In order to address the reproducibility issue that may result in errors, most past studies applied similar control samples for every experiment. Furthermore, its application in genetic studies is rather limited as the reaction conditions, DNA quality, and PCR temperature profiles highly influence the outcomes of using this particular technique (Kumari and Thakur 2014).

Despite that, one of the strengths of using RAPD marker lies in its appropriateness to evaluate the genetic variation of underexplored species, as this particular technique does not require genomic information of the organism under study. In addition, the use of electrophoresis gel and the scoring that is subjected to the presence (or the absence) of the band, simplify the overall technical process of using RAPD marker. The use of an extensive range of universal primers also contributes to its robust analytical power. There are many reports on the application of RAPD marker in assessing genetic diversity of germplasm and commercial varieties in mango (Schnell et al. 1995; Jayasankar et al. 1998; Xu et al. 1998; Kumar et al. 2001; Karihaloo et al. 2003; Xie et al. 2005; Rahman et al. 2007; Rajwana et al. 2008; Matallana et al. 2009; Jena et al. 2010; Bhargava and Khorwal 2011; Majumder et al. 2011; Ramessur and Ranghoo-Sanmukhiya 2011; Souza et al. 2011; Mansour et al. 2014; Kumar and Sreekala 2016; Pruthvish and Chikkaswamy 2016; Galal et al. 2017; Dao 2019).

5.4.2 Amplified Fragment Length Polymorphism (AFLP) Marker

Zabeau and Vos (1991) discovered this particular technique of using the AFLP marker and the previously patented by Keygene NV (Wageningen, The Netherlands) (Blears et al. 1998). Similar to the RAPD marker, the identification of genetic polymorphisms of AFLP is established based on the presence or the absence of DNA fragments (bands) after the amplification and restriction of the genomic DNA takes place. Through the PCR process with the aid of the restriction enzyme, the AFLP marker selectively amplifies the restriction fragments from the genomic DNA. Fundamentally, restriction enzymes with two different forms, specifically with four-base pair recognition site and six-base pair recognition site, slice the total genomic DNA into fragments prior to the ligation of these fragments through the recognition site-specific adapter oligonucleotides. Following that, a subset of approximately between 50 and 100 restriction fragments is selectively amplified in the PCR process, which are subsequently resolved and separated through electrophoresis on high-resolution polyacrylamide gels or Li-Cor or any electrophoresis that is based on the genotyping platform.

One of the strengths of using AFLP marker lies in its capacity to disclose numerous polymorphic bands in a single lane. Unlike other existing PCR-based techniques, this particular technique is highly efficient given its capacity to concurrently analyze a huge number of generated bands on a gel and a substantial amount of loci for polymorphism. In other words, its detection rate for the number of polymorphism per reaction is significantly higher. Besides that, studies have proved the superior performance of this particular technique, specifically on the amount of amplified sequences per reaction and reproducibility. The produced results appeared to be more reliable and reproducible across different laboratory settings. The nature of every segregated AFLP fragment is dominant type DNA marker. Moreover, this particular technique does not require any genomic information, which explains its appropriateness for samples of any origin or level of complexity.

AFLP represents a valuable and robust tool for identification, characterization, genetic diversity analysis, and genetic relationships assessment among mango species or cultivars or rootstocks (Eiadthong et al. 1999a; Kashkush et al. 2001; Yamanaka et al. 2006; Gao et al. 2010; Shi et al. 2011; Zhang et al. 2014; Luo et al. 2015b). AFLPs are highly reproducible and identifiable, which makes it an attractive technique for identifying polymorphisms and for determining linkages by analyzing individuals from a segregating population (Mohan et al. 1997). AFLPs have been employed efficiently for detailed molecular linkage maps in *Mangifera* species or cultivars (Fang et al. 1999; Chunwongse et al. 2000; Kashkush et al. 2001; Yamanaka et al. 2006; Gálvez-López et al. 2009, 2010; Ying 2012; Li 2016; Dang and Chen 2017).

5.4.3 Inter-simple Sequence Repeat (ISSR) Marker

The development of inter-simple sequence repeat (ISSR) marker serves to address the limitations of other markers, such as the reproducibility problem (RAPD marker), high operating cost (AFLP marker), and the absence of genomic sequence information (SSR marker). The use of ISSR marker is one of the PCR-based techniques that amplify the DNA segment at an amplifiable distance between two identical (but opposing directions) microsatellite repeat regions. The use of ISSR markers involves microsatellites that typically range between 16 bp and 25 bp (in length) as single arbitrary primers to aim at several genomic loci for the amplification of the inter-SSR sequences with different sizes. As primers, these microsatellite repeats can be di-, tri-, tetra-, or even penta-nucleotide, which can be either unanchored (Meyer et al. 1993; Gupta et al. 1994) or anchored (which is more commonly used) at 3' or 5' ends with one to four degenerate bases that are stretched into flanking sequences (Zietkiewicz et al. 1994).

Basically, ISSR markers are mainly segregated as dominant markers with respect to the simple Mendelian inheritance (Gupta et al. 1994; Ratnaparkhe et al. 1998) but, in several cases, the use of ISSR marker can be segregated as codominant markers; thus, allowing the differentiation between homozygotes and heterozygotes (Akagi et al. 1996; Sankar and Moore 2001). Unlike the previously discussed markers, such as RAPD marker (only about 10 mers), ISSR marker possesses higher reproducibility due to their longer primers (only between 16 and 25 mers) considering that a longer primer is linked to high annealing temperature that enhances its stringency. Moreover, the irreproducibility of only the faintest bands across diverse experimental settings was successfully demonstrated in several past studies, which reaffirmed the reproducibility of ISSR marker. Several other studies used polyacrylamide gel and demonstrated high reproducibility of about 92% to 95% of the scored fragments for different DNA samples of similar cultivar and different PCR runs (Fang and Roose 1997). Considering its strengths, this particular technique is widely applied across diverse genetic fields for plants, such as mango (Godwin et al. 1997; Krishna and Singh 2007) and can be particularly beneficial for areas of genetic diversity, gene tagging, genome mapping, phylogenetic studies, and evolutionary biology in a diverse range of crop species (Reddy et al. 2002). There are some reports of the application of ISSR marker to evaluate the genetic diversity of mango worldwide (Gonzalez et al. 2002; He et al. 2005, 2007; Pandit et al. 2007; Bajpai et al. 2008; Samant et al. 2010; Gajera et al. 2011; Luo et al. 2011; Tomar et al. 2011; Damodaran et al. 2012; Rocha et al. 2012; Samal et al. 2012; Dang and Chen 2017).

5.4.4 Simple Sequence Repeat (SSR) Marker

SSR marker or also known as microsatellite represents the repeat sequences of one to six found throughout the genome (Buschiazzo and Gemmell 2006). It is broadly distributed throughout the genome (Phumichai et al. 2015) and mainly present in the noncoding area (as compared to the coding region) of the genome (e.g., introns, intergenic spaces, and untranslated regions). The presence of SSR marker in the coding or gene region is generally lesser given its high mutation rate that affects the gene expression.

Meanwhile, a motif is basically a repeat nucleotide sequence. Its length represents the number of nucleotide bases in the repeat motif (where n refers to the repeat frequency): (1) mononucleotide, [A]n; (2) di-nucleotide, [AG]n; (3) tri-nucleotide, [AAG]n; (4) tetra-nucleotide, [CCAT]n; (5) penta-nucleotide, [GAATT]n; (6) hexa-nucleotide, [AAGGTA]n (Scribner and Pearce 2000). Referring to the repeat motif pattern, microsatellites are grouped as (1) perfect with continuous repeat of a single motif (AGAGAGAGAGAGAGAG), (2) imperfect with a base pair disruption between repeats (AGAGAGAGAGTAGAGAG), and (3) compound with multiple types of repeat motif (ATATATATCACACAATATATATCACACA). Basically, the development of microsatellites can be either from the genomic DNAs (gSSRs) or expressed sequenced data, resulting in the formation of EST-SSRs (result of the expressed sequence tags [ESTs]) that are said to be more superior given its link to the expressed genes that directly contribute to phenotype. Meanwhile, when it comes to plants, SSRs that are developed from the nuclear DNA (nuSSRs) can be grouped as nuclear SSRs (nuSSRs) whereas SSRs that are developed from the chloroplast DNA can be grouped as chloroplast SSRs (cpSSRs).

The progression of the next-generation sequencing (NGS) technologies has significantly affected the development of microsatellites given their capacity to identify and develop thousands of SSR markers from the generated reads by the sequencer (Ekblom and Galindo 2011). Notably, the highly informative and polymorphic attributes, codominance in the Mendelian inheritance, and multi-allele genetic markers that experimentally reproducibility and transferability across closely related species have contributed to the extensive use of SSR marker for plant genotyping in the genetic fields for the past two decades (Mason 2015). Apart from being used in evolutionary studies, the use of SSR marker has been widely adopted to assess genetic diversity and DNA fingerprinting as well as to develop linkage and quantitative trait loci (QTL) map and conduct parentage studies and marker-assisted breeding.

Despite its strengths, particular for the genetic field, the use of SSR marker can be a disadvantage for certain studies particularly underexplored species, as this particular technique requires the genomic information in order to develop their primers (Selkoe and Toonen 2006). The need for de novo development on the basis of species-by-species often takes place since it may not be possible to retain the primers from closely related taxa (Whitton et al. 1997). The conventional techniques of developing SSR marker involve the generation of a genomic library through microsatellite repeats enrichment, cloning, plasmid isolation, and Sanger sequencing (Zane et al. 2002). Although this process yields about hundreds of sequences, only a small subset is deemed fitting for the ensuing primer design considering that the design of microsatellite primers requires several criteria to be met. Some of the key criteria are that the sequence must contain a desired repeat, appropriate in size (typically ranges from 100 bp to 400 bp), and appropriate sequence length for primer design to target the regions flanking the repeat motif (Squirrell et al. 2003). In addition, the occurrence of null alleles and stutter band also has limited the application of the microsatellite markers. But, using software such as Micro-Checker is able to resolve this issue. There are a lot of reports on application of this type of marker in assessing the genetic diversity and population structure analysis of the mango accessions and varieties (Eiadthong et al. 1999b; Duval et al. 2005; Schnell et al. 2005, 2006; Hirano et al. 2010; Huang 2010,2011; Shamili et al. 2012; Singh et al. 2012; Surapaneni et al. 2013; Tsai et al. 2013; Dinesh et al. 2015; Ravishankar et al. 2015a; Wang et al. 2015; Alves et al. 2016; Azmat et al. 2016; Bajpai et al. 2016; Ganogpichayagrai et al. 2016; Gitahi et al. 2016; Lal et al. 2017; Nazish et al. 2017; Ab Razak et al. 2019; Yamanaka et al. 2019).

5.4.5 Expressed Sequence Tag-Simple Sequence Repeat (EST-SSR) Marker

Conventional methods used for SSR development markers are usually considered time- and cost-consuming, labor intensive, and also with less efficiency (Ellis and Burke 2007). Thus, as an alternative approach, the source of SSRs development is the development of the expressed sequence tag- (EST-) based SSRs. EST-SSRs are transferable across the closely related species which make it valuable compare to neutral SSR markers (Chapman et al. 2009). EST-SSRs are noticeably linked to the expressed genes, and hence, denote potentially functional markers for anticipated results. According to Bandopadhyay et al. (2004), about 80–90% of the EST-SSRs are characteristically found to be polymorphic in nature. Identification and development of SSR markers were gradually increased as a lot of mango genes and EST sequences were deposited into the database. Additionally, advancement in computational approach especially for primer designing also has eased the researchers to work with SSR markers (De Keyser et al. 2009).

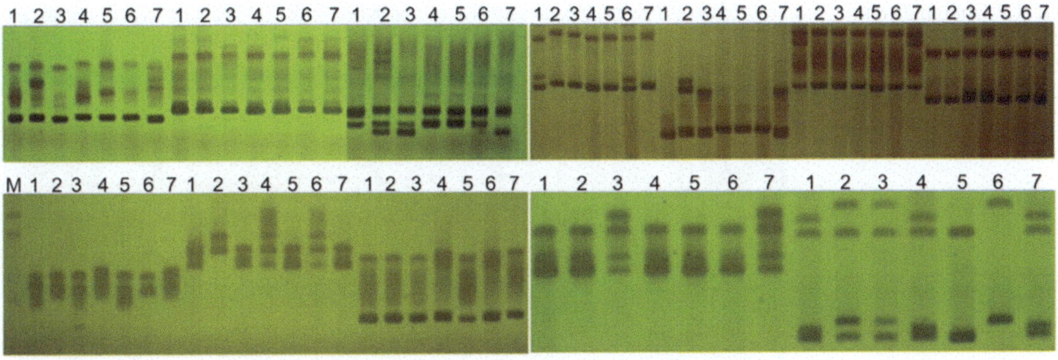

Fig. 5.1 Representative examples of PCR products amplified using several EST-SSR primer pairs and separated by electrophoresis on 6% denaturing polyacrylamide gels. (Luo et al. 2015a)

Dillon et al. (2014) studied 32 *M. indica* cultivars and related species with EST-SSRs in Australia, and out of it, 24 EST-SSRs showed polymorphisms and could distinguish between all of the *Mangifera* studied cultivars/selections. Luo et al. (2015a) developed 230 pairs EST-SSR primers, of which 218 primer pairs for seven mango genotypes showed stable amplification and 93 of the primer pairs yielded polymorphic products (Fig. 5.1). Zhou et al. (2019) analyzed the genetic diversity and relationship of 43 mango varieties by EST-SSRs. Among the 43 varieties, 26 pairs of primers amplified 140 bands, of which 128 were polymorphic, with the polymorphism rate of 90.79%. Therefore, EST-SSRs have tremendous potential to be used for identification and genetic diversity analysis of mango germplasm and other *Mangifera* species (Khan et al. 2015).

5.4.6 Start Codon Targeted (SCoT) Marker

The use of SCoT marker involves the use of a single primer as both forward and reverse primer in the PCR reaction, which is similar to the use of RAPD and ISSR markers (Collard and Mackill 2009). Unlike these two previously discussed techniques, the ATG start codon of the genes on both DNA strands becomes the focus of this particular technique that uses SCoT marker. The SCoT PCR primers were designed from the short conserved region flanking the start codon (ATG) that is conserved for all genes. Their PCR product will be resolved on the standard agarose gel electrophoresis, which make it simple and easy to handle. These attributes have contributed to the appropriateness of this particular marker across numerous studies. However, this particular marker has a major limitation that lies in its inheritance in a dominant manner despite being reported as being codominant in the case of insertion–deletion mutation in the region—nevertheless, such cases of codominance (e.g., RAPD marker and ISSR marker) are deemed insignificant (Zhang et al. 2015). Unlike other dominant markers or random DNA markers, SCoT marker is a gene-specific marker that yields critical insights that are related to biological traits. Moreover, Luo et al. (2010, 2011, 2012) and Gajera et al. (2014) also successfully proved the polymorphic nature and efficiency of the use of SCoT marker in the assessment of genetic diversity for mango. The genetic relatedness and variability among 168 mango (*M. indica* L.) germplasm resources were studied with SCoT markers (Fig. 5.2). Forty-five SCoT primers generated unambiguous and reproducible bands and a total of 337 fragments, 244 bands were polymorphic. These results demonstrate that SCoT markers are useful for cultivar identification and genetic diversity analyses of mango cultivars (Zhou et al. 2020).

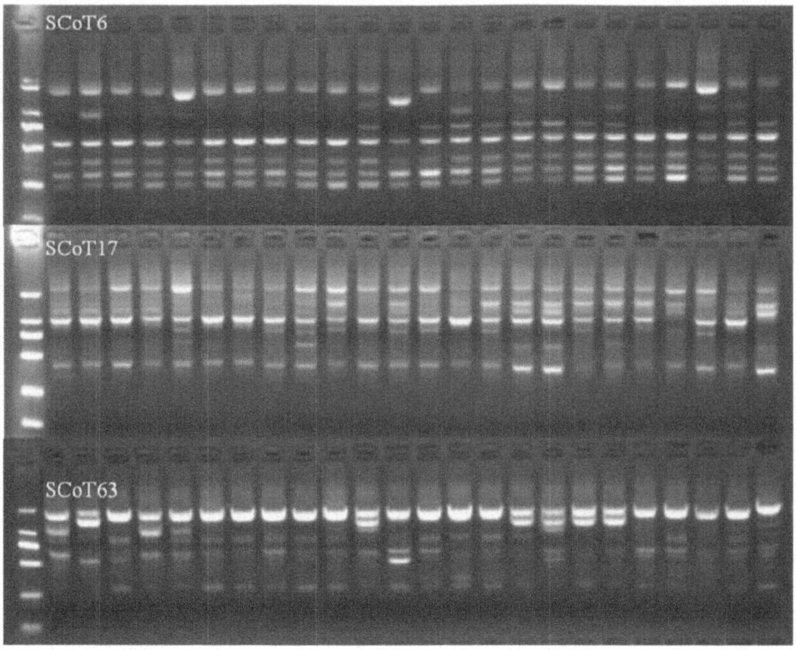

Fig. 5.2 Amplification products generated by SCoT6, SCoT17, and SCoT63 with portion mango varieties (Zhou et al. 2020)

5.4.7 Sequence-Related Amplified Polymorphism (SRAP) Marker

Sequence-related amplified polymorphism (SRAP) marker is a straightforward and low-cost PCR-based marker that exhibits high levels of reproducibility and versatility. The use of this particular marker was first introduced for gene tagging in *Brassica oleracea* L., specifically for the amplification of the coding regions of the genome with ambiguous primers, targeting GC-rich exons (forward primers), AT-rich promoters, introns, and spacers (reverse primers). For this, these primers comprised 17 to 18 nucleotides and core sequences from 13 to 14 bases. The "filler" sequences with no specific constitution were positioned at the first 10 to 11 bases at the 5′ end. The CCGG (forward) sequence or –AATT (reverse) sequence and three other random selective nucleotides at the 3′ end (Li and Quiros 2001) were next. Subsequently, the output of PCR was resolved in standard agarose gel electrophoresis.

Numerous comparative studies reported that SRAP marker exhibited comparatively similar extent of variation to AFLP marker but with markedly lower technical effort and cost when it comes to the same band-pattern variability and reproducibility (Li and Quiros 2001; Levi and Thomas 2007). Besides that, up to 20% of the tested SRAP markers demonstrated codominance pattern (Li and Quiros 2001).

Accordingly, the most typical approach to scoring the band fragment for this dominant marker involves the binary system with the presence (1) or absence (0) of the band. Nevertheless, certain studies modified their approach by opting for fluorescent labeling at the forward primers when it comes to the capillary electrophoresis-based system (Alghamdi et al. 2012). Likewise, SRAP marker is similar to other dominant markers that demonstrate the capacity to detect the variation in genetic at different levels of taxonomy (Uzun et al. 2009). The use of this marker is typically applied for interspecific and intraspecific hybrid analyses (Liu et al. 2007), linkage, and QTL analyses which are highly valuable when it comes to crop improvement (Zhang et al. 2009; Levi et al. 2011). Zhang et al. (2016) have used SRAP markers to evaluate the genetic diversity and relationship of 20 mango germplasms. Liu et al.

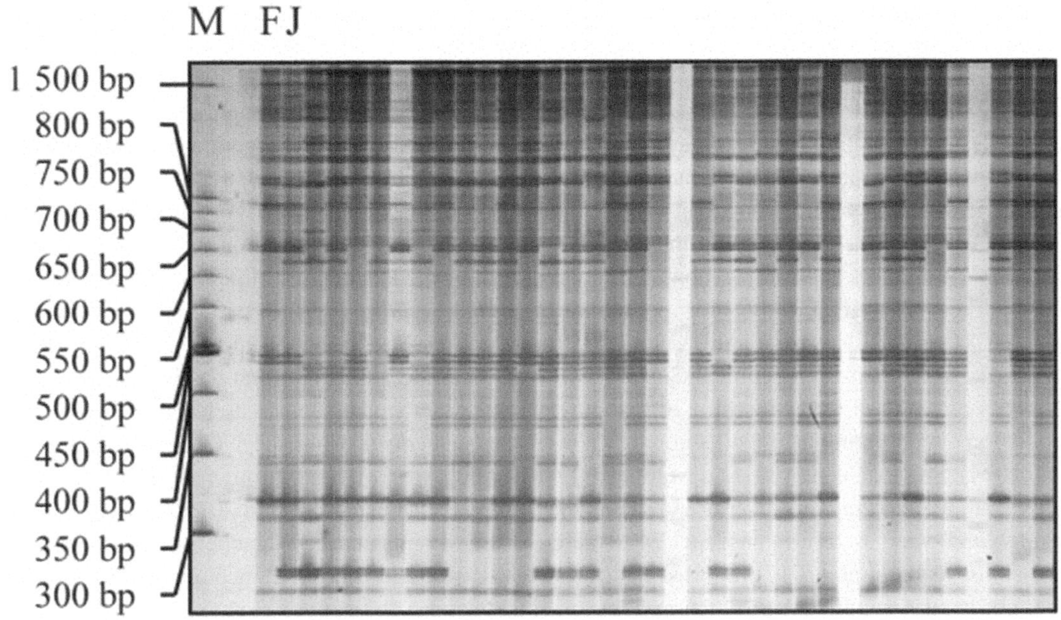

Fig. 5.3 Amplification of SRAP primer pairs M11E9 in mapping populations (Li et al. 2016)

(2019) assessed the genetic diversity of 13 mango germplasm resources by SRAP marker. A population of Guifei mango, Jinhuang mango, and their hybrids were used to construct a linkage map of mango and analyze their genetic diversity by SRAP markers. The genetic map constructed by Joinmap4.0 covered 801.6 cM with an average marker interval of 9.32 cM, containing 20 linkage groups and 113 markers (Fig. 5.3) (Li et al. 2016).

5.4.8 Single Nucleotide Polymorphism (SNP) Marker

The most abundant form of genetic variation in the genome is the single nucleotide polymorphism (SNP) marker (Mammadov et al. 2012). High-density SNP markers were found at specific regions of maize (with 1 SNP per 104 bp) and barley genomes (1 SNP per 189 bp) (Tenaillon et al. 2001; Kanazin et al. 2012). The abundance of SNP markers in the genome with its responsiveness is to be multiplexed up to a substantial amount (based on the genotyping platform) per reaction (Xiong and Jin 1999), resulting in a lower cost, establish this particular marker as the preferred choice.

Accordingly, SNP marker is closely linked to a single base variation in a DNA sequence. This particular marker displays a specific attribute where the least frequent allele for a base position with an alternative base in the genomic DNA sequence should account for at least 1%. Theoretically, SNP marker, in most cases, is bi-allelic for the automation practices considering the presence of any four possible nucleotide bases at every position of a sequence stretch. Certain studies regarded one base pair indels (insertion or deletion) as SNP marker despite the manifestation of different mechanisms. In other words, SNP marker reflects the existence of polymorphism between DNA samples with respect to the single base.

The SNP mutation may modify the characteristics of phenotype due to the variation in amino acid. Non-synonymous SNP reflects functional polymorphism. Despite the bi-allelic nature of SNP marker, which makes this marker less informative than multi-allelic marker (e.g., microsatellite), its amenability toward high-throughout multiplexing and abundance in the genome enables the use of more loci in research. Unlike other forms of markers, especially microsatellites, SNP marker displays less mutable marker system, resulting in higher evolutionary stability (Coates et al. 2009). This particular marker yields superior performance in evaluating complex genetic traits and grasping the genomic evolution. With that, the simple application of SNP marker is deemed fitting for population studies. 381,535 SNPs were identified in 28,959 transcripts from 24 mango cultivars, and these SNPs have been used for genotyping of mapping populations and germplasm collections (Kuhn et al. 2014). Sherman et al. (2015) developed 239 SNPs from mango cultivars 'Keitt' and 'Tommy Atkins' and was used to genotype 74 Israeli accessions. Kuhn et al. (2017) reported that they have observed a significant association between trait and SNP markers for the vegetative trait of branch habit and the fruit traits of bloom, ground skin color, blush intensity, beak shape, and pulp color. Recently, 272 SNP markers were used for identifying over 520,000 mango genotypes (Kuhn et al. 2019).

5.5 Next Generation Sequencing (NGS) of Mango

Essential insights on the population structure and diversity of species under study can be acquired based on the dataset of genomic-scaled DNA polymorphism. Based on these essential insights, the relationships between phenotypic variations and genetic polymorphisms, particularly using genome-wide association study (GWAS), can be effectively determined. A whole-genome sequence and a well-annotated genome are significant in exploring useful genes, determining their physical location in the genome, and predicting their biological functions.

The development of NGS technologies has introduced several radical innovations to interpret species diversity using a large dataset. The proposed NGS technologies have significantly bargained the cost and time required in order to obtain a substantial amount (up to gigabases) of information of nucleotide sequence. Besides that, the integration of NGS technologies and bioinformatics techniques also contributes to the advancement of high-throughput genetic markers and genome-scale investigation of genotypes for crops.

As of December 15, 2019, 533 species of land plants have undergone sequencing as deposited in the NCBI database (http://www.ncbi.nlm.nih.gov/genome/browse/). In particular, even though *M. indica* or commonly known as mango is not listed in the NCBI list (that has been sequenced) but certain studies demonstrated that this species was sequenced using the NGS platform for the development of high-throughput SNP and SSR markers (Ravishankar et al 2015b; Sherman et al. 2015). For instance, Warschefsky and Wettberg (2019) applied the NGS technology to address the genetic diversity of mango. Adding to that, a web database for mango (MiSNPDb) was developed which deposited with 1.25 M of SNP. Notably, the use of SNP marker benefits the genetic improvement programs and management of mango germplasm in terms of genome-assisted breeding, linkage and QTL mapping, traceability or parentage studies, varietal differentiation and purity assessment.

5.6 Genetic Diversity Analysis from Molecular Data

It is crucial to understand the various analysis methods of the data generated by molecular biological techniques before their applications in diversity analyses. Generally, there are two main types of analysis that are followed: (i) genetic relationship analysis and (ii) parameter calculation about population genetics. The analysis of genetic relationships among samples initiates with the construction of a matrix, sample × sample pairwise genetic distance (or similarities) based on the amplification bands (Govindaraj et al. 2015).

Genetic distance and/or similarity between different populations, genotypes, or individuals may be calculated by different statistical calculations depending on the dataset. The common measures of genetic similarity (GS) or genetic distance (GD) are (i) Nei and Li's (1979) coefficient (GDNL), (ii) Jaccard's coefficient (GDJ) (Jaccard 1908), (iii) simple matching coefficient (GDSM) (Sokal and Michener 1958), and (iv) modified Rogers' distance (GDMR). Mohammadi and Prasanna (2003) reviewed about different GD measures in detail. There are two main methods of analyzing the similarity (or distance) matrix, namely, principal coordinate analysis (PCA) and dendrogram (or clustering). Various algorithms were employed for clustering, but some common methods include unweighted pair group method with arithmetic averages (UPGMA), neighbor-joining method, and Ward's method (Karp et al. 1997). Govindaraj et al. (2015) carried on a brief description of common or basic statistical methods and its principle with the advantage and disadvantage of each method for analyzing genetic diversity. Therefore, the features and approaches of measures were not much discussed separately in the text.

A lot of software programs are available for evaluating genetic diversity in the postgenomic era, such as Arlequin, DnaSP, DARwin, PowerMarker, NTSYSpc, MEGA, PAUP, STRUCTURE, fastSTRUCTURE, ADMIXTURE, fineSTRUCTURE, POPGENE, GENEPOP, and GenAIEx, most of them are freely available through Internet. Many of these perform same tasks, with the main differences being in the platform, type of data input and output, and user interface. Therefore, choosing which to use depends largely on personal preferences (Govindaraj et al. 2015).

5.7 Conclusion

The manifold plant genetic resources are priceless treasures for humankind. The existence of genetic variability in crops is essential for breeding new varieties and hybrids. This can be acquired through phenotypic, biochemical, and molecular characterization of plant genetic resources. Molecular markers are indispensable means for analyzing the diversity of plant species. The application of molecular markers is one of the best strategies for construction of mango genomic map, which can be employed for determination of genetic relationship between exact heritable and desired characters. Mango genetic resources for future exploitation in breeding and correct assessment of mango varieties or cultivars are great demands of the present time. Major goal of breeding is to obtain quick fruit-bearing pattern, high yield, excellent fruit quality, and strong resistance cultivars against biotic and abiotic stresses. Morphological, biochemical, and molecular markers are very useful not only in assessment and documentation, but also extremely necessary concerning mango germplasm conservation and management for future strategies to broaden the scope of mango industry in the world. Regarding future scenarios, exploitation and utilization of various markers for uncultivated wild mango germplasm and variety identification diagrams are much imperative. Availability of the whole mango genome may be very helpful in breeding excellent quality, high-yielding and good-resistance cultivars. Next generation sequencing reduced the time and cost required for sequencing the whole genome. With the development of high-throughput molecular biological technologies ensuring speed and quality of data generated, it is possible to analyze the larger number of germplasm with limited time and genetic resources. Many software packages are available for evaluating phenotype and genetic diversity that enhanced the efficiency of mango germplasm curators, breeders, and producers to accelerate the mango improvement and effective utilization of genetic resources.

References

Ab Razak S, Azman NHEN, Ismail SN, Yusof MFM, Ariffin MAT, Sabdin ZHM, Abdullah N (2019) Assessment of diversity and population structure of mango (*Mangifera indica* L.) germplasm based on microsatellite (SSR) markers. Aust J Crop Sci 13 (2):315–320

Ahmed S (2004) Origin, history and distribution. In: Ahmed S (ed) Mangoes in Pakistan. The Horticultural Foundation of Pakistan, Islamabad, Pakistan, pp 1–26

Akagi H, Yokozeki Y, Inagaki A, Nakamura A, Fujimura T (1996) A codominant DNA marker closely linked to the rice nuclear restorer gene, *Rf-1*, identified with inter-SSR fingerprinting. Genome 39(6):1205–1209

Alghamdi S, Al-Faifi S, Migdadi H, Khan M, EL-Harty E, Ammar M, E (2012) Molecular diversity assessment using sequence related amplified polymorphism (SRAP) markers in *Vicia faba* L. Intl J Mol Sci 13 (12):16457–16471

Ali S (2013) Morphological, physico-chemical characterization and evaluation of mango (*Mangifera indica* L.) germplasm in Multan (Pakistan). MSc Thesis, Institute of Horticultural Sciences, University of Agriculture, Faisalabad, Pakistan

Alves E, Lima-Neto FP, Santos CAF, Ribeiro ICNS, Melo CAF, Holanda ISA, Souza AP, Corrêa RX (2016) Genetic diversity of mango accessions (*Mangifera indica*) using new microsatellite markers and morphological descriptors. Aust J Crop Sci 10 (9):1281–1287

Azmat MA, Khan AA, Khan IA, Rajwana IA, Cheema HMN, Khan AS (2016) Morphological characterization and SSR based DNA fingerprinting of elite commercial mango cultivars. Pak J Agri Sci 53 (2):321–330

Bajpai A, Srivastava N, Rajan S, Chandra R (2008) Genetic diversity and discrimination of mango accessions using RAPD and ISSR markers. Indian J Hort 65 (4):377–382

Bajpai A, Muthukumar M, Ahmad I, Ravishankar KV, Parthasarthy VA, Sthapit B, Rao R, Verma JP, Rajan S (2016) Molecular and morphological diversity in locally grown non-commercial (heirloom) mango varieties of North India. J Environ Biol 37(2):221–228

Bandopadhyay R, Sharma S, Rustgi S, Singh R, Kumar A, Balyan HS, Gupta PK (2004) DNA polymorphism among 18 species of Triticum-Aegilops complex using wheat EST-SSRs. Plant Sci 166:349–356

Barholia AK, Yadav S (2014) Divergence for fruit characters in mango (*Mangifera indica* L.). Afr J Agri Sci Technol 2:65–67

Bhargava R, Khorwal R (2011) Molecular characterization of *Mangifera indica* by using RAPD marker. Indian J Fund Appl Life Sci 1:47–49

Bhat ZA, Dhillon WS, Rashid R, Bhat JA, Dar WA, Ganaie MY (2010) Role of molecular markers in improvement of fruit crops. Not Sci Biol 2:22–30

Blears MJ, De Grandis SA, Lee H, Trevors JT (1998) Amplified fragment length polymorphism (AFLP): a review of the procedure and its applications. J Ind Microbiol Biotechnol 21(3):99–114

Buschiazzo E, Gemmell NJ (2006) The rise, fall and renaissance of microsatellites in eukaryotic genomes. BioEssays 28(10):1040–1050

Cao D, Oard JH (1997) Pedigree and RAPD-based DNA analysis of commercial US rice cultivars. Crop Sci 37 (5):1630–1635

Chapman MA, Hvala J, Strever J, Matvienko M, Kozik A, Michelmore RW, Tang S, Knapp SJ, Burke JM (2009) Development, polymorphism, and cross-taxon utility of EST-SSR markers from safflower (*Carthamus tinctorius* L.). Theor Appl Genet 120:85–91

Chen WY, He HM (2000) Study on peroxidase isoenzyme of mango. Chin Wild Plant Resour 19(5):52–55

Chunwongse J, Phumichai C, Chungwongse C, Sukansawan S, Bonreungrawd R, Babprasert C (2000) Molecular mapping of mango cultivars 'Alphonso' and 'Palmer'. Acta Hort 509:193–206

Coates BS, Sumerford DV, Miller NJ, Kim KS, Sappington TW, Siegfried BD, Lewis LC (2009) Comparative performance of single nucleotide polymorphism and microsatellite markers for population genetic analysis. J Hered 100(5):556–564

Collard BC, Mackill DJ (2009) Start codon targeted (SCoT) polymorphism: a simple, novel DNA marker technique for generating gene-targeted markers in plants. Plant Mol Biol Rep 27(1):86

Daga A, Gazit S, Eisenstein D, El-Batsri R, Degani C (1999) Effect of the male parent on pericarp and seed weights in several Floridian mango cultivars. Sci Hort 82:325–329

Dang ZG, Chen YY (2017) Construction of a genetic linkage map of mango based on SRAP, AFLP and ISSR Markers. Agri Biotechnol 6(6):9–12, 16

Dao Y (2019) RAPD analysis of genetic diversity of mango at Panzhihua region in Sichuan. Jian Agri Sci 47(13):35–38

Degani M, Cohen O, Reuveni RF, Basstri, x (1993) Frequency and characteristics of zygotic seedling from polyembryonic mango cultivars determined using isozyme as genetic markers. Acta Hort 341:256–259

De Keyser E, de Rick J, van Bockstaele E (2009) Discovery of species-wide EST-derived markers in Rhododendron by intron-flanking primer design. Mol Breed 23:171–178

Desai AR, Dhandar DG (2000) Variation in physico-chemical and morphogenetic characters of some mango varieties of Goa. Acta Hort 509:243–252

Dillon NL, Innes DJ, Bally ISE, Wright CL, Devitt LC, Dietzgen RG (2014) Expressed sequence tag-simple sequence repeat (EST-SSR) marker resources for diversity analysis of mango (*Mangifera indica* L.). Diversity 6:72–87

Diaz-Matallana M, Schuler-Garcia I, Ruiz-Garcia M, Jaramillo EHD (2009) Analysis of diversity among six populations of Colombian mango (*Mangifera indica* L. cvar. Hilacha) using RAPDs markers. Elec J Biotechnol 12(3):1–8

Dinesh MR, Ravishankar KV, Sthapit B, Parthasarathy VA, Sandya BS, Nischita P, Lavanya B (2015) Genetic diversity studies in certain Indigenous mango (*Mangifera indica* L) varieties. Indian J Plant Genet Resour 28(1):153–160

Duval MF, Bunel J, Sitbon C, Risterucci AM (2005) Development of microsatellite markers for mango (*Mangifera indica* L.). Mol Ecol Notes 4:824–826

Damodaran T, Kannan R, Ahmed I, Srivastava RC, Rai RB, Umamaheshwari S (2012) Assessing genetic relationships among mango (*Mangifera indica* L.) accessions of Andaman Islands using inter simple sequence repeat markers. NZ J Exp Agri 40(4):229–240

Eiadthong W, Yonemori K, Kanzaki S, Sugiura A, Utsunomiya N, Subhadrabandhu S (1999a) Amplified fragment length polymorphism (AFLP) analysis for studying the genetic relationships among *Mangifera* species in Thailand. J Am Soc Hort Sci 125:160–164

Eiadthong W, Yonemoni K, Kanzaki S, Sugiura A, Utsunomiya N, Subhadrabandhu S (1999b) Identification of mango cultivars of Thailand and evaluation of their genetic variation using the amplified fragments by simple sequence respect (SSR) anchored primers. Sci Hort 82:57–66

Ekblom R, Galindo J (2011) Applications of next generation sequencing in molecular ecology of non-model organisms. Heredity 107(1):1–15

Ellis JR, Burke JM (2007) EST-SSRs as a resource for population genetic analyses. Heredity 99:125–132

Fan WG, Luo Y, Wu SF, Ge HM (2012) Morphological Characteristics and distribution of wild germplasm resources of *Mangifera indica* in South and North Pan River Valley. SW Chin J Agri Sci 25(6):2244–2247

Fang JG, Liu DJ, Zhang Z, Hillel J, Lavi U (1999) The construction of the fingerprinting of two mango cultivars using AFLP. J Nanjing Agri Univ 22:25–27

Fang DQ, Roose ML (1997) Identification of closely related citrus cultivars with inter-simple sequence repeat markers. Theor Appl Genet 95(3):408–417

Fukuoka S, Hosaka K, Kamijima O (1992) Use of random amplified polymorphic DNAs (RAPDs) for identification of rice accessions. Jpn J Genet 67(3):243–252

Gajera HP, Tomar RS, Patel SV, Viradia RR (2011) Golakiya BA (2011) Comparison of RAPD and ISSR markers for genetic diversity analysis among different endangered *Mangifera indica* genotypes of Indian Gir forest region. J Plant Biochem Biotechnol 20(2):217–223

Gajera HP, Bambharolia RP, Domadiya RK, Patel SV, Golakiya BA (2013) Molecular characterization and genetic variability studies associated with fruit quality of indigenous mango (*Mangifera indica* L.) cultivars. Plant Syst Evol 300(5):1011–1020

Galal OA, Galal HA, Aboulila AA (2017) Genetic variability and molecular characterization of some local and imported mango cultivars in Egypt. Egyp J Genet Cytol 46(1):121–138

Gálvez-López D, Hernández-Delgado S, González-Paz M, Becerra-Leor EM, Salvador-Figueroa M, Mayek-Pérez N (2009) Genetic analysis of mango landraces from Mexico based on molecular markers. Plant Genet Resour 7(3):244–251

Gálvez-López D, Salvador-Figueroa M, Becerra-Leor EN, González-Paz M, Hernández-Delgado S, Mayek-Pérez N (2010) Molecular diversity and genetic relationships of mango germplasm from Chiapas. Mexico. Agrociencia 44(8):907–915

Ganogpichayagrai A, Rungsihirunrat K, Palanuvej C, Ruangrungsi N (2016) Characterization of *Mangifera indica* cultivars in Thailand based on macroscopic, microscopic, and genetic characters. J Adv Pharmaceut Technol Res 7(4):127–133

Gao AP, Chen YY, Jonathan HC, Zhu M, Huang JF, Luo HY, He YH (2010) AFLP analysis on the genetic diversity of two hundred accessions of mango in China. IX International Mango Symposium 992:295–307

Godwin ID, Aitken EA, Smith LW (1997) Application of inter simple sequence repeat (ISSR) markers to plant genetics. Electrophoresis 18(9):1524–1528

Gonzalez A, Coulson M, Brettell R (2002) Development of DNA markers (ISSRs) in mango. Acta Hort 575:139–143

Govindaraj M, Vetriventhan M, Srinivasan M (2015) Importance of genetic diversity assessment in crop plants and its recent advances: an overview of its analytical perspectives. Genet Res Intl. https://doi.org/10.1155/2015/431487

Gupta M, Chyi YS, Romero-Severson J, Owen JL (1994) Amplification of DNA markers from evolutionarily diverse genomes using single primers of simple-sequence repeats. Theor Appl Genet 89(7–8):998–1006

Gitahi R, Kasili R, Kyallo M, Kehlenbeck K (2016) Diversity of threatened local mango landraces on smallholder farms in Eastern Kenya. Forests Trees Livelihoods 25(4):239–254

Hartless AC (1913) The flowering of mango. Agri J India 8:9–12

He XH, Li YR, Guo YZ, Tang ZP, Li RB (2005) Genetic analysis of 23 mango cultivar collection in Guangxi Province revealed by ISSR. Mol Plant Breed 3 (6):829–834

He XH, Guo YZ, Li YR, Ou SJ (2007) Assessment of the genetic relationship and diversity of mango and its relatives by cpISSR marker. Agri Sci China 6:137–142

Hirano R, Oo TH, Watanabe KN (2010) Myanmar mango landraces reveal genetic uniqueness over common cultivars from Florida, India, and Southeast Asia. Genome 53(4):321–330

Huang LF (2010) Study on genetic diversity of rootstock germplasm resources from main mango producing areas by SSR marker. MSc Thesis, Hainan University, Haikou, China

Huang LF, Yan L, Fan R, Yao QS, Liu Y (2011) Genetic Diversity Analysis of Seedling Resources of Mango (*Mangifera indica* L.) by SSR Markers. Chin J Trop Crops 32(10):1828–1832

Human CF (2008) Production areas. In: deVilliers EA, Joubert PH (eds) The Cultivation of Mango. ARC Institute for Tropical and Subtropical Crops, Florida, USA, pp 5–64

Ibrahim M (1952) A study of the distinguishing vegetative, floral and fruit characteristics of Punjab Mangoes. MSc Thesis, University of Punjab, Lahore, Pakistan

IPGRI (2006) Descriptors for Mango (*Mangifera indica* L.). International Plant Genetic Resources Institute, Rome, Italy

Iyer CPA, Degani C (1997) Classical breeding and genetics. In: Litz RE (ed) The mango, botany, production, and uses. CAB International, Wallingford, Oxon, UK, pp 49–68

Jaccard P (1908) Nouvelles researches sur la distribution florale. Bulletin de la Société Vaudoise des Sciences Naturelles 44:233–270

Jayasankar S, Litz RE, Schnell RJ, Hernandez A (1998) Embryogenic mango cultures selected for resistance to colletotrichum gloeosporioides culture filtrate show variation in random amplified polymorphic DNA (RAPD) markers. Vitro Cell Dev Biol-Plant 34 (2):112–116

Jena RC, Samal KC, Chand PK, Das BK (2010) Molecular characterization of 12 mango germplasm using RAPD markers. Plant Tiss Cult Biotechnol 20(1):91–99

Jintanawongse S, Changtragoon S (2000) Identification of cultivars and certification of hybrids in mango (*Mangifera indica* L.) by isozymes gene markers. Acta Hort 509:177–184

Kanazin V, Talbert H, See D, DeCamp P, Nevo E, Blake T (2002) Discovery and assay of single-nucleotide polymorphisms in barley (*Hordeum vulgare*). Plant Mol Biol 48(5–6):529–537

Karibasappa GS (1995) Morpho-phenological, fruit physico-chemical and isozymes characterization of (*Mangifera indica* L) germplasm, using principal component analysis. MSc Thesis, Dharwad University of Agricultural Sciences, Dharwad, Karnataka, India

Karihaloo JL, Dwivedi YK, Archak S, Gaikwad AB (2003) Analysis of genetic diversity of Indian mango cultivars using RAPD markers. J Hort Sci Biotechnol 78(3):285–289

Karp A, Kresovich S, Bhat KV, Ayad WG, Hodgkin T (1997) Molecular tools in plant genetic resources conservation: a guide to the technologies. IPGRI Technical Bulletin 2, International Plant Genetic Resources Institute, Rome, Italy

Kashkush K, Jinggui F, Tomer E, Hillel J, Lavi U (2001) Cultivar identification and genetic map of mango. Euphytica 122:129–136

Kaur A, Ha CO, Jong K, Sands VE, Chan HT, Soepadmo E, Ashton PS (1980) Apomixis may be widespread among trees of the climax rain forest. Nature 271:440–442

Khan AA (1960) Relation of growth to fruit bearing in mangoes. Punjab Fruit J. 23:117–140

Khan SK, Ali S, Khan IA (2015) Morphological and molecular characterization and evaluation of mango germplasm: An overview. Sci Hort 194:353–366

Knight RJ Jr, Campbell RJ, Maguire I (2009) Important mango cultivars and their descriptors. In: Litz RE (ed) The Mango, Botany, Production and Uses. CAB International, Boca Raton, Florida, USA, pp 43–52

Krishna H, Singh SK (2007) Biotechnological advances in mango (*Mangifera indica* L.) and their future implication in crop improvement. Biotechnol Adv 25:223–243

Krishnapillai N, Wijeratna RSW (2016) Morphometric analysis of mango varieties in Sri Lanka. Aust J Crop Sci 10(6):784–792

Kuhn D, Dillon N, Innes D, Wu LS, Mockaitis K (2014) Development of single nucleotide polymorphism (SNP) markers from the mango (*Mangifera indica*) transcriptome for mapping and estimation of genetic diversity. XXIX International Horticultural Congress on Horticulture: Sustaining Lives, Livelihoods and Landscapes (IHC2014): IV 1111. 315–322

Kuhn DN, Bally ISE, Dillon NL, Innes D, Groh AM, Rahaman J, Ophir R, Cohen Y, Sherman A (2017) Genetic Map of mango: a tool for mango breeding. Front Plant Sci 8:577

Kuhn DN, Dillon N, Bally I, Groh A, Rahaman J, Warschefsky E, Freeman B, Innes D, Chambers AH (2019) Estimation of genetic diversity and relatedness in a mango germplasm collection using SNP markers and a simplified visual analysis method. Sci Hort 252:156–168

Kumar H, Narayanaswamy P, Prasad T, Mukunda GK, Sondur S (2001) Estimation of genetic diversity of commercial mango (*Mangifera indica* L.) cultivars using RAPD markers. J Hort Sci Biotechnol 76(5): 529–533

Kumari N, Thakur SK (2014) Randomly amplified polymorphic DNA-A brief review. Amer J Anim Vet Sci 9(1):6–13

Kumar KL, Sreekala AK (2016) Genetic diversity study, by using RAPD techniques among the selected natural varieties of *Mangifera Indica* L. occurring at Nedumangadu Taluk of the Thiruvananthapuram District in Kerala State of India. Imper J Interdiscipl Res 2(5): 1123–1126

Lakshminarayana S (1980) The mango. In: Nagy S, Shaw PE (eds) Tropical and subtropical fruits: composition, properties and uses. AVI Publishers, West port, Connecticut, USA, pp 184–257

Lal S, Singh AK, Singh SK, Srivastav M, Singh BP, Sharma N, Sing NM (2017) Association analysis for pomological traits in mango (*Mangifera indica* L.) by genic-SSR markers. Trees 31(5):1391–1409

Levi A, Thomas CE (2007) DNA markers from different linkage regions of watermelon genome useful in differentiating among closely related watermelon genotypes. HortScience 42(2):210–214

Levi A, Wechter P, Massey L, Carter L, Hopkins D (2011) An extended genetic linkage map for watermelon based on a testcross and a BC2F2 population. Amer J Plant Sci 2(2):93–110

Li G, Quiros CF (2001) Sequence-related amplified polymorphism (SRAP), a new marker system based on a simple PCR reaction: its application to mapping and gene tagging in *Brassica*. Theor Appl Genet 103:455–461

Li GP, Zhang LH, Xie DH, Yu YC, Chen YF, Chen HR, Wamg MC, Ni ZG (2010) Analysis of the correlations between fruit traits and diversity of quality traits within mango (*Mangifera indica* L.) germplasm resource in Nujiang hot-dry valley. SW China J Agri Sci 23(4):1225–1229

Li ZQ (2016) Construction of molecular genetic map and genetic diversity analysis in mango hybrid F1. MSc Thesis, Hainan University, Haikou, China

Li ZQ, Dang ZG, Zhao ZC, Huang JF, Gao AP, Chen YY (2016) Genetic diversity analyze of mango's F1 hybrids and construction of genetic map. Mol Plant Breed 14(4):953–958

Liu L, Liu G, Gong Y, Dai W, Wang Y, Yu F, Ren Y (2007) Evaluation of genetic purity of F1 hybrid seeds in cabbage with RAPD, ISSR, SRAP, and SSR markers. HortScience 42(3):724–727

Liu R, Gong DY, Liu QG, Huang H, Fan JX (2019) Genetic diversity analysis of thirteen mango germplasm resources based on SRAP molecular markers. Chin J Trop Crops 40(1):87–91

Litz RE (ed) (2003) The mango: botany, production and uses. Tropical Research and Education Centre, University of Florida, USA

Litz RE (2004) Biotechnology and mango improvement. Acta Hort 645:85–92

Luo C, He XH, Chen H, Ou SJ, Gao MP (2010) Analysis of diversity and relationships among mango cultivars using Start Codon Targeted (SCoT) markers. Biochem Syst Ecol 38(6):1176–1184

Luo C, He XH, Chen H, Ou SJ, Gao MP, Brown JS, Tondo CT, Schnell RJ (2011) Genetic diversity of mango cultivars estimated using SCoT and ISSR markers. Biochem Syst Ecol 39:676–684

Luo C, He XH, Chen H, Hu Y, Ou SJ (2012) Genetic relationship and diversity of *Mangifera indica* L.: Revealed through SCoT analysis. Genet Resour Crop Evol 59(7):1505–1515

Luo C, Wu HX, Yao QS, Wang SB, Xu WT (2015a) Development of EST-SSR and TRAP markers from transcriptome sequencing data of the mango. Genet Mol Res 14(3):7914–7919

Luo HY, Ying DS, Huang JF, Peng SN, Hong JW, Wang M, Chen YY (2015b) AFLP analysis of the genetic diversity of 78 mango germplasms resources. Chin J Trop Crops 36(12):2130–2137

Mammadov J, Aggarwal R, Buyyarapu R, Kumpatla S (2012) SNP markers and their impact on plant breeding. Intl J Plant Genom 2012, doi:10.1155/2012/728398

Mansour H, Mekki LE Hussein MA (2014) Assessment of genetic diversity and relationships among Egyptian mango (*Mangifera indica* L.) cultivers grown in Suez Canal and Sinai region using RAPD markers. Pak J Biol Sci 17(1):56–61

Mason AS (2015) SSR genotyping. In: Batley J (eds) Plant genotyping. Methods in molecular biology (Methods and Protocols), vol 1245. Humana Press, New York, pp 77–89

Matallana DM, Garcia SI, Garcia RM, Jaramillo DEH (2009) Analysis of diversity among six populations of Colombian mango (*Mangifera indica* L. cv. Hilacha) using RAPDs markers. Elec J Biotechnol 12:1–8

Majumder DAN, Hassan L, Kabir MA (2011) Genetic diversity of mango (*Mangifera indica* L. Anacardiaceae) detected by RAPD markers. Intl J Agri Environ Biotechnol 4(1):45–51

Meyer W, Mitchell TG, Freedman EZ, Vilgalys R (1993) Hybridization probes for conventional DNA fingerprinting used as single primers in the polymerase chain reaction to distinguish strains of Cryptococcus neoformans. J Clin Microbiol 31(9):2274–2280

Mo R, Luo YH, Zhou SM, Liu JP (2005) Polyembryony in mango (*Mangifera indica* L.) and genetic analysis. J Trop Subtrop Bot 13(6): 475– 479

Mohammadi SA, Prasanna BM (2003) Analysis of genetic diversity in crop plants–salient statistical tools and considerations. Crop Sci 43(4):1235–1248

Mohan M, Nair S, Bhagwat A, Krishna TG, Yano M, Bhatia CR, Sasaki T (1997) Genome mapping, molecular markers and marker-assisted selection in crop plants. Mol Breed 3(2):87–103

Mukherjee SK, Rao MM (1943) Studies on blossom biology and pollination in mangoes (*Mangifera indica* L.). Indian J Hort 1:107–119

Mukherjee SK (1997) Introduction: botany and importance. In: Litz RE (ed) The mango: botany, production and uses. CAB International, Boca Raton, Florida, USA, pp 1–19

Mukherjee SK, Litz RE (2009) Introduction: Botany and Importance. In: Litz RE (ed) The mango, 2nd edn: botany, production and uses. CAB International, New York, USA, pp 1–18

Mussane CRB (2010) Morphological and genetic characterization of mango (*Mangifera indica* L.) varieties in Mozambique. MSc Thesis, University of the Free State, Bloemfontein, South Africa

Nazish T, Shabbir G, Ali A, Sami-ul-Allah S, Naeem M, Javed M, Seher R (2017) Molecular diversity of Pakistani mango (*Mangifera indica* L.) varieties based on microsatellite markers. Genet Mol Res 16(2): 2–8

Nei M, Li WH (1979) Mathematical model for studying genetic variation in terms of restriction endonucleases. Proc Natl Acad Sci USA 76(10):5269–5273

Pandit SS, Mitra SM, Giri AP, Pujari KH, Patil BP, Jambhale ND, Gupta VS (2007) Genetic diversity analysis of mango cultivars using inter simple sequence repeat markers. Curr Sci 93(8):1135–1141

Phumichai C, Phumichai T, Wongkaew A (2015) Novel chloroplast microsatellite (cpSSR) markers for genetic diversity assessment of cultivated and wild Hevea rubber. Plant Mol Biol Rep 33(5):1486–1498

Prasad A, Prasad A (1972) Performance studies on polyembryonic varieties of mango (*Mangifera indica* L.). Acta Hort 24:31–35

Pruthvish R, Chikkaswamy BK (2016) Genetic diversity and relationships among mango varieties using RAPD molecular markers. Intl J Curr Microbiol Appl Sci 5 (1):778–787

Rahman ML, Rabbani MG, Siddique MNA, Rahman MA, Garvey EJ, Rahaman EHMS (2007) Molecular characterization of 28 mango germplasm using RAPD. Plant Tiss Cult Biotechnol 17(1):71–77

Rajwana IA, Tabbasam N, Malik AU, Malik SA, Rahman M, Zafar Y (2008) Assessment of genetic diversity among mango (*Mangifera indica* L.) genotypes using RAPD markers. Sci Hort 117:297–301

Rajwana IA, Khan IA, Malik AU, Saleem BA, Khan AS, Ziaf K, Anwar R, Amin M (2011) Morphological and bio-chemical markers for varietal characterization and quality assessment of potential indigenous mango (*Mangifera indica* L.) germplasm. Intl J Agri Biol 13:151–158

Ramessur AD, Ranghoo-Sanmukhiya VMS (2011) RAPD Marker-assisted identification of genetic diversity among mango (*Mangifera indica*) varieties in Mauritius. Intl J Agri Biol 13:167–173

Ramírez F and Davenport TL (2010) Mango (*Mangifera indica* L.) flowering physiology. Sci Hort 126 (2):65–72

Ramírez F and Davenport TL, F (2016) Mango (*Mangifera indica* L.) pollination. A review. Sci Hort 203:158–168

Ratnaparkhe MB, Tekeoglu M, Muehlbauer FJ (1998) Inter-simple-sequence-repeat (ISSR) polymorphisms are useful for finding markers associated with disease resistance gene clusters. Theor Appl Genet 97(4):515–519

Ravishankar KV, Bommisetty P, Bajpai A, Srivastava N, Mani BH, Vasugi C, Rajan S, Dinesh MR (2015a) Genetic diversity and population structure analysis of mango (*Mangifera indica*) cultivars assessed by microsatellite markers. Trees 29(3):775–783

Ravishankar KV, Dinesh MR, Nischita P, Sandya BS (2015b) Development and characterization of microsatellite markers in mango (*Mangifera indica*) using next-generation sequencing technology and their transferability across species. Mol Breed 35(93):1–13

Raza SA, Khan AS, Khan IA, Rajwana IA, Ali S, Khan AA, Rehman A (2017) Morphological and physico-chemical diversity in some indigenous mango (*Mangifera indica* L.) germplasm of Pak J Agric Sci 54(2):287–297; 2017

Reddy MP, Sarla N, Siddiq EA (2002) Inter simple sequence repeat (ISSR) polymorphism and its application in plant breeding. Euphytica 128(1):9–17

Rhodes MA, Campbell G, Malo SE, Camer SG (1970) A numerical taxonomic studies of the mango (*Mangifera indica* L.). J Am Soc Hort Sci 95:252–256

Rocha A, Salomão LCC, Salomao TMF, Cruz CD, Siqueira DL (2012) Genetic diversity of 'Uba'mango tree using ISSR markers. Mol Biotechnol 50(2):108–113

Rymbai H, Laxman RH, Dinesh MR, John Sunoj VS, Ravishankar KV, Jha AK (2014) Diversity in leaf morphology and physiological characteristics among mango (*Mangifera indica*) cultivars popular in different agro-climatic regions of India. Sci Hort 176:189–193

Sachar RC, Chopra RN (1957) A study of the endosperm and embryo in *Mangifera indica* L. Indian J Agric Sci 27:219–228

Samal KC, Jena RC, Swain SS, Das BK, Chand PK (2012) Evaluation of genetic diversity among commercial cultivars, hybrids and local mango (*Mangifera indica* L.) genotypes of India using cumulative RAPD and ISSR markers. Euphytica 185(2):195–213

Samant D, Singh AK, Srivastav M, Singh NK (2010) Assessment of genetic diversity in mango using inter simple sequence repeat markers. Indian J Hort 67:1–8

Sandip M, Sandip AN, Barad AV, Nawade BD (2015) Physiology of flowering - the case of mango. Intl J Appl Res 1(11):1008–1012

Sankar AA, Moore GA (2001) Evaluation of inter-simple sequence repeat analysis for mapping in Citrus and extension of the genetic linkage map. Theor Appl Genet 102(2–3:206–214

Schnell RJ, Brown JS, Olano CT, Meerow AW (2006) Mango genetic diversity analysis and pedigree inferences for Florida cultivars using microsatellite markers. J Am Soc Hort Sci 131:214–224

Schnell RJ, Ronning CM, Knight RJ (1995) Identification of cultivars and validation of genetic relationships in *Mangifera indica* L. using RAPD markers. Theor Appl Genet 90(2):269–274

Schnell RJ, Olano CT, Quintanilla WE, Meerow AW (2005) Isolation and characterization of 15 microsatellite loci from mango (*Mangifera indica* L.) and cross-species amplification in closely related taxa. Mol Ecol Resour 5(3):625–627

Selkoe KA, Toonen RJ (2006) Microsatellites for ecologists: a practical guide to using and evaluating microsatellite markers. Ecol Lett 9(5):615–629

Semagn K, Bjørnstad Å, Ndjiondjop MN. An overview of molecular marker methods for plants. Afr J Biotechnol 5(25):2540–2568

Sennhenn A, Prinz K, Gebauer J, Whitbread A, Jamnadass R, Kehlenbeck K (2014) Identification of mango (*Mangifera indica* L.) landraces from Eastern and Central Kenya using a morphological and molecular approach. Genet Res Crop Evol 61:7–22

Shamili M, Fatahi R, Hormaza JI (2012) Characterization and evaluation of genetic diversity of Iranian mango (*Mangifera indica* L. Anacardiaceae) genotypes using microsatellites. Sci Hort 148:230–234

Shenoy SB, Rajagopala A (2016) Morphological and anatomical characters related to dwarfness in mango (*Mangifera indica* L.). J Trop Agric 54 (1):84–86

Sherman A, Rubinstein M, Eshed R, Benita M, Ish-Shalom M, Sharabi-Schwager M, Ophir R (2015) Mango (*Mangifera indica* L.) germplasm diversity based on single nucleotide polymorphisms derived from the transcriptome. BMC Plant Biol 15(1):277

Shi SY, Wu HX, Wang SB, Liu LQ, Wang YC, Ma WH (2011) Genetic diversity of mango germplasm based on morphological characters and AFLP markers. Acta Hort Sin 38(3):449–456

Singh LB (1968) The mango botany, cultivation and uses. Leonard Hill Books, London, UK
Singh A, Singh AK, Singh SK (2012) SSR markers reveal genetic diversity in closely related mango hybrids. Indian J Hort 69(3):299–305
Sokal RR, Michener CD (1958) A statistical method for evaluating systematic relationships. Univ Kansas Sci Bull 38:1409–1438
Souza IGB, Valente SES, Britto FB, de Souza VAB, Lima PDC (2011) RAPD analysis of the genetic diversity of mango (*Mangifera indica*) germplasm in Brazil. Genet Mol Res 10(4):3080–3089
Squirrell J, Hollingsworth PM, Woodhead M, Russell J, Lowe AJ, Gibby M, Powell W (2003) How much effort is required to isolate nuclear microsatellites from plants? Mol Ecol 12(6):1339–1348
Sturrock TT (1968) Genetics of mango polyembryony. Proc Fla State Hort Soc 82:318–321
Surapaneni M, Vemireddy LR, Begum H, Reddy BP, Neetasri C, Nagaraju J, Anwar SY, Siddiq EA (2013) Population structure and genetic analysis of different utility types of mango (*Mangifera indica* L.) germplasm of Andhra Pradesh state of India using microsatellite markers. Plant Syst Evol 299:1215–1229
Tenaillon MI, Sawkins MC, Long AD, Gaut R L, Doebley J F, Gaut BS (2001) Patterns of DNA sequence polymorphism along chromosome 1 of maize (*Zea mays* ssp. mays L.). Proc Natl Acad Sci USA 98(16):9161–9166
Tisserat B, Esen EB, Murashige T (1979) Somatic embryogenesis in angiosperms. Hort Rev 1:1–72
Tomar RS, Gajera HP, Viradiya RR, Patel SV, and Golakiya, BA (2011) Phylogenetic relationship among Mango (*Mangifera indica* L.) landraces of Saurashtra based on DNA fingerprinting. J Hort For 3(13):379–385
Tsai CC, Chen YKH, Chen CH, Weng IS, Tsai CM, Lee SR, Lin YS, Chiang YC (2013) Cultivar identification and genetic relationship of mango (*Mangifera indica)* in Taiwan using 37 SSR markers. Sci Hort 164:196–201
Uzun AYDIN, Yesiloglu T, Aka-Kacar Y, Tuzcu O, Gulsen O (2009) Genetic diversity and relationships within Citrus and related genera based on sequence related amplified polymorphism markers (SRAPs). Sci Hort 121(3):306–312
Warschefsky EJ, von Wettberg EJ (2019) Population genomic analysis of mango (*Mangifera indica*) suggests a complex history of domestication. New Phytol 222(4):2023–2037
Wang M, Ying DS, Wang QF, Li LP, Zhang RL (2015) Genetic diversity analysis and fingerprint construction for mango cultivars in China. J Southern Agric 46 (7):1154–1159
Whitton J, Rieseberg LH, Ungerer MC (1997) Microsatellite loci are not conserved across the Asteraceae. Mol Biol Evol 14(2):204–209
Xie DH, Zhang GX, Zhang YC, Ni ZG,Chen YF, Zhang FM, Bai TQ, Hao XH (2018) Fruit trait diversity of mango(*Mangifera indica* L.)germplasm resources in Lancang River basin. J Southern Agric 49 (2):214–221
Xie JH, Liu CM, Ma WH, Lei XT (2005) Analysis of genetic relationship among mango germplasm by RAPD markers. J Fruit Sci 22(6):649–653
Xiong M, Jin L (1999) Comparison of the power and accuracy of biallelic and microsatellite markers in population-based gene-mapping methods. Am J Hum Genet 64(2):629–640
Xu BY, Jin ZQ, Peng SQ, Xu SP, Zheng XQ (1998) RAPD Analysis of genomic DNA in mango cultivars in Hainan Island. Chine J Trop Crops 19(3):33–36
Xu WT, Wu HX, Li L, Liang QZ, Luo C, Yao QS, Wang SB (2018) Genetic differences and cluster analysis of pollen quantity of 53 mango cultivars. Chin J Trop Agric 38(10):26–30
Yamanaka N, Hasran M, Xu DH, Tsunematsu H, Idris S, Ban T (2006) Genetic relationship and diversity of four *Mangifera* species revealed through AFLP analysis. Genet Resour Crop Evol 53:949–954
Yamanaka S, Hosaka F, Matsumura M, Onoue-Makishi Y, Nashima K, Urasaki N, Yamamoto T (2019) Genetic diversity and relatedness of mango cultivars assessed by SSR markers. Breed Sci 69:332–344
Ying DS (2012) Construction of 78 mango germplasm fingerprinting. MSc Thesis, Hainan University, Haikou, China
Yu K, Pauls KP (1992) Optimization of the PCR program for RAPD analysis. Nucleic Acids Res 20(10):2606
Zane L, Bargelloni L, Patarnello T (2002) Strategies for microsatellite isolation: a review. Mol Ecol 11(1):1–16
Zhang CX, Ni ZG, Chen HR, Chen YF, Xie DH, Long YQ, Zhang FM (2014) Genetic diversity of mango (*Mangifera indica* L) in Nujiang Dry-hot valley revealed by morphological characters and AFLP marker. J Plant Genet Resour 15(4):753–758
Zhang CX, Xie DH, Bai TQ, Luo XP, Zhang FM, Ni ZG, Chen YF (2020) Diversity of a large Collection of natural populations of mango (*Mangifera indica* Linn.) revealed by Agro-Morphological and quality traits. Diversity 12:27
Zhang J, Xie W, Wang Y, Zhao X (2015) Potential of start codon targeted (SCoT) markers to estimate genetic diversity and relationships among chinese *Elymus sibiricus* accessions. Molecules 20(4):5987–6001
Zhang Y, Huang G, Huang Q (2016) Analysis on genetic relationship of some mango (*Mangifera indica* L.) germplasms by SRAP markers and ISSR markers. SW China J Agric Sci 29(9):2045–2051
Zhang ZS, Hu MC, Zhang J, Liu DJ, Zheng J, Zhang K, Wan Q (2009) Construction of a comprehensive PCR-based marker linkage map and QTL mapping for fiber

quality traits in upland cotton (*Gossypium hirsutum* L.). Mol Breed 24(1):49–61

Zhou L, Luo C, He TX, Yu HX, He XH (2019) Analysis of Genetic Diversity and Relationship of Mango Varieties based on EST-SSR Markers. Mol Plant Breed http://kns.cnki.net/kcms/detail/46.1068.S.20190513.1420.002.html

Zhou L, He XH, Yu HX, Chen MY, Fan Y, Zhang XJ, Fang ZB, Luo C (2020) Evaluation of the genetic diversity of mango (*Mangifera indica* L.) seedling germplasm resources and their potential parents with start codon targeted (SCoT) markers. Genet Resour Crop Evol 67:41–58

Zietkiewicz E, Rafalski A, Labuda D (1994) Genome fingerprinting by simple sequence repeat (SSR)-anchored polymerase chain reaction amplification. Genomics 20(2):176–183

Alternate Flowering in Mango

6

Hutchappa Ravishankar, Nimisha Sharma, and V. K. Singh

Abstract

Alternate bearing is the tendency of fruit trees to produce a heavy crop in 1 year (on-year), followed by a light crop or no crop production in the next (off-year). This phenomenon in perennial fruit crops appeared to be governed by positive and negative regulatory factors and their interplay decided the fate of meristems under optimum environmental cues. Mango production dynamics in different parts of the world in the recent years are gripped with considerable uncertainties arising due to the occurrence of weather extremes, inadequacies of water, and nutrient management, and impacts of altered pests and disease prevalence. In mango especially, in which the issue though has been investigated extensively yet the problem eludes tangible solutions, the choice varieties in the Indian sub-continent continue to be gripped with the alternate bearing rhythm resulting in low productivity and economics of commercial orcharding. Greater and in-depth understanding of mechanisms involved in floriferous condition of mango fruit crop is essential to strategize appropriate orchard protocols to minimize or smoothen alternate bearing rhythms. Further, genomics and transcriptome approaches along with physiological events, elucidate the complete understanding of bearing and non-bearing tendency of mango crop.

Abbreviations

AB	Alternate bearing
ABI	Alternate bearing index
ABI	Alternate Bearing Intensity
ABA	Abscisic acid
AFL2	Apple floricula/lfy
AP	Apetala
BBI	Biennial bearing index
BFT	Brothers of ft
CH	Carbohydrates
CiFT	Citrus flowering locus T
CO	Constans
DEGs	Differentially expressed genes
FBD	Fruit bud differentiation
FI	Flower induction
FLC	Flowering locus c
FUL	Fruitful
GAs	Gibberellins

H. Ravishankar · V. K. Singh
ICAR-Central Institute for Subtropical Horticulture, Lucknow 227107, UP, India
e-mail: drhravishankar@gmail.com

V. K. Singh
e-mail: vksingh.icar@gmail.com; singhvk_cish@rediffmail.com

N. Sharma (✉)
Division of Fruits and Horticultural Technology, ICAR-Indian Agricultural Research Institute, New Delhi 110012, India
e-mail: nims17sharma@gmail.com

C. Kole (ed.), *The Mango Genome*, Compendium of Plant Genomes,
https://doi.org/10.1007/978-3-030-47829-2_6

GUs	Growth units
LFY	Leafy
MABI	Modified alternate bearing index
miRNA	Micro RNA
NGS	Next-generation sequencing
SAMs	Shoot apical meristems
SoC1	Suppressor of over expression of constants
SPL5	Squamosa promoter binding
SVP	Short vegetative phase
TFL1	Terminal flower1

6.1 Introduction

Alternate bearing (synonymously termed as a biennial or uneven bearing) is the tendency of fruit trees to produce a heavy crop in 1 year (on-year), followed by a light crop or no crop production in the next (off-year). This phenomenon is widespread among fruit trees, including both deciduous and evergreen (Monselise and Goldschmidt 1982).In mango especially, in which the issue though has been investigated extensively, the problem eludes tangible solutions, the choice varieties in the Indian sub-continent continue to be gripped with the alternate bearing rhythm resulting in low productivity and poor economics of commercial orcharding. The International Mango Research Network targets to achieve mango productivity of 40 tons ha^{-1} in order to make the world mango industry sustainable. The prevailing scenario of mango productivity world over is not more than 17–18 tons ha^{-1} (Brazil) while in India it being 7–8 tons ha^{-1}. Flowering of plants is a complicated developmental process that involves a series of morphological and physiological stages under the control of a number of external signals and internal factors. Genes have been identified as genetically governing floriferous identity of meristems and fruit development. The expression of these genes is reportedly regulated by photoperiod and hormones. Alternate bearing though found triggered by a variety of factors, the direct cause of alternate bearing is the lack of floral buds in the '*Off*' year rather than lack of fruit set. Greater and in-depth understanding of the mechanisms involved in floriferous condition of fruit trees is essential to strategize appropriate orchard protocols to minimize or smoothen alternate bearing rhythms. In the ensuing review, though based on a few fruit crops, we have tried to critically present the status on understanding of the problem, the prevailing gaps, the challenges ahead, and the way forward.

6.2 The Alternate Bearing Phenomenon

Flowering is an important event in the plant life cycle, as blooming is the key factor in fruit production. Under favorable growth conditions the timing and intensity of flowering greatly determine where and how much fruits are produced. However, in several times there are very contrasting phenomena where prolific flowering takes place without appreciable fruit set. Many important details about flowering are becoming clearer, especially in annual/herbaceous plants at the physiological, biochemical, and molecular level, but this complex biological phenomenon is less understood in perennial fruit crops. Flowering is promoted by photoperiod, temperature or autonomous factors, or some combinations thereof (Léchaudel et al. 2005), suggesting that flowering is triggered environmentally and genetically. Thus, to gain further insights into environmental adaptation, it is important to understand the regulation of seasonal flowering in fruits like mango, litchi, guava exhibiting different flowering characteristics in varying agro climates both in the tropics and subtropics. Fruits like mango, when grown in the subtropics (latitude 23°–30°) where substantial seasonal temperature changes occur, show floral induction from exposure to night temperatures of 10–15 °C (Davenport 2007). The flowering response to temperature occurs in mangoes growing in subtropical latitudes where cool temperature is the dominant induction factor. The effects of temperature and water relations in determining vegetative and reproductive growth of mango have been addressed by several research groups (Davenport 2000; Kulkarni 2004). Many cultivars flower erratically in the low-latitude tropics, providing continuously warm temperatures with high soil and atmospheric moisture. Under such conditions, the age of stems is the dominant inductive factor (Ravishankar et al. 1979; Batten and Mc-Conchie 1995; Davenport 2007), and occasional cool night temperatures in the upper latitude tropics have a positive moderating effect (Davenport 2003). In tropics, the age of the last leaf was recently considered as dominant factor regulating the flowering (Davenport 2011).

6.2.1 Quantification of Alternate Bearing

Alternate bearing (AB) phenomenon refers to the tendency of a fruit tree to produce a high yield (*'On'* year) followed by a low yield (*'Off'* year). This results in the operation of fundamental hypothesis that yield in 1 year affects yield in the subsequent year, however necessitating then need to quantify the extent to which this serial dependence takes place (Rosen stock et al. 2010). Evaluation of Alternate Bearing Intensity (ABI) has been a matter of intense deliberations by many researchers of fruit crops in the recent past as it exerted severe negative economic impacts on the commercial fruit industry. The occurrence of AB is also highly variable among the mango (*Mangifera indica* L.) varieties, across mango producing agro-ecologies impacted by diverse factors. Different researchers have attempted to address this problem from many angles in the past but without appreciable success (Ravishankar 1987; Chacko 1991). However, a chemical manipulation of this problem had been successfully demonstrated (Singh and Singh 2003) which, however, could have serious consequences on the tree and environment health if pursued on a continuum besides posing the problem of consistency of results obtained across varieties and geographical areas. Despite existence of voluminous information on understanding of AB phenomenon in mango, no study till date statistically outlined the AB in this species that captured the true behavior of the crop/variety. There existed extensive biological/horticultural literature on AB, although they are crop-specific with different issues dealt with being relevant for the respective crops. These broadly though highlighted different explained and unexplained factors of AB, many of them however, failed to come to grips with the precise approaches for quantification of ABI including mango. Evaluation of ABI itself has been a matter of considerable interest to researchers of perennial fruit trees for quite some time. Based upon an exhaustive survey of literature available on the subject, we strongly believe that the precise evaluation of degree of AB has either been underestimated or seriously misestimated in majority of the studies. For example, Hoblyn et al. (1936) proposed two parameters *'B'* and *'I'* to quantify alternate bearing tendency, with *'B'* expressing bienniality as a percentage of occasions (pairs of successive years) where each trend of increase or decrease in yield is reversed in calculations. This was subsequently re-evaluated by Pearce and Doberšek-Urbanc (1967) who concluded that *'B'* is very insensitive, as alternation of positive and negative signs could continue by chance for 8 years so that cent percent value became significant at the 5% level, only if obtained after 9 consecutive years. Full bienniality as per this approach, however, may not accrue in actuality due to its very nature of simplicity, parameters used for calculation of bienniality over long period of years. The other index *'I'* is a measure of intensity of deviation in yield in successive years which is estimated by adopting the procedure outlined by Pearce and Doberšek-Urbanc (1967).

The *'I'* value was applied for the first time in mango to quantify ABI, following the application of different treatments to manage alternate bearing (Ravishankar 1987) in 'Alphonso' mango by applying the method of Pearce and Doberšek-Urbanc(1967). Huff (2001) reported that $I = 1$, showed an extreme alternate bearing pattern in which no crop being produced in *'Off'* years and $I = 0$, showed an absence of influence between successive crops, rather than complete absence of variation. It was opined that '*I*', a descriptive statistics had a limited value in the absence of a test for significance (Huff 2001).

The absence of statistical testing, the long prevailing bias of '*I*', and the differential behavior of mango varieties across agro-ecologies/years generated questions about the predominance of AB problem in mango. Singh et al. (2014) statistically tested AB occurrence using one hundred physiologically matured 25 years old 'Langra' mango trees grafted on seedling rootstock with 5 years of production data under uniform cultural conditions. The study tested the null hypothesis that yields were random patterns against the alternative hypothesis that yield of subsequent year is dependent on the yield of previous year, along standing premise to define AB phenomenon by adopting time series methods which could be very useful for documenting and predicting the magnitude of yield variations across years based on historical yield data. Their study could possibly assist researchers based on at least 10 years yield data in the development of predictive models of yield in perennial fruit crops as a component of decision support to empower growers for optimizing inputs management for sustained tree health despite being gripped with alternate bearing rhythms.

6.3 Mango (*Mangifera indica* L.)

Different varieties of mango exhibited distinct patterns of alternate bearing tendency manifesting '*On*' and '*Off*' phases in fruit production. An ideal mango variety is described as the one that had characters of regularity in bearing; dwarf tree size; precocity, high yield (40 MT ha^{-1} with 40 percent 'A' grade (>250 g) fruits); quality (attractive color; 21° Brix, firm pulp consistency, freedom from spongy tissue/internal breakdown: improved shelf-life; for export less sweet); resistance to pests and diseases. Till date, the world mango industry is yet to harness such an ideal variety.

Flowering is an important event in the plant life cycle, as it is the key factor of fruit production. Under favorable growth conditions, the timing and intensity of flowering greatly determined where and how much fruits were produced. However, several times there existed very contrasting phenomena where prolific flowering took place without appreciable fruit set. Many important details about flowering are now becoming clearer, especially in annual/herbaceous plants at the physiological, biochemical, and molecular levels, but this complex biological phenomenon continued to remain less understood in perennial tree fruit crops. Flowering was shown promoted by photoperiod, temperature or autonomous factors, or some combinations thereof (Léchaudel et al. 2005; Léchaudel and Joas 2007), suggesting that flowering was triggered environmentally and genetically. Thus, to gain further insights into environmental adaptation, it is important to understand the regulation of seasonal flowering in fruits like mango, litchi, guava exhibiting different flowering characteristics in varying agro climates both in tropics and subtropics. Fruits like mango, when grown in the subtropics (latitude 23°–30°) where substantial seasonal temperature changes occurred, showed floral induction from exposure to night temperatures of 10–15 °C (Ravishankar 1979; Davenport 2007). The flowering response to temperature occurred in mangoes growing in subtropical latitudes where cool temperature played the dominant induction factor. The effects of temperature and water relations on determining vegetative and reproductive growth of mango have been fairly well addressed by different researchers (Davenport 2000; Kulkarni 2004). Many cultivars flowered erratically in the low-latitude tropics, providing continuously warm temperatures with high soil and atmospheric moisture. Under such conditions, the age of stems was the dominant inductive factor (Ravishankar et al. 1979; Batten and Mc-Conchie 1995; Davenport 2007; Ramirez and Davenport 2012), and occasional cool night temperatures in the upper latitude tropics had a positive moderating effect (Davenport 2003). In the tropics, age of the last leaf was recently considered as the dominant factor-regulating flowering (Davenport 2011).

Floral stimulus, most commonly photoperiod or temperature, leads to floral initiation. The juvenile phase was the first which lasted for several years during which time no flowering or fruiting occurred. Thereafter, the interactions between vegetative growth, flowers, and fruit of the previous year on floral initiation in the current year, affected production through phenomenon such as alternate bearing. Verreynne and Lovatt (2009) reported that for the 'Pixie' mandarin in California, the alternate bearing cycles appeared to be a crop load-dependent inhibitory effect of fruits on bud-break.

Two major differences reportedly existed between tropical and temperate deciduous fruit trees with respect to floral initiation. The tropical species such as mango initiated flowers in response to an environmental stimulus, while temperate deciduous species, such as apple, initiated flowers autonomously. On the other hand, temperate deciduous fruit trees underwent, a period of dormancy between floral initiation and anthesis. While in the tropical species, including mango, floral development was continuous from floral induction to anthesis. Induction of flowering in the subtropics was found primarily governed by chilling temperatures from passing cold fronts during winter-spring months. The age of the previous flush modified the cool-temperature-induced floral response, with older stems exhibiting a higher probability of a floriferous response and younger stems displaying a higher probability of a vegetative response. In the tropics, however, the age of the last flush is the dominant factor-regulating flowering (Chen and Huang 2001). Several researchers have averred that stems must be in rest for sufficient time, generally about 4–5 months to be induced to flower in the absence of chilling temperatures. This extended rest period occurred naturally as trees increased in architectural configuration, but it could also be achieved by mild plant water stress or ensuring low nitrogen fertility. Some studies also indicated that moderately cool temperatures that often reached deep into tropical dry and high elevation locations provided additional stimulus to flower in stems of a given age. Effect of different parameters and their functions are presented in Table 6.1.

Armed with the basic information provided here, growers could be empowered to manage flowering responses to occur at any desired week of the year. Local environmental conditions may, however, alter the expected responses, but it is hoped that scrutiny of all of the factors in decision support mode to develop management strategies should bring consistent success.

No single hypothesis could, however, explain the problem of alternate bearing or irregular flowering. A number of theories have been proposed by the researchers from time to time to provide answer to this problem, but very few of them have met with some success and the puzzle continues.

6.3.1 Environmental Control of Floral Initiation

With respect to floral initiation, mango has been more extensively researched than any other tropical or subtropical tree species. The juvenile period for mango varied with cultivars but could be approximately 3 years. Floral initiation occurred during late autumn and early winter; flower panicles emerged from the terminal and subterminal buds and grew continuously until anthesis occurred in the spring. Phenomenon of '*cauliflory*' also occurred. Mango flowering occurred during the coolest months of the year, and required 4–6 weeks of shoot dormancy and cool night temperatures to trigger floral induction of the terminal shoot apical meristems (SAMs). The absolute temperature needed for floral induction varied among cultivars and climates, but night temperatures between 8 and 15 °C (46–59°F) with day temperatures around 20 °C (68° F) were typically needed. Better flowering was seen in trees growing in the subtropics where the seasonal temperature differences were stronger

Table 6.1 Impact of hormones, carbohydrate metabolism, and genetic factors on alternate bearing in mango

Factor	Summary	References
Hormone	Autonomous GA pathways play an important role in both biennial bearing and flowering control in mango. Hormone study	Nakagawa et al. (2012), Kachru et al. (1972), Tongumpai et al. (1991), Nuñez-Elisea and Davenport (1991), (1998), Blaikie et al. (2004), Davenport (2007, 2009), Bangerth et al. (2004), Bangerth (2009)
Flowering	Flower bud phenomenon Studied florigenic promoter, required to induce flowering in mango	Blaikie and Kulkarni (2002), Ramirez et al. (2010), Ramírez and Davenport (2010, 2012), Tiwari et al. (2018)
Carbohydrate Metabolism	Source-sink relationship Biochemical changes in the leaf of bearing and non-bearing trees of some mango (*Mangifera indica* L.) varieties Studies on changes in carbohydrate metabolism in regular bearing and 'Off' season bearing cultivars of mango during flowering. Fruit load study	Chacko and Randhawa 1971; Kalayanaruk et al. (1982), Ravishankar (1987), (Whiley et al. 1989; Stitt and Quick 1989; Chacko (1991), Urban et al. (2004), Urban Alphonsout (2007), Singh and Rajan (2009), Singh et al. (2011), Jyothi et al. (2010), Shivu Prasad et al. (2014), Urban et al. (2004))
Differential Gene Expression Studies	Differential expression of flowering control genes like *FT*, *LFY*, *AP1*, *TFL*, and *miR156*-regulated *SPL5* was seen due to the effect of fruit load in leaves and buds of mango. Differential pathways studies in regular and irregular mango varieties. Deep RNA sequencing of regular and irregular varieties of mango.	Nakagawa et al. (2012), Sharma et al. (2015, 2017, 2019, 2020), Mahato et al. (2016)
Association Mapping Studies	Association mapping studies were done in 60 mango genotypes using 100 genic SSR markers. Phylogenetic analysis showed that mostly apple and olive genes have shown close association with genus *Mangifera*.	Lal (2016), Lal et al. (2017), Sharma et al. (2020)

and more reliable than in the hot tropics. In Hawaii, the main flowering was found between December and April (Bally 2011).

Floral induction in mango as mentioned above occurred in response to cool temperatures perceived by mature leaves (*L*), which were necessary for floral initiation. Flower panicles (*FP*) originated from terminal or subterminal buds of the most recent vegetative flush. Many mango cultivars flowered irregularly. This has been largely attributed to variation in flowering and fruit retention in subtropical areas. Variations in the amount of flowering may be within trees from year to year, between trees in the same year, and between branches on the same tree besides variation in flowering behavior within the varieties (Ravishankar 1979). Thus, even in bearing trees, there are non-bearing units which can be attributed to the differential potentiality of the shoots to form flower buds (Singh et al. 2011). Some Indian cultivars are reportedly alternate bearing, having distinct '*On*' and '*Off*' years and many exotic cultivars grown in different countries, flowered irregularly but without predictable '*On*' and '*Off*' years (Blaikie and Kulkarni 2002). Under subtropical conditions also mango flowered in response to cool temperatures (Whiley et al.1988, 1989; Nunez-Elisea and Davenport 1995). The suitable day/night temperature in majority of the cultivars was found 15/10 °C and at 20/15 °C. However, floral-inductive temperatures varied between cultivars (Sukhvibul et al. 2000). Temperature also has a negative effect on inflorescence size (Dambreville et al. 2013) and on the number of flowers per inflorescence (Sukhvibul et al.1999). It is possible that the larger diurnal temperature difference appeared a significant factor for flowering in mango.

Both light and temperature are important factors for photosynthesis besides CO_2 concentration (Searle and Coupland 2004). Increase in the levels of these factors would have a positive effect on photosynthesis. However, some studies have indicated that higher temperatures and CO_2 concentrations have counterbalancing effects because they increase both photosynthetic assimilation and respiratory losses. Extreme temperatures (>45 °C) or high levels of light intensity provoke damages to the photosynthesis machinery. The warmer climates during flowering and the warmer and wetter climates during the season of vegetative rest will probably lead to a lower floral induction. On the opposite, the hot and wet climates during fruit growth and vegetative growth will promote good fruit growth and important vegetative growth after harvest. But fruit quality could be reduced and pests and diseases problems and other disorders could increase. Provided nutrient and water are not limiting, the increasing temperature will lead to more rapid growth rhythms (shorter time-lag between successive vegetative flushes) and development of growth units and leaves and to the delay or the avoidance of growth cessation at the beginning of winter.

Reports on the duration of cold temperature needed for floral initiation vary from 4 days to 2 weeks in mango cultivar 'Haden' and up to 35 days in 'Tommy Atkins' and 'Keitt' (Yeshitela et al. 2004). In mango and litchi, cool temperatures may be sensed in the leaves because mature leaves appear to be the source of an essential floral stimulus (Ying and Davenport 2004).

The prediction of the effects of climate change on mango production and cultivation requires, therefore, a complete mango crop model. Several partial models, simulating specific processes of the mango tree, are available: a biochemical model of photosynthesis (Urban et al. 2007), a model of stomatal conductance (Damour et al. 2010), a model of fruit growth and quality build-up (Léchaudel et al. 2005; Lechaudel et al. 2007), and thermal time models (Dambreville et al. 2013). But there is no complete mango crop as of now available. Climate and weather play critical roles in the economic success or failure of tropical fruit tree species including commercial mango production. Air temperatures and rainfall patterns could influence vegetative and phenological phases in mango and are two of the most important factors determining suitability or otherwise of an area's climate for successful mango production. Weather-related changes have already brought widespread changes in the flowering and fruiting patterns of mango in different mango producing regions of the country during the current decade. This is adversely affecting fruit production in some areas. But rising temperature in areas previously too cold for mango production is now making them more suitable for mango production. For instance, an increase in temperature during the coldest month has made mango cultivation possible in the valley areas of Himachal Pradesh and Uttarakhand previously thought not possible. In several parts of the globe, increasing temperatures are likely to offer opportunities for mango production in new areas.

6.3.2 Importance of Stress in Flowering

In the tropics of India, floral induction of mango tree is mainly driven by stress conditions during the period preceding fruit bud differentiation (FBD) (Ravishankar 1979) and any intervening precipitation would predispose growth units (*GUs*) to predominantly, manifesting vegetative phenology. The type of irrigation adopted for mangoes, hence, assumes an important consideration if flowering management is desired. Majority of the growers in the tropics, with irrigation infrastructure provided water to the trees through the typical 6-month annual dry season, used furrow irrigation along the tree rows. Row-furrow irrigation has the disadvantage of providing water periodically around the base of trees. The major problem, however, was many roots outside the limits of the irrigation ditches never got hydrated during the dry season; hence, water moved away from roots located in or near the irrigation ditches to not only the canopy but also out to the dry roots in response to water potential gradients in the root system (Boyer 1985; Passioura 1988; Canny 1995). This backwards xylem flow towards root tips reportedly prevented upward xylem movement of shoot initiating hormones (cytokinins) that are synthesized in the root tips (Mok 1994) and the cytokinins, instead, accumulated in the tips of the roots. Active water uptake by these roots when the first rains set in moved the accumulated cytokinins to stem buds in the canopy, thus providing the stimulus to initiate an undesired flush of vegetative growth in insufficiently physiologically mature stems (Davenport 2008). In some areas, the first rains of the rainy season stimulated flowering in those stems that had attained sufficient time in rest and attained physiological maturity due to lack of flushes during the dry season. This underscored the importance of initiation of vegetative phenology well in advance so that it has a reasonable time to physiology mature to provide fruiting wood. Evidently, this aspect formed the crux of the problem of alternate bearing in perennial fruit trees. Drought and flooding reportedly have both negative effects on the vegetative development of mango trees as they potentially reduce tree growth.

6.3.3 Physiological Aspects

6.3.3.1 Source-Sink Relationships

A nutritional concept was given by Goldschmidt et al. (1985). This theory explained how the developing fruit provides a strong sink for photo-assimilates and further explains the mechanism of depletion of photo-assimilates, especially carbohydrates from the bud that prevents flower induction. Producing and exporting organs in the plant (typically mature leaves) are known as sources, while non-photosynthetic organs (fruits, roots, and tubers) and immature leaves are known as sinks (Taiz and Zeiger 2006). Marschner (2012) pointed out that the source (supply of photosynthates)—sink (utilization of photosynthates) limitations were strongly affected by the interactions between genotype and environment. Photosynthesizing organs, known as sources, mainly mature leaves, produced photosynthates, mainly carbohydrates, translocated by the sieve tubes of the phloem to non-photosynthetic organs (fruits, roots, and tubers) and immature leaves, known as sinks. High, yearly consistent yields, and fruit quality required an adequate leaf-fruit ratio (number of leaves, certain leaf area per fruit, or fresh weight unit). Fischer et al. (2012) based on their studies remarked that fruit production and quality depended upon adequate source-sink relationships. They observed carbohydrates (CH) translocated from leaves or reserve organs were the most important for the growth and development of sink organs (mainly fruits) and it is also paradoxical that up to 60% of CH produced daily could be lost through respiration. Carbohydrates constituted over 65% of the dry matter of tree crops. Increasing the leaf-fruit ratio generally increased fruit growth and CH content. They further observed photosynthesis increased with fruit load and the leaves next to fruits were the strong sources for CH. It was surmised, leaf-fruit ratio is species, cultivar, and geographic location dependent. Ravishankar (1987) studied the pattern of ^{14}C-Sucrose translocation and assimilation patterns in distinctly alternate (‘*Alphonso*’) and regular (‘*Totapuri*’) bearing mango varieties during pre-FBD and post-FBD stages. Based on ^{14}C-Sucrose translocation, assimilation, and metabolic labeling pattern, he provided evidence on the possible nature of interrelationships between the shoot apex and other parts of the mango tree, viz., stem portion and leaves and the presumptive role of root system in the regulation of translocation and assimilation in flowering of mango. His studies also pointed out to the strong possibilities of regular bearing feature of ‘Totapuri’ attributable to higher metabolic turnover, energy transducing mechanism probably under up/down regulation of genes at specific stages of fruit bud differentiation in response to environmental cues via hormonal route since, Paclobutrazol mediated flower induction in mango have been widely and successfully practiced globally. Further, the increasing concentration of atmospheric CO_2, which is a main driving force of climate change, has to be considered for appraising consequences of climate change on mango production too, because CO_2 is involved in such key process as photosynthesis; besides impacting temperature dynamics. Empirical studies have indicated that probably more dry matter would be allocated to the roots under increasing CO2 concentrations to some limits.

6.3.4 Hormonal Regulation

Flowering time is mainly influenced by important chemical constitutes like plant hormones (Domagalska et al. 2010). How hormones influence the plants, mainly governed by three core factors like (i) time of release, (ii) target tissue susceptibility, and (iii) concentration of the hormone (Achard et al. 2006). Studying the mechanism of hormones influencing the flowering time and their interactions with the other plant metabolites is useful to infer the bearing phenomenon. Mango flowering involved hormonal regulation of shoot initiation and induction events resulting in reproductive shoot formation. A balance or ratio of endogenously regulated phytohormones, thought to be auxins from leaves and cytokinins from roots, appeared to govern the initiation cycle independently from inductive influences (Fig. 6.1). Induction of reproductive or vegetative shoots is thought to be governed by the ratio of a temperature-regulated florigenic promoter and an age regulated vegetative promoter at the time of shoot initiation. Management of off-season flowering in mango trees is being accomplished in the tropics by successfully synchronizing shoot initiation through tip pruning and use of nitrate sprays coupled with management of the stem age to induce flowering such that it could be accomplished during any desired week of the year (Davenport 2006, 2007).

A direct inhibitory role of GA in mango floral initiation was reported (Davenport 2008; 2010). The concentration of GA in terminal bud decreased before panicle emergence in fruit trees including mango that subsequently flowered, and increased over the same period that remained vegetative (Tongumpai et al. 1991; Upreti et al. 2013). It suggested the direct inhibitory role of

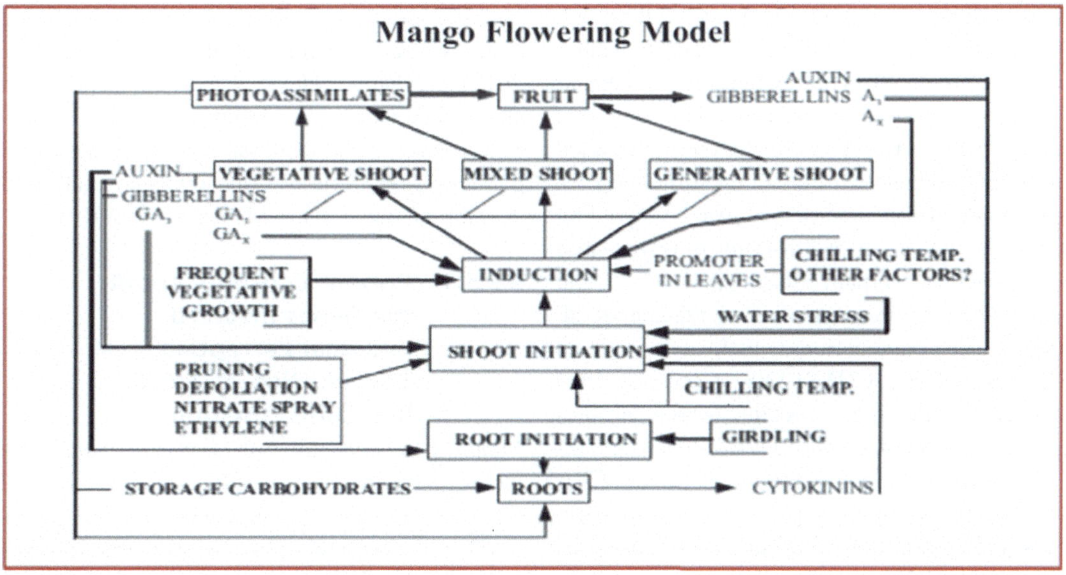

Fig. 6.1 Conceptual flowering model of mango. The model summarizes the proposed roles for various phytohormones in the initiation of shoot growth and in defining the vegetative or reproductive outcome of that growth (induction). Single lines in the scheme are promotive and double lines are inhibitory (Davenport 2006)

GA in mango floral initiation. Exogenous applications of GA may indirectly regulate mango flowering by delaying bud release. This may be because applied GA delayed bud release and did not inhibit flowering so long as bud release occurred under florally inductive conditions (Nunez-Elisea and Davenport 1998). GA biosynthesis inhibitors such as Paclobutrazol (Rademacher 1995) both hastened and increased the flowering intensity and reduced vegetative vigor of mango (Blaikie et al. 2004). So Paclobutrazol directly promoted flowering or acted indirectly by increasing the likelihood of bud release during floral-inductive conditions. Therefore, GA could inhibit floral initiation endogenously or acted indirectly by influencing the timing of bud release.

6.3.5 Molecular Approaches

In plants phase transition from vegetative to reproductive is a complex mechanism (Huijser and Schmidt 2011). This leads to explore the regulatory mechanisms to understand the phase transitions. Recently, the development of next-generation sequencing (NGS) tool, genomic, and transcriptome has contributed to a thorough understanding of the metabolic and molecular processes involved in floral biology. Molecular biology of flowering in the facultative, long-day, model plant, *Arabidopsis thaliana* provided insight into the nature of the floral stimulus (*FP*). A network of four interacting genetic signaling pathways may result in flowering in response to photoperiodic, vernalization, gibberellin, and autonomous environmental cues (Zhang et al. 2005). The photoperiodic pathway involved activation of the *CONSTANS* (*CO*) gene that encoded a zinc-finger protein, which in turn induced expression of the *FLOWERING LOCUS T* (*FT*) gene in the phloem tissue of leaves. *FT* is the terminal, integrating gene of the four pathways regulating flowering in Arabidopsis. Its transcribed mRNA was initially thought to be the *FP* that is transported in phloem to buds (Huang et al. 2005); however, evidence indicated that the translated protein product of *FT* is translocated to Arabidopsisbuds (Corbesier et al. 2007). Similar mechanisms are likely to exist in mango. Davenport et al. (2006) isolated a *CONSTANS* like gene (*MiCOL*) from mango leaf DNA. The mango ortholog has 79%, 76%, and 62% homology with two apple *CO* genes, *MdCOL2* and *MdCOL1*, and the Arabidopsis *CO* gene (*AtCO*), respectively. Studies with mango indicated that a *FP* is synthesized in leaves during exposure to cool, floral-inductive temperatures, and moved to buds to induce flowering (Davenport et al. 2006). The putative, temperature-regulated *FP* is reportedly short-lived in situ. Davenport et al. (2006, 2007) demonstrated the quantitative movement of mango *FP* with photo-assimilates in phloem from donor leaves to buds in the receiver stems. The conditions controlling floral transition were determined by certain complex growth correlations (Tan and Swain 2006; Nocker and Gardiner 2014) and are difficult to know about the certainty of branch that produced flowering buds that posed main hurdles in understanding the mechanism of flowering in fruit trees. However, recent studies have indicated that epigenetic modification, alternative splicing, antisense *RNA*, and chromatin silencing regulatory mechanisms played an important role in this process, by regulating related flowering gene expression (Kumar et al. 2013; Khan et al. 2014). Dynamic changes between chromatin states facilitating or inhibiting DNA transcription, regulated the expression of floral induction pathways, in response to environmental and developmental signals (Meijon et al. 2010).

6.3.5.1 Present Concepts About the Genes Involved in Signal Perception, Transduction on Flower Bud Differentiation

Major pathways to flowering in fruit trees included environmental induction through photoperiod, vernalization, gibberellins, autonomous floral initiation, and aging by sequentially operating *miRNAs7* (typically *miR156* and *miR172*) responding to endogenous cues. The balance of signals from these pathways is reportedly integrated by a common set of genes (*FLOWERING*

LOCUS C, FLOWERING LOCUS T, LEAFY, and *SUPPRESSOR OFOVEREXPRESSION OF CONSTANS 1*) that determined the flowering time. Gene regulation studies during vegetative to flowering and fruiting transition at transcription and post-translational level are required to understand the 'On' and 'Off' mechanism operating in perennial fruit trees (Khan et al. 2014; Sharma et al. 2015). Differential gene expression studies were carried out in various fruit crops for example in leaves of mango few genes, viz., *FT, AP1,* and *LFY* were up-regulated during the flower induction period. As a principle, the expression of flower control genes in mango is induced in leaves, buds, and stems in coordinate with the onset of the flower induction period (November–December) in regular bearing varieties, and in alternate bearers during the 'Off' crop year, while GA treatment decreased their expression (Nishikawa et al. 2007; Muñoz-Fambuena et al. 2011; Shalom et al. 2012; Goldberg-Moeller et al. 2013). Biennial bearing is an apparently genetic trait that is governed both by multigene as well as environmental cues and cultural factors (Guitton et al. 2011). Despite the fact, flowering genes may not directly regulate biennial bearing, their control by phytohormones might be one of the mechanisms leading to biennial bearing tendencies in perennial fruit crops. It has, however, been observed that *MdFT* and *MdTFL1* flowering genes in *Malus* are not responsible for biennial bearing (Mimida et al. 2009; Kotoda et al. 2010). However, in biennial bearing mango trees *FLOWERING LOCUS T*-like and gibberellins metabolism genes were identified by Nakagawa et al. (2012). The *FP* appeared to be up-regulated during the exposure of leaves to cool temperatures and down-regulated to a constant basal level when temperatures were warm (Davenport 2000, 2003, 2008). For example, floral induction of initiating shoots during exposure of trees to cool, subtropical temperature conditions is considered to be the result of elevated levels of up-regulated *FP*. High *VP* levels in young stems, however, may counterbalance the impact of *FP*, such that exposure to cool temperatures at the time of shoot initiation caused vegetative shoot induction despite the presence of cool temperatures (Davenport 2000). Shalom et al. (2012) study highlighted among genes induced in '*Off*'-crop trees was one homologous to *SQUAMOSA PROMOTER BINDING-LIKE* (*SPL*), which controlled juvenile-to-adult and annual phase transitions, regulated by *miR156*. The expression pattern of *SPL*-like, *miR156,* and other flowering control genes suggested that fruit load affected bud fate, and therefore development and metabolism, a relatively long time before the flowering induction period. Their results shed light on some of the metabolic and regulatory processes that are altered in '*On*' and '*Off*' buds. This data could be utilized. Recently, Bouché and colleagues provided FLOR-ID, the flowering interactive database to the public domain, which is freely available at www.flor-id.org (Bouché et al. 2016). Interestingly, FLOR-ID is a simple and effective tool (UniProt, www.uniprot.org) and PubMed (www.ncbi.nlm.nik.gov/pubmed) database finder combined with information taken from other sources (Wellmer 2017). Furthermore, orthologs and co-orthologous detection were done to study the evolutionary relationship between alternate bearing genes of different fruit plant species like avocado, litchi, apple, mango, olive, pistacia, prunus, etc. (Sharma et al. 2019). A total of 1,282 nucleotide sequences were compared with all-against-all BLASTP utilizing the data from NCBI (ESM1) and phylogenetic tree (Fig. 6.2) was constructed using Mega (version 7.0, https://www.megasoftware.net) Kumar et al. (2016). Recently Sharma et al. (2020) studied the alternate bearing phenomenon in contrasting regular (Neelam) and irregular (Dashehari) mango genotypes using differential gene approach. Deciphering the differentially expressed genes (DEGs) and potential candidate genes associated with alternate bearing, hormone, and carbohydrate metabolism pathways will help for illustrating the molecular mechanisms underlying the bearing tendencies in mango. In mango, the availability of the draft genome sequence is expected to greatly assist the candidate gene approach to solve the mystery of alternate bearing at gene level (Singh et al. 2016). The improved understanding of mango

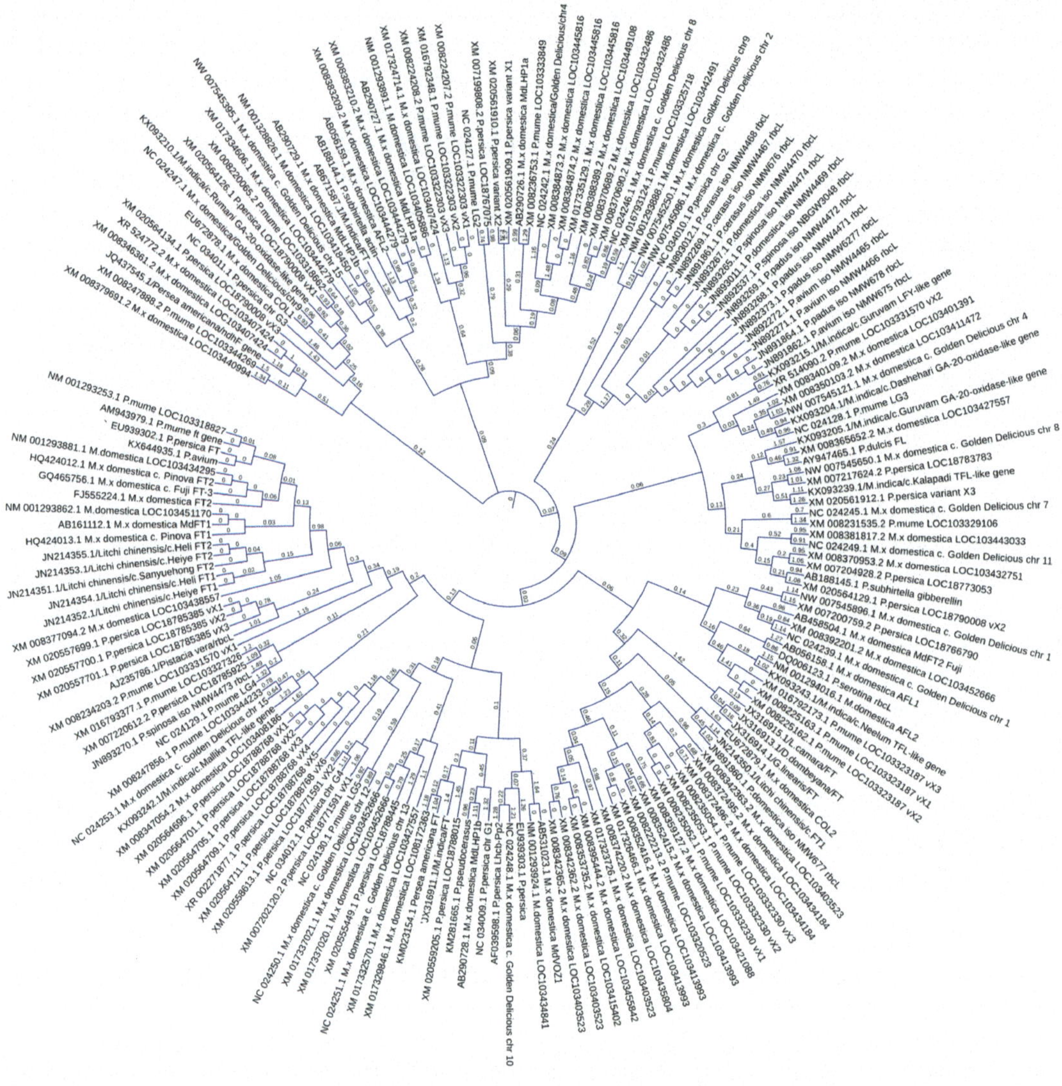

Fig. 6.2 Phylogenetic tree represents the evolutionary relationship among gene sequences of different alternate bearing fruit crops (Sharma et al. 2019)

genetics and adoption of molecular genetics in breeding programs is enabling breeders to greatly improve their breeding efficiency through improved selection of parents and progeny. This along with research in mango genomics and gene discovery will identify genes and markers associated with traits of interest and improve the breeding outcomes.

6.4 Conclusions

Presently, it may not be possible to draw a general scenario for India in the absence of information on the extent, magnitude, and complexion of regional and seasonal variability and different mango cultivar responses as there is no precise scientific information available on response of the mango tree to each weather variable and interactions between variables with reference to the effect of the tree phenological stage, cultivar effects, and optimal threshold values of these variables. Future researches need to articulate on mining of genes of climate resilience among the large variability of genetic resources available in the sub-continent and integrate them into commercial cultivars in order to fit mango in the emerging paradigms of weather dynamics. Need for elucidating weather dynamics and behavior of mango cultivars, developing dwarf ideotypes conforming to 'ideal mango' variety (regularity in bearing; precocity; dwarf tree architecture; good fruit size; high yield and quality and resistance to key pests and diseases) and rootstock selection (low vigor, but with strong dynamic root systems having greater rhizosphere competence) in specific, through international networking and understanding of the same in the context of the Indian sub-continent is underscored. Researchers have indicated that genetic control of the target genes as of now is elusive in perennial crops system. Till they are deciphered, judicial use of chemical(s) in floral regulation as a means of overcoming negative regulators is inevitable.

Acknowledgements Authors are thankful to DST-SERB and Director, ICAR-IARI.

References

Achard P, Cheng H, De Grauwe L, Decat J, Schoutteten H, Moritz T, Van Der Straeten D, Peng J, Harberd NP (2006) Integration of plant responses to environmentally activated phytohormones signals. Science 311(5757):91–94. https://doi.org/10.1126/science.1118642

Bally ISE (2011) Advances in research and development of mango industry. Rev Bras Fruit Jabot cabal-SP, Volume Especial, E. 057–063, Outubro

Bangerth F, Naphrom D, Sruamsiri P, Hegele M, Boonplod N, Manochai P (2004) Hormonal changes in various tissues of mango trees during flower induction following cold temperature. Acta Hort 645:453–457. https://doi.org/10.17660/Acta Hortic.2004.645.58

Bangerth KF (2009) Floral induction in mature, perennial angiosperm fruit trees: Similarities and discrepancies with annual/biennial plants and the involvement of plant hormones. Sci Hort 122:153–163. https://doi.org/10.1016/j.scienta.2009.06.014

Batten DJ, McConchie CA (1995) Floral induction in growing buds of litchi (*Litchi chinensis*) and mango (*Mangifera indica*). Aust J Plant Physiol 22:783–791

Blaikie SJ, Kulkarni VJ, Müller WJ (2004) Effects of morphactin and Paclobutrazol flowering treatments on shoot and root phenology in mango cv. Kensington Pride. Sci Hort 101:51–68. https://doi.org/10.1016/j.scienta.2003.09.009

Blaikie SJ, Kulkarni V (2002) Manipulating flowering in mango, cv, Kensington Pride. Acta Hort 575:791–796

Bouché F, Lobet G, Tocquin P, Perilleux C (2016) FLOR-ID: An interactive database of flowering-time gene networks in *Arabidopsis thaliana*. Nucleic Acids Res 44:D1167–D1171. https://doi.org/10.1093/nar/gkv1054

Boyer JS (1985) Water transport. Annu Rev Plant Physiol 36:473–516

Canny MJ (1995) Apoplastic water and solute movement: new rules for an old space. Annu Rev Plant Physiol Plant Mol Biol 46:215–236

Capelli M, Lauri PE Normand (2016) the Costs of Reproduction in Mango—Vegetative Growth Matters. Front Plant Sci. https://doi.org/10.3389/fpls.2016.01531

Chacko EK, Randhawa GS (1971) Towards an understanding of the factors affecting flowering in mango (*Mangifera indica* L.). Andhra Agri J 18:226–36

Chacko EK (1991) Mango flowering - still an enigma. Acta Hort 291:12–21

Chen HB, Huang HB (2001) China litchi industry: development, achievements and problems. Acta Hort 31–39. https://doi.org/10.17660/ActaHortic.2001.558.2

Corbesier L, Vincent C, Jang S, Fornara F, Fan Q, Searle I, Giakountis A, Farrona S, Gissot L, Turnbull C, Coupland G (2007) FT protein movement contributes to long-distance signalling in floral induction of *Arabidopsis*. Science 316:1030–1033. https://doi.org/10.1126/science.1141752

Dambreville A, Normand F, Lauri PE (2013) Plant growth co-ordination in natura: a unique temperature-controlled law among vegetative and reproductive organs in mango. Funct Plant Biol 40:280–291

Damour G, Simonneau T, Cochard H, Urban L (2010) An overview of models of stomatal conductance at the leaf level. Plant Cell Environ 33:1419–1438

Davenport TL (2003) Management of flowering in three tropical and subtropical fruit tree species. HortSci 38:1331–1335

Davenport TL, Ying ZT, Kulkarni V, White TL (2006) Evidence for a translocatable florigenic promoter in mango. Sci Hort 110:150–159

Davenport TL (2009) Reproductive physiology. In: (ed) Litz RE The Mango: Botany, Production and Uses, 2nd edn. CAB International, Wallingford, UK, pp 97–169

Davenport TL (2007) Reproductive physiology of mango. Braz J Plant Physiol 19(4):363–376. https://doi.org/10.1590/S1677-04202007000400007
Davenport TL (2010) Mango: reproductive physiology. In: DaMatta F (ed) Ecophysiology of tropical tree crops. Nova Science Publishers, New York, pp 217–234
Davenport TN (2011) Mango reproductive physiology, challenges and opportunities. In: Ravishankar H, Misra AK, Singh VK (eds) Souvenir, global conference on mango. ISHS, Belgium and SDSH, Lucknow, India, pp 58–71
Davenport TL (2000) Leaves are not necessary for citrus floral induction. In: Albrigo LG (ed) Proceedings of the International Society of Citriculture IX Congress, pp 660–661. http://www.crec.ifas.ufl.edu/societies/ISC/, 690 p
Domagalska MA, Sarnowska E, Nagy F, Davis SJ (2010) Genetic analyses of interactions among gibberellin, abscisic acid, and brassinosteroids in the control of flowering time in *Arabidopsis thaliana*. PLoS One 5. https://doi.org/10.1371/journal.pone.0014012
Fischer G, Almanza-Marchan PJ, Ramirez F (2012) Source-sink relationships in fruit species. A review. Rev Colom Cien Hort Colas 6(2):238–253
Goldberg-Moeller R, Shalom L, Shlizerman L, Samuels S, Zur N, Ophir R, Blumwald E, Sadka A (2013) Effects of gibberellin treatment during flowering induction period on global gene expression and the transcription of flowering-control genes in citrus buds. Plant Sci 198:46–57. https://doi.org/10.1016/j.plantsci.2012.09.012
Goldschmidt EE, Aschkenaki N, Herzano Y, Schaffer AA, Monselise SP (1985) A role for carbohydrate levels in the control of flowering in citrus. Sci Hort 26:159–166
Guitton B, Kelner JJ, Velasco R, Gardiner SE, Chagne D, Costes E (2011) Genetic control of biennial bearing in apple. J Exp Bot 1–19
Hoblyn TN, Grubb NH, Painter AC, Wates BL (1936) Studies in biennial bearing.Intl J Pomol Hort Sci 14:39–76
Huang T, Böhlenius H, Eriksson S, Parcy F, Nilsson O (2005) The mRNA of the Arabidopsis gene *FT* moves from leaf to shoot apex and induces flowering. Science 309:1694–1698
Huff A (2001) A significance test for biennial bearing using data resampling. J Hort Sci Biotechnol 76:534–535
Huijser P, Schmidt M (2011) The control of developmental phase transitions in plants. Development 138:4117–4129. https://doi.org/10.1242/dev.063511
Jyothi MH, Ekbote S (1998) Biochemical changes in the leaf of bearing and non bearing trees of some mango (*Mangifera indica* L.) varieties/hybrids. Adv Agric Res India 10:17–23
Kachru RB, Singh RN, Chacko EK (1972) Inhibition of flowering in *Mangifera indica* L. by gibberellic acid. Acta Hort 24:206–209
Kalayanaruk S, Subhadrabandhu S, Baabprasert C (1982) Total nonstructural carbohydrate and total nitrogen content in leaves and terminal shoots of mango (*Mangifera indica* L.) cv. Nam Dok Mai throughout the year. Kaset J Nat Sci 16: 41
Khan MRG, Ai XY, Zhang JZ (2014) Genetic regulation of flowering time in annual and perennial plants. Wiley Interdiscipl Rev: RNA 5(3):347–359. https://doi.org/10.1002/wrna.1215
Kotoda N, Hayashi H, Suzuki M, Igarashi M, Hatsuyama Y, Kidou SI, Igasaki T, Nishiguchi M, Yano K, Shimizu T, Takahashi S, Iwanami H, Moriya S, Abe K (2010) Molecular characterization of flowering *LOCUS t*-like genes of apple (*Malusdomestica*Borkh.). Plant Cell Physiol 51:561–575. https://doi.org/10.1093/pcp/pcq021
Kulkarni VJ (2004) The tri-factor hypothesis of flowering in mango. Acta Hort 645:61–70
Kumar S, Garrick DJ, Bink MC (2013) Whitworth C, Chagne D and Volz RK (2013) Novel genomic approaches unravel genetic architecture of complex traits in apple. BMC Genom 14:393. https://doi.org/10.1186/1471-2164-14-393
Kumar S, Stecher G, Tamura K (2016) MEGA7: molecular evolutionary genetics analysis version 7.0 for bigger datasets. Mol Biol Evol 33(7):1870–1874. https://doi.org/10.1093/molbev/msw054
Lal S (2016) Identification of genomic regions for alternate bearing and fruit quality traits in mango (*Mangifera indica* L.). PhD thesis, Fruits and Horticultural Technology, PG School, ICAR- Indian Agriculture research Institute, New Delhi, India
Lal S, Singh AK, Singh SK, Srivastava M, Singh BP, Sharma N, Singh NK (2017) Association analysis for pomological traits in mango (*Mangifera indica* L.) by genic-SSR markers. Trees Struct Funct 31:1391. http://https://doi.org/10.1007/s00468-017-1554-2
Léchaudel M, Génard M, Lescourret F, Urban L, Jannoyer M (2005) Modelling effects of weather and source–sink relationships on mango fruit growth. Tree Physiol 25:583–597
Lechaudel M, Vercambre G, Lescourret F, Normand F, Génard M (2007) An analysis of elastic and plastic fruit growth of mango in response to various assimilate supplies. Tree Physiol 27:219–230
Léchaudel M, Joas J (2007) An overview of preharvest factors influencing mango fruit growth, quality and postharvest behaviour. Braz J Plant Physiol 19:287–298
Marschner P (ed) (2012) Marschner's mineral nutrition of higher plants, 3rd edn. Elsevier, Oxford, UK
Meijón M, Feito I, Valledor L, Rodríguez R, Cañal MJ (2010) Dynamics of DNA methylation and Histone H4 acetylation during floral bud differentiation in azalea. BMC Plant Biol 10, 10.https://doi.org/https://doi.org/10.1186/1471-2229-10-10
Mimida N, Kotoda N, Ueda T, Igarashi M, Hatsuyama Y, Iwanami H, Moriya S, Abe K (2009) Four *TFL1/CEN*-Like genes on distinct linkage groups show different expression patterns to regulate vegetative and reproductive development in apple (*Malusdomestica* Borkh.). Plant Cell Physiol 50:394–412. https://doi.org/10.1093/pcp/pcp001

Mok MC (1994) Cytokinin and plant development-An overview. In: Mok DWS, Mok MC (eds) Cytokinin: chemistry, activity and function. CRC Press, Boca Raton, FL, USA, pp 155–166

Monselise SP, Goldschmidt EE (1982) Alternate bearing in fruit trees. Hort Rev 4:128–173

Muñoz-Fambuena N, Mesejo C, Carmen González-Mas M, Primo-Millo E, Agustí M, Iglesias DJ (2011) Fruit regulates seasonal expression of flowering genes in alternate-bearing "Moncada" mandarin. Ann Bot 108 (3):511–519. https://doi.org/10.1093/aob/mcr164

Nakagawa M, Honsho C, Kanzaki S, Shimizu K, Utsunomiya N (2012) Isolation and expression analysis of *FLOWERING LOCUS T*-like and gibberellins metabolism genes in biennial-bearing mango trees. Sci Hort 139:108–117. https://doi.org/10.1016/j.scienta.2012.03.005

Nishikawa F, Endo T, Shimada T, Fujii H, Shimizu T, Omura M, Ikoma Y (2007) Increased *CiFT* abundance in the stem correlates with floral induction by low temperature in Satsuma mandarin (*Citrus unshiu* Marc.). J Exp Bot 58:3915–3927. https://doi.org/10.1093/jxb/erm246

Nocker SV, Gardiner SE (2014) Breeding better cultivars, faster: applications of new technologies for the rapid deployment of superior horticultural tree crops—Mini review. Hort Res 1:1–8. https://doi.org/10.1038/hortres.2014.22

Nuñez-Elisea R, Davenport TL (1991) Flowering of 'Keitt' mango in response to deblossoming and gibberellic acid. Proc Fl State Hort Soc 104:41–43

Nunez-Elisea R, Davenport TL (1998) Gibberellin and temperature effects on dormancy release and shoot morphogenesis of mango (*Mangifera indica* L.). Sci Hort 77:11–21. https://doi.org/10.1016/S0304-4238(98):00158-7

Nunez-Elisea R, Davenport TL (1995) Effect of leaf age, duration of cool temperature treatment, and photoperiod on bud dormancy release and floral initiation in mango. Sci Hort 62:63–73

Passioura JB (1988) Water transport in and to roots. Annu Rev Plant Physiol Plant Mol Biol 39:245–265

Pearce Doberšek-Urbanc (1967) The measurement of irregularity in growth and cropping. J Hort Sci 442:295–305

Rademacher W (1995) Growth retardants: biochemical features and applications in horticulture. Acta Hort 394:57–73

Ramírez F, Davenport TL (2010) Mango (*Mangifera indica* L.) flowering physiology. Sci Hort 126:65–72. https://doi.org/10.1016/j.scienta.2010.06.024

Ramírez F, Davenport TL (2012) Reproductive biology (physiology)- The case of mango. In: Valavi SG, Rajmohan K, Govil JN, Peter KV, Thottappilly G (eds) Mango: Volume 1 Production and processing technology, Edition: 1, Chapter: Studium Press, Houston, Texas, USA, ISBN: 1-933699-93-0 Series ISBN: 1-933699-92-2 pp 56-81

Ramírez F, Davenport TL, Fischer G (2010) The number of leaves required for floral induction and translocation of the florigenic promoter in mango (*Mangifera indica* L.) in a tropical climate. Sci Hort 123:443–453. https://doi.org/10.1016/j.scienta.2009.10.005

Ravishankar H (1987) Studies on physiological and biochemical aspects into the causes and control of alternate bearing in mango (*Mangifera indica* L.). PhD Thesis, University of Agricultural Sciences, Dharwad, Karnataka, India 310 p

Ravishankar H, Rao MM, Bojappa KM (1979) Fruit-bud differentiation in mango "Alphonso" and "Totapuri" under mild tropical rainy conditions. Sci Hort 10:95–99. https://doi.org/10.1016/0304-4238(79):90073-6

Rosen stock TS, Rosa UA, Plant RE, Brown PH (2010) A re-evaluation of alternate bearing in pistachio. Sci Hort 124:149–152. https://doi.org/10.1016/j.scienta.2009.12.007

Searle I, Coupland G (2004) Induction of flowering by seasonal changes in photoperiod. EMBO J 23:1217–1222

Shalom L, Samuels S, Zur N, Shlizerman L, Zemach H et al (2012) Alternate bearing in citrus: changes in the expression of flowering control genes and in global gene expression in ON- versus OFF-crop trees. PLoS One7(10):46930. http:// https://doi.org/10.1371/journal.pone.0046930

Sharma N, Singh SK, Mahato AK, Ravishankar H, Dubey AK, Singh NK (2019) Physiological and molecular basis of alternate bearing in perennial fruit crops. Sci Hort 243:214–225. https://doi.org/10.1016/j.scienta.2018.08.021

Sharma N, Singh SK, Prakash J, Mahato AK, Srivastava M, Singh A et al (2017) Flowering gene and genomic region in fruit crops: a tool for future breeding. Intl J Genomes Data Mining J108

Sharma N, Singh SK, Singh NK, Srivastava M, Singh BP, Mahato AK et al (2015) Differential gene expression studies: a possible way to understand bearing habit in fruit crops. Transcriptome: Open Access 3:110. http:// https://doi.org/10.4172/2329-8936.1000110

Sharma N, Singh AK,, Singh SK, Mahato AK, Srivastava M, Singh NK (2020) Comparative RNA sequencing based transcriptome profiling of regular bearing and alternate bearing mango (*Mangifera indica* L.) varieties reveals novel insights into the regulatory mechanisms underlying alternate bearing. Biotechnol Lett. https://doi.org/10.1007/s10529-020-02863-8

Shivu Prasad SR, Reddy YTN, Upreti KK, Rajeshwara AN (2014) Studies on changes in carbohydrate catabolism in regular bearing and "Off" season bearing cultivars of mango (*Mangifera indica* L.) during flowering Intl J Fruit Sci 14(4):437–459. https://doi.org/10.1080/15538362.2014.897891

Singh A, Singh VK, Ravishankar H, Shukla GK (2014) Time series application for quantification of Alternate Bearing Intensity (ABI) in mango (*Mangifera indica*) cv Langra. Indian J Agri Sci84(1): 137–141, January 2014/Article

Singh NK, Mahato AK, Jayaswal PK, Singh A, Singh S, Singh N et al (2016) Origin, diversity and genome sequence of mango (*Mangifera indica* L.) Indian J Hort Sci 51 (2.2):355–368

Singh VK, Singh A (2003) Effect of Paclobutrazol on regularity of bearing in mango (*Mangifera indica* L.). Physiol Mol Biol Plants 9(2):239–48

Singh VK, Ravishankar H (2011) Photosynthetic aptitude of leaves on flowering and non-flowering terminals in mango (*Mangifera Indica* L.) cv. Dashehari. In: Singh B, Sengar RS, Umrao, Vijay K, Naresh RK, Chauhan, Pankaj, M, Sunil, Singh V, Veer K, Goswami, Verma A, Nidhi P, Yogesh (eds) International conference on issues for climate change, land use diversification and biotechnological tools for livelihood security (ICLDBT-2011), 08–10 October, 2011, Sardar Vallabhbhai Patel University of Agriculture & Technology, Meerut, India, pp 174–179

Singh VK, Rajan S (2009) Changes in photosynthetic rate, specific leaf weight and sugar contents in mango (*Mangifera indica* L.). Open Hort J 2:34–37

Singh VK, Singh A, Rajan S (2011) Physiological diversity and its significance in flowering of mango (*Mangifera indica* L.). Progr Hort 43(1):61–65

SinghVK, Pooja Saxena, Rajan S, Upreti KK, Ravishankar H (2014) Changes in the levels of endogenous phytohormones associated with flowering in mango (*Mangifera indica* L.) cultivars. In: Ravishankar H, Singh VK, Misra AK, Mishra M(ed) Souvenir, National Seminar-cum-Workshop on Physiology of Flowering in Perennial Fruit Crops, Director, CISH, Lucknow, India, p 29

Stitt M, Quick WP (1989) Photosynthetic carbon partitioning, its regulation and possibilities for manipulation. Physiol Plant 77:633–664

Sukhvibul N, Whiley AW, Smith MK, Hetherington SE, Vithanage V (1999) Effect of temperature on inflorescence development and sex expression of mono- and poly-embryonic mango (*Mangifera indica* L.) cultivars. J Hort Sci Biotechnol 74:64–68

Sukhvibul N, Hetherington SE, Vithanage V, Whiley AW, Smith MK (2000) Effect of temperature on inflorescence development and floral biology of mango (*Mangifera indica* L.). Acta Hort 509:601–607

Taiz L, Zeiger E (2006) Plant Physiology. 4th edn. Sinauer Associates, Sunderland, MA, p 690

Tan FC, Swain SM (2006) Genetics of flower initiation and development in annual and perennial plants. Physiol Plant 128(1):8–17. https://doi.org/10.1111/j.1399-3054.2006.00724.x

Tiwari DK, Patel VB, Pandey AK (2018) Floral induction in mango: physiological, biochemical and molecular basis. Int J Chem Studies 6(1):252–259

Tongumpai P, Jutamanee K, Sethapakdi R, Subhadrabandhu S (1991) Variation in level of gibberellin like substances, during vegetative growth and flowering of mango cv. Khiew Sawoey Acta Hort 291:105–108. https://doi.org/10.17660/ActaHortic.1991.291.13

Upreti KK, Reddy YTN, Shivu Prasad SR, Bindu GV, Jayaram HL, Rajan S (2013) Hormonal changes in response to Paclobutrazol induced early flowering in mango cv, Totapuri. Sci Hort 150:414–418

Urban L, Alphonsout L (2007) Girdling decreases photosynthetic electron fluxes and induces sustained photo protection in mango leaves. Tree Physiol 27 (3):345–52

Urban L, Léchaudel M, Lu P (2004) Effect of fruit load and girdling on leaf photosynthesis in *Mangifera indica* L. J Exp Bot 55:2075–2085. https://doi.org/10.1093/jxb/erh220

Verreynne JS, Lovatt CJ (2009) The effect of crop load on bud break influences return bloom in alternate bearing "Pixie" mandarin. J Am Soc Hort Sci 134:299–307

Wellmer F (2017) Flowering highlights a database for all things flowering. J Exp Bot 68(23):1–20. https://doi.org/10.1093/jxb/erx087

Whiley AW, Saranah JB, Rasmussen TS, Winston EC, Wolstenholme BN (1988) Effect of temperature on 10 mango cultivars with relevance to introduction in Australia. In: Proceedings of the 4th Australian Conference on Tree and Nut crops. ACOTANC, Lismore, Australia, pp 176–185

Whiley AW, Rasmussen TS, Saranah JB, Wolstenholme BN (1989) Effect of temperature on growth, dry matter production and starch accumulation in ten mango (*Mangifera indica* L.) cultivars. J Hort Sci 64:753–765

Yeshitela T, Robbertse PJ, Stassen PJC (2005) Effects of pruning on flowering, yield and fruit quality in mango (*Mangifera indica*). Aust J Exp Agri 45:1325–1330

Ying Z, Davenport TL (2004) Leavesrequired for floral induction of lychee. Plant Growth Reg Soc Am Q 32:132–137

Zhang X, Garreton V, Chua NH (2005) The AIP2 E3 ligase acts as a novel negative regulator of ABA signaling by promoting ABI3 degradation. Genes Dev 19:1532–1543

Classical Genetics and Breeding

7

M. Sankaran, M. R. Dinesh, and K. V. Ravishankar

Abstract

Mango (*Mangifera indica* L.) is believed to be originated as an allopolyploid most probably amphidiploid with chromosome number, 2n = 40 with the size of 0.2–2.0 μm. As revealed by cytological studies, these chromosomes undergo regular pairing and disjunction in 20 bivalents during meiosis. The mango (*Mangifera indica L.)* has been originated from north-eastern India covering the Indo Myanmar border region and Bangladesh, where still wild mango trees with very small fruits are found. *Mangifera* genus has more than 60 species with Myanmar-Siam-Indochina or Malay Archipelago being its primary center of diversity comprising of highest diversity where as the Sunda Islands (Java, Sumatra, Bornco)–the Philippines and Celebes–Banda-Timor are considered as the secondary center of diversity. There are certain species such as *M. odorata*, *M. zeylanica, M. caloneura, M. sylvatica, M. foetida,* and *M. caesia that* are cross compatible with each other owing to the identical number of chromosomes. On accord of its cross pollinated and heterozygous nature, *M. indica* L. and other related species possess very huge variability that is useful for mango improvement programme. More than 1000 mango centres of India. Depending on the number of embryos that a mango fruit produces, mango varieties are classified as monoembryonic and polyembryonic. Systematic hybridization work has resulted more than 50 hybrids across the globe and several of them are being cultivated commercially. Genetical studies conducted have shown Amrapali to be a good combiner as female parent whereas Vanraj, Janardhan Pasand, Sensation, and Tommy Atkins have been found to be good combiners for shelf life, peel color, and sugar: acid blend. Studies on inheritance of traits have been conducted and pre-selection indices for early screening of hybrids have been established. Attempts have been made to introgress genes for biotic and abiotic stress resistance/tolerance from the wild species. Molecular markers such as RAPD, AFLP, SSR, and SNP are being employed for diversity analysis and hybridity confirmation in mango breeding.

M. Sankaran (✉) · M. R. Dinesh · K. V. Ravishankar
ICAR-Indian Institute of Horticultural Research, Hesaraghatta Lake Post, Bengaluru 560089, Karnataka, India
e-mail: kmsankaran@gmail.com; m.sankaran@icar.gov.in

C. Kole (ed.), *The Mango Genome*, Compendium of Plant Genomes,
https://doi.org/10.1007/978-3-030-47829-2_7

7.1 Introduction

Mango being a perennial has inherent characteristics such as outbreeding, heterozygosity, sterility, self-incompatibility and polyembryony, etc. which forces the breeders to raise large populations for effective selections. Fruit breeding in general aims for creation of genetic variability and selection of desired genetic variation is one of the critical objectives. The basic knowledge on classical cytogenetic is very important for a breeder as it gives the information on structure and properties of chromosomes during mitosis and meiosis as well as their influence on phenotype besides the factors causing the chromosomal changes. In the recent past, with advent of molecular cytogenetic techniques, it is possible to delineate genetic marker order, define contig order and gap size, heterochromatin-euchromatin regions and genetic events such as addition, deletion, reversion, and chromosome rearrangement. Cytogenetic studies are also important to reduce the risk of genetic erosion and to broaden the genetic base in cultivated types/varieties, economically important genes could be introgressed from wild relatives in fruit crops (e.g., Mango). Tropical fruit species like mango are highly suitable for arid and semi-arid regions and hence, increase in genetic diversity of these crops will in turn lead to their increased adaptability in the changing climatic scenario. Furthermore, as an important export commodity, the commercial cultivation of mango is limited to only a few cultivars in order to meet the export criteria and consumer demand. The continuous neglect of available germplasm and increased practice of monoculture may lead to genetic erosion, loss of the wild crop relatives which are reservoirs of several important traits and increased threat of diseases or pest (Campbell 2007). The diversely rich gene pool of mango comprising of wild, semi-cultivated species are under the threat of extinction and hence, several in situ and ex situ measures are being taken to conserve the available germplasm. In this chapter, an attempt has been made to compile all the works pertaining to genetics and breeding of mango.

7.2 Cytology

Mango (*M. indica* L.) was found to be a diploid (2x) with somatic chromosome number (2n) of 40 (Mukherjee 1950; Roy and Visweswaraiya 1951) and a polyembryonic variety, Vellaikolamban was as tetraploid with $2n = 4x = 80$ (Roy and Visweswaraiya 1951). This report was contradicted by Akbar Hussain (1983) (Pandey 1984). Later, it was confirmed to be a diploid with $2n = 2x = 40$ by Majumder and Sharma (1990) The reason for this anomaly was the choice of plant material which had supposedly undergone chromosome doubling but was unavailable for further studies (Singh 1960). Mango chromosomes show regular pairing and disjunction into 20 bivalents suggesting a normal meiotic division. The chromosomes are of small size, i.e., 0.2–2.0 μm and can be classified into eight types based on the size and presence or absence of primary and secondary constrictions and satellites. Eleven common chromosomes have been reported in *M. indica* L. and *M. sylvatica* Roxb. and they differ from one another primarily in the assortment of these eleven chromosomes(Mukherjee 1950). The wild relatives of mango are supposed to have originated through allopolyploidy and most probably through amphidiploidy (Mukherjee 1950). Mango genotypes are divided into two distinct groups on the basis of their origin such as monoembryonic (Indian types) and polyembryonic (Southeast Asian types). The seeds of monoembryonic mangoes contain a zygotic embryo, and the fruit skin is highly colored (mixes of red, purple, and yellow), while the seeds of polyembryonic mangoes have several embryos originated from nucellar tissue and the skin is soft or pale in color (green to light green to yellow) (Iyer and Degani 1997; Viruel et al.

2005). The present day mango varieties are mainly a result of spontaneous mutation, selection, and hybridization. Mukherjee (1950) reported mango to be allopolyploid based on the characteristics like more chromosome number, high number of nucleolar chromosomes, secondary association of bivalents, regular pairing, absence of multivalent formation, and good pollen fertility. Haploid genome size of mango is 0.91 pg, which is three times larger than *Arabidopsis thaliana* (L.) Heynh. and comparable to that of rice (Arumuganathan and Earle 1991). Owing to the presence of secondary associations at meiotic metaphase stage, Mukherjee (1950) suggested that the basic chromosome number of *Mangifera* is 8. In addition, the high number of somatic chromosomes and the correspondingly high number of nucleolar chromosomes led him to conclude that mango is an allopolyploid. Molecular markers can be used to identify true hybrids in polyembryonic species such as Indochinese type mango and some citrus species. Furthermore, MAS allows early selection of major genes controlling traits and thus reduces the time and space needed for growing out seedlings. The molecular markers such as microsatellite markers are available in mango (Viruel et al. 2005; Duval et al. 2005; Schnell et al. 2005; Honsho et al. 2005; Ukoskit 2007; Ravishankar et al. 2011; and Suprapaneni et al. 2013). Microsatellite, or simple sequence repeat (SSR), markers are polymerase chain reaction (PCR) based markers which detect differences in the copy numbers of short stretch repetitive DNA sequences. There were several reports on applications of mango microsatellite markers, including genetic diversity (Duval et al. 2005; Suprapaneni et al. 2013), cultivar identification (Eiadthong et al. 1999), and pedigree analysis (Olano et al. 2005). With the vast germplasm in the Indian sub-continent and Southeast Asia (Bompard and Schnell 1997) would help for better understanding and utilization of genetics controlling agronomic and quality traits and subsequent introgression. In order to utilize marker technology to its full potential, more markers are needed to construct a high-density mango genetic linkage map. The saturated map with whole genome coverage will be the basis for many genetic analyses such as quantitative trait loci (QTL) analysis, map-based gene cloning, comparative genomics studies, and genome-wide association studies (Liu et al. 1996; Milbourne et al. 1998; Fukino et al. 2008; Huang et al. 2010; Li et al. 2013). With the advances in whole genome sequencing technology and EST sequencing projects in several crop species such as barley (Pasam et al. 2012), maize (Li et al. 2013), and rice (Huang et al. 2010), genome wide association studies of mango become feasible.

7.3 Genetic Studies

One of the main problems in mango breeding has been the lack of information on inheritance pattern of traits, high heterozygosity, and formation of single seed per crossing which in turn is highly prone to fruit drop have further compounded the problem. However, some data generated though not conclusive can be used as indicator in the breeding programme. Genetical studies revealed that the traits such as dwarf, regular bearing, and precocity are recessive genes wherein, regular bearing was tightly linked with precocity and biennial bearing was found to dominant over regular bearing habit Sharma and Majumder (1988a). Flesh color was reported to be governed by additive genes (Sharma 1987) whereas, Iyer (1991) found light yellow colour was dominant over orange-yellow in the progenies of Alphonso x Neelum cross. In a another study carried out by Dinesh (2003) using half-sib population revealed that traits such as fruit weight, TSS and pulp percentage were found to be governed by non-additive factors with low heritability. Selection of parents based on phenotypic characters is not recommended as the progeny performance is highly unpredictable in such cases (Lavi et al. 1998). With regard to peel colour, it was found that hybridization between red with green colored varieties resulted in progenies with gradation of colour indicating that peel colour is governed by a several loci (Sharma 1987; Iyer and Subramanyam 1987). The presence of beak on the fruit was dominant when 'Totapuri' (with prominent beak) was used as one of the parents in hybridization, all the resulting progenies were having beaked fruit (Iyer and Subramanyam 1979). Similarly, bunch bearing habit was found to be dominant over single fruiting (Sharma et al. 1972). Lavi et al. (1989) observed that in an open-pollinated population, there is no female parent effect on juvenile period and fertility; however, it was observed on harvest season and fruit color and there was a slight effect on fruit size and taste. Resistance to bacterial canker was reported to be of cytoplasmic inheritance since all the progenies were bacterial canker susceptible when Neelum was used as female parent irrespective of the male parent (Sharma and Majumder 1988a). Iyer (1991) observed that recessive genes govern the internal breakdown (Spongy tissue). Sharma and Majumder (1988a) found that susceptibility to mango malformation is governed by dominant genes, as hybrid progenies of Bhadauran (resistant variety) lacked resistance.

Use of improved crossing technique, i.e., use of few flowers from large number of panicles for hybridization and covering the fertilized flowers with polythene bags instead of crossing large number of flowers in single panicle and use of muslin clothes for covering has increased the fruit set percentage in mango. Additionally the information on gene donors and crossing between the varieties with desired traits has further improved the efficiency of hybridization and resulted in saving of resources on indiscriminate hybridization between varieties. Work carried out at various centres has revealed that the varieties Totapuri, Neelum, Banganapalli, and Lazzat Baksh are good source regular bearing. The variety of Neelum can also be used as a source for extending crop duration. However, Totapuri whenever has been used, the progenies were found to have beak in fruits which is undesirable. The varieties Creeping, Kerala Dwarf, Rumani, and Mohammada Vikarabad were observed to be good source for dwarfness; work on evaluation at various places has shown that varieties like Totapuri, Red Small Sensation, Lafra, Kalapadi, Manipur, and the hybrid variety, Amrapali can have good potential for dwarfing. Evaluation of germplasm for good skin color has shown that several varieties have very good skin color. The cultivars, Janardhan Pasand, Suvarna Rekha, Fernandina, Rosa, and Sindhura can be good gene donors. The exotic varieties, Tommy Atkins, Kensington, Irwin, and Sensation have also been found to have good potential as a donor source for skin color. Till date the works carried out to find a resistance source for some of the pests and diseases have yielded no satisfactory result. The varieties, which were initially observed to be resistant, have shown susceptibility in later stages and using them in breeding programme based on earlier findings resulted in susceptible progenies. The main bottlenecks in mango breeding arise from the inherent nature of this species such as long juvenile phase, high heterozygosity, and single seed per fruit. Inheritance pattern of traits in mango is as follows:

1. Upright tree habit is dominant over spreading and spreading is dominant over dwarf.
2. Dwarfness is governed by recessive genes.
3. Bearing habit and fruit quality are strongly linked traits.
4. Biennial bearing is dominant over regular bearing.
5. Regular bearing is controlled by recessive genes which is linked with precocity.
6. Bunch bearing habit is dominant over single fruit bearing.
7. Presence of beak on fruits is a dominant trait.
8. Heterosis and transgressive segregation for fruit size was observed in the F1 hybrids.
9. Red peel color was found to be dominant and gradations in skin color of F1 have suggested the role of duplicate genes, same is the case with flesh color.
10. Floral malformation appears to be controlled by recessive genes.
11. Bacterial canker is governed by cytoplasmic genes.
12. Spongy tissue is controlled by recessive genes.

It was reported that many fruit quality traits such as fruit weight, fruit shape, ground skin color, fruit width, and pulp depth have high heritability, and thus, can be improved by breeding programme. Fruit flavour was relatively high in progenies derived from the combinations involving Kensington Pride. Subsequent studies also confirmed that blush colour and fruit flavour have a strong genetic component and high frequency of hybrids with red or burgundy blush can be obtained from crosses where one parent has an intense red blush colour. High variation between phenotypic coefficient of variation (PCV) with high genetic advance was observed for yield per plant and fruit weight in 31 chance seedlings of mango (Singh et al. 2004).

7.3.1 Inheritance of Quantitative Traits

While selecting the parents for hybridization programme, the quantitative characters such as fruit weight, fruit length, fruit thickness, fruit diameter, and pulp recovery should be considered as all of them have high heritability. Iyer and Subramanyam (1987) observed the transgressive segregation for fruit size whereas Sharma and Majumdar (1988a) observed it to be controlled by additive genes. Prabhuram (1998) reported that the heritability, expected genetic advance, and total genetic variance for fruit weight was influenced by environmental conditions whereas the high degree of broad sense heritability for fruit length, fruit weight, peel weight, and stone weight in different mango varieties was observed by Rajan et al. (2009). Fruit length was suggested to be a polygenic trait as it shows transgressive segregation and is highly influenced by environmental conditions (Pandey 2012). Additionally, total beta carotenoid pigments and TSS content was also found to have transgressive segregation (Sharma and Majumder (1988)). In the progenies of Alphonso x Neelum, light yellow pulp colour appeared to be dominant over orange-yellow (Iyer 1991).

7.3.2 Inheritance of Qualitative Characters

Out of 30 qualitative characters studied in 400 germplasm, the fruit shape distribution was observed to be oblong (67.5%),.elliptic (1.50%), roundish (30.5%), ovoid (0.25%), and obovoid (0.25%), fruit attractiveness, fibre free, pulp recovery (>70%), pulp aroma (mild 55.50 (%), intermediate (28.75%), and strong (15.75%)); TSS and eating quality (poor (16.75%), good (63.50%), very good (15.50%) and excellent (4.25%) are very important for commercial point of view (Sankaran et al. personal communication).

Sharma (1987) observed that flesh colour is governed by additive genes but light yellow pulp colour was found to be dominant over orange-yellow in the progenies of Alphonso x Neelum (Iyer 1991). The non-additive factors and low heritability governs the fruit weight, TSS, and pulp percentage traits (Dinesh 2003). The progenies of Amrapali (yellow peel colour) and Sensation (red peel colour) were observed to have various peel colour thus suggesting that red peel colour is not dominant over yellow peel colour. Cross between red peeled varieties Totapari Red Small and Sensation and yellow-peeled varieties Dashehari and Amrapali yielded progenies with varying shades of pink blush and few hybrids with green peel colour suggesting that red peel colour is heterozygous dominant and is governed by duplicate genes. Iyer and Subramanium (1987) reported fruit peel colour to be governed by a number of loci.. Prabhuram (1998) observed fruit pulp colour is highly heritable and controlled by additive genes. On the basis of study conducted on Amrapali and Mallika, fibre content in pulp was found to have high heritability and high genetic advance suggesting additive nature of genetic variance. The pulp aroma was found to have high heritability coupled with genetic advance which suggest the contribution of both additive and non-additive genetic variance while fruit taste was governed by non-additive genes.

7.4 Diversity Studies

Mango (*Mangifera indica* L.) is originated from the Indo-Burma region and so far 69 species have been reported under the genus *Mangifera* from the regions of Malayan Peninsula, Borneo, and Sumatra having the highest diversity (Bompard 1993). Mango has known to be under cultivation for almost 4000 years and more than 1000 varieties are reported to be in cultivation (Mukherjee 1953). Most of the varieties in present day cultivation are selection from open-pollinated seedlings or hybrids of unknown parentage (Singh. 1963). Centers of mango variability in India are (i) Subtropical humid region (south Assam, Manipur, Tripura and Mizoram), (ii) Chhota-Nagpur Plateau (Madhya Pradesh, Odisha and Bihar), (iii) Santhal Paragana, (iv) Southern Madhya Pradesh (tribal area) adjoining Odisha and Andhra Pradesh, (v) Dhar Plateau of Madhya Pradesh adjoining South Rajasthan and Gujarat, (vi) Tropical humid region of southern peninsular India and (vii) Andaman and Nicabar group of islands (Yadav and Rajan 1993).

7.4.1 Characterization and Cataloguing of Mango

Earliest scientific description of mango was done by Watt (1891). Maries (1902) collected more than 500 varieties of mango across the globe including Indian mangoes and described them with botanical terminology. Wood house (1909) described 40 mango varieties based on the key characters. Burns and Prayag (1927) described 89 mango varieties collected from the Bombay Presidency. Popenoe (1932) described 300 varieties collected from across the world. Mukherjee (1948) has described 72 varieties of Bengal, Bihar, and Uttar Pradesh. Simultaneously, Naik and Gangolly (1950) have described 335 varieties of South India. Apart from the fruit characters, they have also laid great stress on the vegetative characters. Huge variability has been recorded in cultivated mango as well as wild types and cultivars Southeast Asia. Pandey and Dinesh (2010) revised and updated the names mango varieties and the updated list now contains the names of 1682 cultivars. They also identified the donors for traits such as dwarfness, regular bearing, fruit weight, fruit shoulder colour, high pulp recovery, high TSS, long shelf-life, precocity and lateness in fruit maturity and good processing quality. Dinesh et al. (2012) characterized and catalogued 223 genotypes of mango using Bioversity International Descriptors and barcodes were developed for the same through molecular characterization. These include some polyembryonic varieties and 34 pickling varieties known as Appemidi from Western Ghats. Dinesh et al. (2017) characterized and catalogued another 177 genotypes of mango including the appemidi types and three *Mangifera* species.

Pandey and Dinesh (2010) grouped mango varieties based on the fruit weight as below:

S. no	Group	Varieties
1	Very small (up to 99 g)	Bhowani Chowras (50–70 g)
2	Small (100–149 g)	Amrapali (130 g) and Rataul (107 g)
3	Medium large (150–299 g)	Dashehari, Alphonso Bombay, Pairi, Langra, Pusa Arunima, Himsagar, Totapari, Neeleshwari, Neelum, Suvarnarekha, Ambika, Zardalu, Bombay Green, Arka Puneet, Arka Anmol (India), Tommy Atkins, Eldon and Sensation (Florida), Aroumanus and Golek (Indonesia), Kyo Savoy and Okrong (Thailand), Vallenato of Colombia and Momi-K (Momi Kinney) (Hawaii).
4	Large (300–500 g)	Arka Aruna, Mallika, Alfazli, A.U. Rumani, Manjeera, Mulgoa, Rajapuri, Samarbehist Chowsa and Sukul (India), Amelie (West Africa), Bourbon (Brazil);

(continued)

S. no	Group	Varieties
		Golden Brooks, Florigon, Davis Haden, Zrifin, Glenn and Van Dyke (Florida, USA), Kensington (Australia); Nam Doc Mai, Nuwun Chan, Namtal and Pak-Kra Bok (Thailand) and Tahar and Naomi (Israel)
5	Very large (more than 500 g)	SufedMulgoa (1813 g), HimamPasand (536 g), Anardana (717 g), Balakondapari (647 g), Tenneru (944 g), Cowasji Patel (713 g), Sora (1166 g), Gaddemar (713 g), Hathijhool, Pansera and Fazli of India, Manzanillo Nunez of Mexico, Osteen, Keitt, Anderson and Haden of Florida (U.S.A) and R2E2 of Australia.

Out of 61 mango varieties of U.S.A., 19 are reported to bear large to very large fruits (Brooks and Olmo 1972). Mango cultivars show great variations in fruit quality and climatic requirements. Varate Giduga is one of the unique genotypes from Sirsi region of Karnataka and it has got potential for commercial cultivation (Dinesh et al. 2019).

7.5 Resistant Sources for Biotic Stress Tolerance

There are several mango cultivars that possess the tolerance/resistance against biotic and abiotic stresses like 'Golden Brooks' has cold tolerance, both 'Bhadauran' and 'Elaichi' are resistant to mango malformation (Pandey 1984). RamNath et al. (1987) reported that 'Bhadayandaula','Samar Bahist Rampur', and 'Mian Sahib' were found to be free from malformation in field conditions whereas the cultivars such as Dashehari, Langra, Samarbehist Chowsa, Bombay Green, and Rataul of North and Central India, Gulab Khas, KishenBhog and Zardalu from Bihar and Alphonso, Pairi and Kesar of Maharashtra and Gujarat are susceptible to malformation. The mango hybrid 'Amrapali' is highly susceptible to floral malformation.

Most of the mango varieties of South India like Mulgoa Neelum, Bangalora, Banganpalli, Suvarnarekha, etc., are found to be free from malformation in south India; however, they exhibited malformation in North India. All the commercial mango varieties are susceptible to fungal and bacterial diseases and hence, proper management practices are important to minimize the loss. The varieties such as 'Parish', 'Fairchild', and 'Tommy Atkins' are moderately resistant to fungal disease, anthracnose, caused by *Colletotrichum gloeosporioides* Penz (Yee 1958) whereas the Janardhan Pasand, Neelum, and Zardalu show tolerance to powdery mildew (*Oidium mangiferae* Berthet) and 'Bombay Green' is found to be tolerant to bacterial canker caused by *Xanthomonas campestris* pv. *Mangiferae indicae* (Gupta 1976; Om Prakash and Srivastava 1987).

7.6 Problems in Breeding

There are several problems in mango breeding which have hindered the progress. Some of the important ones are discussed hereunder with possible solutions.

7.6.1 Fruit Drop

Inspite of the large number of flowers being crossed, because of heavy fruit drop, the number of progenies generated is very low. Several studies have been conducted to minimize fruit drop using growth regulators; however, their use in the breeding programme is very low owing to the less flowers remaining in panicle. Use of embryo rescue technique has been recommended by Iyer and Subramanyam (1972) for culturing more hybrid progenies. The performance of exotic cultivars of mango under South Indian conditions has also shown that some of the cultivars like Kensington had higher percentage of bisexual flowers. Probably these varieties if used in the breeding programme may help in developing varieties having higher fruit retention.

7.6.2 Long Juvenile Phase

Mango seedlings generally take 3 to 10 years for first fruiting. The identification and multiplication of selected progenies on a desirable rootstocks can shorten juvenility. Iyer (1991) observed the significant differences among seedlings and grafted plants of the same genotype for fruit size, quality, and even colour. Singh (1969) reported that mango seedlings can be induced to flower and fruit if defoliated and girdled scions are grafted into comparable shoots of a bearing tree few days before flowering. Ethephon has been shown to induce flowering (Chacko et al. 1974). Paclobutrazol was also observed to induce flowering in the variety Dashehari (Sant Ram 1996).

7.6.3 Polyembryony

Polyembryony comes in the way of using good polyembryonic varieties as female parents as isolation of zygotic seedlings from nucellar ones becomes difficult. The use of polymorphic enzyme systems (isozymes) to identify zygotic seedlings (Degani et al. 1990, 1992, 1993, Schnell and Knight 1992; Truscott 1992) is based on the fact that nucellar seedlings should possess the same isozyme alleles as that of the female parent. The difference at a locus coding for an enzyme suggests that the plant has originated zygotic embryo. The zygotic seedlings arising from self-pollination can be distinguished from nucellar seedlings by being homozygous at one or more loci as the female parent is heterozygous. Similarly, zygotic seedlings from cross-pollination are distinguishable from those resulting from self-pollination if they express an allele not carried by the female parent. The degree of the occurrence of zygotic seedlings varies among the polyembryonic cultivars (Degani et al. 1993; Schnell and Knight 1992). Sturrock (1968) suggested that the polyembryonic condition in mango is a recessive factor and is governed by a pair of genes. Contrarily, Aron et al. (1998), and Brettell et al. (2004) reported that polyembryony is controlled by dominant genes. However, they also observed that polyembryony varied with the site. Cordeiro et al. (2006) determined in the mango cv. Rosinha using RAPD markers the polyembryonic origin of platelets.

Globally, the maximum traded mangoes are polyembryonic varieties but in India all the commercial cultivars are monoembryonic in nature. There is a less effort or no efforts have been made to breed the polyembryonic scion varieties except the screening of some them for their rootstock traits. The mango varieties such as Bappakai, Bellary, Chandra Karan, Goa, Goa Kasargod, Kurukkan, Mazagaon, Mylepilian, Muvandan, Nekkare, Nileswar Dwarf, Olour, Peach, and Salem are polyembryonic in India whereas Cambodiana, Carabao, Corazon, Edward, Palo, 13–1, Pahutan, Pico, Senora, Starch, Strawberry, and Florigon in other countries (Singh 1960; Shukla et al. 2004). Lopez-Valenzuela et al. (1997) differentiated mango cultivars by their embryo type and identified their geographical origin using RAPD marker. The studies on genetic relatedness based on RAPD markers using genomic DNA and chloroplast DNA, RFLP analysis, and lineage among polyembryonic and monoembryonic varieties indicated that these two types of Indian mango cultivars have a different genetic base (Ravishanker et al. 2004; Abirami et al. 2008). Polyembryonic varieties were observed to have been introduced in India from other parts of Southeast Asia (Ravishankar et al. 2004). The important rootstocks such as 'Kurukkan'of India and 13-1 of Israel are polyembryonic that are tolerant to salt while Olour as rootstock has been found to have dwarfing effect on 'Himsagar' and 'Langra' in West Bengal. 'Bappakai' rootstook was found to be superior to 'Olour' rootstock for tree vigour, spread, and productivity for Neelum variety at Coimbatore conditions. Trials on polyembryonic varieties as rootstocks for tree vigour in high-density plantation and biotic and abiotic stresses must be continued for a longer time before coming to the conclusion for their recommendation on commercial scale.

7.7 Breeding

7.7.1 Objectives

The objectives in mango breeding are set separately for scion and rootstock breeding wherein the objectives may vary from place to place within the country and between the countries. The performance varieties are specific to a particular region (e.g., Alphonso suffers from the internal break down 'spongy tissue' in western parts of India). Hence, the breeding programme undertaken in these parts has as one of the objectives in screening against this malady. With the expansion of export market, demands for coloured mango varieties are increasing making it one of the most important objectives of scion breeding especially in Bangladesh where majority of varieties are green in colour. Besides colour, development of dwarf, and regular bearing mango hybrids is one of the major breeding objectives in India in a view of increasing the productivity per unit area. Till date no polyembryonic rootstock has shown consistent dwarfing effect on the scion and true to type rootstocks cannot be raised using monoembryonic seeds, hence, transfer of character is important. Rootstock breeding programme in Israel are mainly concerned with developing rootstocks for problematic soils in addition to varieties with better yield and quality parameters (Lavi et al. 1993). In Australia, the major breeding objective is to improve the cv. 'Kensington', that is susceptible to anthracnose and bacterial black spot and has poor shelf life by using Floridian and Indian mango cultivars as parents through hybridization. The Brazilian mango breeding aims at developing varieties with improved fruit yield and quality, regular bearing, and small tree size (Pinto and Byrne 1993). Hence, objectives in mango breeding vary from country to country as well as with growing states within a country. In general, the scion-breeding objectives are to develop dwarf, regular bearing, precocity bearing, medium to large size fruits with better fruit quality, shipping quality couple with resistant diseases and pests (Iyer 1991). With regard to rootstock breeding, the main desirable features are polyembryony, dwarfing, tolerance to adverse soil conditions (high pH, calcareous soil, etc.), and good scion-compatibility. However, combining all the desired traits in a single variety is difficult owing to high heterozygosity and long juvenile period (Singh 1977). Hence breeding objectives have to be defined for specific purposes (Sharma 1988).

7.7.2 Methods of Mango Breeding

7.7.2.1 Selection

Most of the commercial cultivars of mango are the results of selections made from chance seedlings. Many of the present-day cultivars in Florida were selected from open-pollinated seedling progenies. The variety 'Haden' was found to be a seeding of 'Mulgoba' a Indian variety. The cultivars Edward, Lippens, Tommy Atkins, and Zill were found to have originated from 'Haden'. In Israel, the varieties 'Maya' & 'Nimrod' are a selection from open-pollinated seedlings of Haden (Oppenheimer 1967). Likewise 'Dahar' was found to be a seedling of 'Irwin' (Slor and Gazil 1982) and 'Naomi' is a seedling of 'Palmer' (Tomer et al.1993). Two selections MDCH-1 and STH-1 are dwarf, and flower within 1 year in Manipur, India (Singh 1996).

7.7.2.2 Clonal Selection

It is one of the important tools in the improvement of perennial crops. Asexual propagation helps in the preservation of accumulated mutations over the years. Naik (1948) has reported significant variation due to bud mutation in trees of the same clone in an orchard with respect to fruit shape, size, colour, and quality. Oppenheimer (1956), after a survey of many orchards reported wide variability in the performance of the same variety in the same orchard.

Singh and Chadha (1981) selected four superior clones C24, C18, C35, and C45 from the variety 'Dashehari' at Lucknow. Two high yielding clones have been identified from 'Langra' by Singh et al. (1985). One off-season clonal selection has also been reported by Deshmukh and Budrukkar (1985). 'Paiyur-1' a clonal selection from Neelum has also been made at Tamil Nadu, India. Whiley et al. (1993) have reported the strains of Kensington resistant to bacterial black spot. Rajput et al. (1996) have isolated a high yielding and regular bearing clone Dashehari 51 from Uttar Pradesh, India.

7.7.2.3 Hybridization

The traditional method of pollinating the flowers on a panicle for several days has now been replaced by one-time pollination of about ten flowers per panicle at the most as it is rare that not more than one fruit/panicle is carried to maturity. By doing so fruit set can be up to 3.85% from 0.23 to 1.57% obtained by other methods (Mukherjee et al. 1968; Singh et al. 1980).

Another improvement in crossing has been the use of 24 × 21" perforated polythene covers of 100 gauges instead of muslin bags (Mukherjee et al. 1961). Studies carried out on inheritance of traits such as colour of young leaves, panicles and beaked characteristic of fruit by Sharma et al. (1985) helped in assessing the hybrid nature of F1 seedlings Singh et al. (1980) suggested that once the flower is carefully pollinated since the stigmatic surface is very small, the pollen deposited first has an advantage, there is very rare possibility of contamination from foreign pollen, the fruit set can be as high as 3.85 percent, without covering the flowers.

After the discovery of sporophytic self-incompatibility, a specialized insect-proof cages were used for mango breeding (Sharma and Singh 1970) in which both self-incompatible (female) and male parents planted and freshly reared house flies, or any other suitable pollinators are introduced to effect cross pollination (Sharma et al.1972).

A new off-season crossing technique was suggested by Kulkarni (1986). It involves the induction of flowering in the desired parents by veneer grafting of defoliated shoots on to normal shoots of off-season flowering cv. Royal Special and allowing open pollination between the desired parents. As no other cultivar flowers during this season, this is a safe technique.

Selection of Hybrid Progenies

The hybrid progenies are subjected to primary selection for the traits such as precocity, fruit size, fruit shape, skin colour, high pulp percentage, free from physiological disorders, and fruit quality. The selected hybrids are retained for further screening and also grafted and at least ten grafts per hybrid are planted in row trial and evaluated for yield, regularity in bearing, and for screening against diseases and pests. The selected progenies are compared with local check and national check for yield and other attributes for 3 years before deciding on their release.

Pre-selection Indices

Mango breeding is time consuming as it has long juvenile phase and need more land for evaluation of progenies. Inspite of difficulties, breeding time can be hastened by using pre-selection indices that can help in overcoming the said problem to a greater extent.

Young leaf aroma was found to be directly correlated with fruit aroma (Majumder et al. 1972; Whiley et al. 1993). The emergence of new growth flushes simultaneously with fruiting or immediately after harvest is indicative of regular bearing Sharma et al. (1972). Higher phloem: xylem ratio was found to be associated with dwarfness. If the ratio more than 1.0, they tend to be least vigorous; if it is 0.6–1.0 they have medium vigour and less than 0.6 are most vigorous (Kurian and Iyer 1992). A higher levels of phenol in the apical bud was found to be associated with dwarfing (Iyer 1991) and Majumder et al. (1981) also reported that lower stomata density is an indication of dwarfness which could not be confirmed by other workers (Iyer 1991).

Appemidi mangoes possess a typical pleasant odor due to the presence of Terpenoids in fruit sap mainly monoterpenoids like α and β-Phellandrenes, limonene, β-terpinene, α-Pinene, Camphene, *trans*-Ocimene, and 3-Carene. More odor intensive types have higher content of

Phellandrenes and other odors depend on the relative proportion of the above compounds. The tender mangoes also recorded very high total phenols and flavonoids which were up to 450 and 30 mg/100 g, respectively. This will give them the high anti-oxidant capacity. Some of the genotypes also found to have high vitamin C. This also can be used as pre-selection indices or biochemical markers for hybridity confirmation.

Breeding Achievements

In India, most of the cultivars are results of selections from naturally occurring, open-pollinated seedlings (Singh 1963) With the introduction of inarching technique, the process of selecting mango varieties was initiated during 1650–1820 of the Mughal period (Mukherjee 1953). Breeding work was first initiated by Burns and Prayag (1921) during the year 1911 with the aim of obtaining varieties having regular bearing habit, fruit quality, high yield with resistance or tolerance to pests and diseases. Three hybrids from the cross Cowasji Patel x Pairi were isolated from 153 crosses (Sen et al. 1946). Wagle (1929) made attempts to improve Alphonso by hybridization. Systematic hybridization work was started during 1940 at Anantharajupet with special reference to South Indian varieties (Bhujanga Rao and Rangacharlu 1958) and at Sabour with special reference to north and eastern region cultivars (Sen et al 1946 and). The main objective was to breed regular bearing varieties having good quality. The mango breeding work at Kodur (Andhra Pradesh) resulted in release of four hybrids, namely, Neeleshan (Neelum x Baneshan), Neelgoa (Neelum x Yerramulgoa), Neeluddin (Neelum x Himayuddin), and Swarnajehangir (Chinna-Survarnarekha x Jehangir) (Bhujanga Rao et al. 1963). The breeding work started at Krishnagar, Saharanpur, and Punjab did not result in the release of any hybrid. However, the mango improvement programme carried out at Sabour resulted in two hybrids, viz., Mahmood Bahar and Prabha Shankar, both of the combination Bombay and Kalapady (Mallik 1951).

In 1950s, systematic hybridization work started in several centres in India that resulted more than 50 hybrids across the globe. Several of these hybrids are slowly growing in popularity with the mango growers. At ICAR- Indian Agricultural Research Institute, New Delhi, two hybrids, Mallika, a cross between Neelum and Dashehari with a fruit size of 300–350 g, high T.S.S. high pulp recovery and fibreless pulp and Amrapali, a cross between Dashehari x Neelum, which is dwarf with a precocious bearing habit, good quality fruit rich in vitamin A were developed.

Mango improvement work at ICAR-Indian Institute of Horticultural Research, Bangalore resulted in six hybrids ArkaAruna (Banganpalli x Alphonso), Arka Puneet (Alphonso x Banganpall), Arka Anmol (Alphonso x Janardhan Pasand), ArkaNeelkiran (Alphonso x Neelum), Arka Udaya (Amrapali x Arka Anmol) and Arka Suprabhath (Amrapali x Arka Anmol). ArkaAruna has high pulp recovery with a fruit size of 500 g and is ideal for homestead planting as it is dwarf. Arka Puneet is similar to Alphonso but is free from spongy tissue. Arka Anmol is a late season variety with good sugar-acid blend and good keeping quality (Iyer and Subramanyam 1993). Arka Neelkiran is also a late season hybrid with attractive colour and good keeping quality. Recent hybrids (Arka Udaya and Arka Suprabhath) are having excellent fruit quality, bunch bearing, and late maturing. Currently, work is being done on introgression of peel colour from Vanraj on Amrapali and Alphonso background.

Two hybrids 'Ratna', a hybrid between Neelum and Alphonso and Sindhu (Ratna x Alphonso) have been developed at Fruit Research Station, Vengurla. Ratna has an excellent fruit quality comparable to Alphonso and is free from spongy tissue (Salvi and Gunjate 1988). 'Sindhu' is a parthenocarpic mango variety having good fruit quality and very thin stone (Gunjate and Burondkar 1993).

These hybrid varieties 'Au-Rumani (Rumani x Mulgoa)', Manjeera (Rumani x Neelum), and KMH-1 (Cherukurasam x Khader) have been developed by the Andhra Pradesh Agricultural University. 'Au-Rumani' is a prolific bearer. Manjeera and KMH-1 is dwarf having good quality. In Gujarat, three mango hybrids have been released, namely, Neelphonso (Neelum x

Alphonso), Neeleshan Gujarat (Neelum x Baneshan), and Neeleshwari (Neelum x Dashehari). These hybrids have high TSS, total sugars, and Vitamin C in addition to being dwarf (Sachan et al. 1988).

At Fruit Research Station, Sabour, four mango hybrids developed and released as Sundar Langra (Sardar Pasand x Langra) having Langra quality and regular bearing habit; AlFazli (Alphonso x Fazli) early ripening with Fazli quality; Sabri (Gulabkhas x Bombai) having Bombai fruit shape and colour of Gulabkhas with regular bearing habit and Jawahar (Gulabkhas x Mahmood Bahar) having high pulp recovery, fibreless pulp and early bearing habit have been developed (Hoda and Ramkumar 1993).

In Israel, the seedlings were evaluated on calcareous soils of which 9 were selected based on their performance with regard to rootstock traits (Lavi et al. 1993). A new cultivar 'Naomi' which had originated from the cv. 'Palmer' either by self pollination or by cross pollination with 'Maya', Haden, 'Nimrod' or 'Irwin' was released (Tomer et al. 1993). Another new variety 'Tango', a open-pollinated seedlings selection of the variety 'Naomi' has been released. It has attractive skin colour with good quality. In South Africa, four new cultivars 'Heidi', Neldica 'Neldawn', and 'Ceriese' were identified by mass selection (Marais 1992). The list of hybrids released in India and Australia is given in Table 6.

S. No	**Hybrid**	**Parents**	**Place of release**
1	ArkaAruna	Banganpalli X Aplhonso	ICAR-IIHR, Bangalore
2	Arka Puneet	Alphonso X Bangapalli	ICAR-IIHR, Bangalore
3	Arka Anmol	Alphonso X Janardhan pasand	ICAR-IIHR, Bangalore
4	ArkaNeelkiran	Alphonso X Neelum	ICAR-IIHR, Bangalore
5	Arka Udaya	Amrapali X Arka Anmol	ICAR-IIHR, Bangalore
6	ArkaSuprabhath	Amrapali X Arka Anmol	ICAR-IIHR, Bangalore
7	Swarna Jehangir	Chinnasuvarnarekha x Jehangir	FRS, Kodur
8	Neeluddin	Neelum X Himayuddin	A.P.A.U.
9	Au Rumani	RumanixYerramulgoa	A.P.A.U.
10	Manjeera	RumaniXneelum	A.P.A.U.
11	KMH-1	Cherukurasam X Khader	A.P.A.U.
12	Neelgoa	Neelum X Yerramulgoa	A.P.A.U.
13	Neeleshan	Neelum X Baneshan	A.P.A.U.
14	Neelphonso	Neelum X Aplhonso	GAU, PARIA
15	Neeleshwari	Neelum X Dashehari	GAU, PARIA
16	Neeleshan	Neelum X Baneshan	GAU, PARIA
17	Mahmood Bahar	Bombay X Kalepad	FRS, Sabour
18	Prabhashankar	Bombay X Kalepad	FRS, Sabour
19	Al Fazli	Alphoso X Fazli	FRS, Sabour
20	Sundar langra	Sardar Pasand X Langra	FRS, Sabour
21	Sabri	Gulab Khas X Bombai	FRS, Sabour
22	Jawahar	Gulab Khas X M. Bahar	FRS, Sabour
23	Ratna	Neelum X Alphonso	FRS, Vengurla

(continued)

S. No	Hybrid	Parents	Place of release
24	Sindu	Ratna X Alphonso	FRS, Vengurla
25	PKM-1	ChinnaswarnarekhaXNeelum	Periyakulam,T.N.A.U.
26	PKM-2	Neelum X Mulgoa	Periyakulam,T.N.A.U.
27	Mallika	Neelum X Dashehari	IARI,N.Delhi
28	Amrapali	Dashehari X Neelum	IARI,N.Delhi
29	Pusa Arunima	Amrapali x Sensation	IARI,N.Delhi
30	Pusa Surya	Amrapali x Sensation	IARI,N.Delhi
31	Pusa Pratibha	Amrapali x Sensation	IARI,N.Delhi
32	PusaShreshth	Amrapali x Sensation	IARI,N.Delhi
33	PusaPeetamber	Amrapali x Lal Sundari	IARI,N.Delhi
34	PusaLalima	Dushehari x Sensation	IARI,N.Delhi
35	Ambika	AmrapaliX Janardhan pasand	CISH
36	Arunika	Amrapali x Vanraj	CISH
37	Sai sugandh	Totapuri x Kesar	MPKV, Rahuri
38	CISH-M-2	Dasheharixchausa	CISH
39	CISH–M-1	Amraphalixjanardhanpasand	CISH
40	Konkan Samrat	Alphonso x Tomy Atkins	BSSKKV
41	Suvarna	Alphonso x Neelum	DAPOLI
42	NMBP-1243	Irwin x Kensington Pride	Australia
43	Neeleshan Gujarat	Neelam x Baneshan	AES, Paria
44	Sonpari	Alphonso x Baneshan	AES, Paria
45	NMBP-1201	Irwin x Kensington Pride	Australia
46	NMBP-4069	Van Dyke and Kensington Pride	Australia
47	Hybrid-949	Amrapalli x Vanraj	ICAR-CISH
48	Hybrid-1084	AmrapalixVanardhanpasand	ICAR-CISH
49	GMH-1	Alphonso X Baneshan	GAU
50	Hybrid 45	(Bennet Alphonso x Himayuddin),	ICAR-CISH
51	Hybrid 87	(Kalapady x AlampurBenishan)	ICAR-CISH
52	Hybrid 151	(Kalapady x Neelum)	ICAR-CISH

Sankaran and Dinesh (2018)

7.7.2.4 Wide Hybridization

Wild relatives of mango are the reservoirs of useful traits that can be introgressed on any commercial variety background provided wild species should have edible fruits with other desirable characters and species that could act as gene donors for specific improvements like resistance to pests and diseases (Iyer 1991). Bompard (1993) has discussed the potential use of *M. laurina* as a potential source of resistance against anthracnose. *M. orophila*, a species from Malaysia, and *M. dongnaiensis* from Vietnam, which are restricted to mountain forests 1000–1700 metres above sea level may make growing mango in the high altitude areas of Mediterranean region.

The species *M. magnifica* (free from fibres), *M. Rufocostata and M. swintonioides* (off-season bearing habit), *M. paiang* and *M.foetida* (good quality fruits), *M. caesia* var.Wani from 'Bali & Borneo has a distinctive taste, *M. casturi* occurring in South Kalimantan is a prolific bearer that can be used as gene donors in mango breeding programme (Bompard 1993; Kostermans and Bompard 1993). The other species *M. altissimo* has been reported to be resistant to hoppers, tip borers, and seed borers (Angeles 1991). Strains of *M. sylvatica* which are polyembryonic in nature with good fruit quality have been reported from northeastern Indian States (Yadav 1996).

Hybrids with good quality pollen can be produced by crossing pentamerous mango flowers with the Indian mango having only one fertile stamen (Fairchild 1948). Thirty-two inter-specific hybrids involving *M. odorata* & *M. zeylanica* developed (Bhujanga Rao et al. 1963) suggesting that different *Mangifera* spp. are cross compatible (Mukherjee 1963). However, in order to use these species in mango improvement programme, their thorough evaluation is paramount hence, extensive research is warranted in order to conserve, screen, and utilize them in future breeding programme.

7.7.2.5 Breeding for Biotic Stress Tolerance

M. laurina 'Lombok' has been reported to be resistant to anthracnose which was confirmed through artificial inoculation screening. This particular accession is being used as the male parent in a cross with the advanced breeding line *M. indica* 'NMBP4046' (Bally et al. 2009) to introgress the anthracnose resistance. A total of ten candidate disease resistant genes have been identified from an expressed sequence tag (EST) and EST-SSR markers have been developed for six genes and screened for polymorphism across a selection of 32 accessions identified from the diversity analysis, as representing the major groupings in the Australian National Mango Genebank (ANMG). Further, six EST-SSR putative disease resistance markers have shown to be polymorphic in 32 *Mangifera* accessions were assessed. The number of alleles identified varied from two to nine with an average of 5.3 alleles per locus and these EST-SSR markers will be tested in future to determine their ability to detect post-harvest anthracnose disease resistance by screening them across the segregating disease resistant hybrid population generated between *M. indica* and *M. laurina.* Identification of further putative disease resistant genes and gene markers (SSRs and single nucleotide polymorphisms (SNPs) will be done by cloning of genes from resistant and susceptible accessions to look for sequence differences (Dillon et al. 2013). Molecular markers have also been used to identify or fingerprint individual mango varieties by unique patterns of marker alleles. DNA fingerprinting data have been used to support plant breeders' rights and patents (IP AUSTRALIA 2008) and in maker assisted selection (MAS). In U.S.A, although several varieties have shown moderate resistance to anthracnose so far no commercial cultivar is resistant to be produced in humid areas without fungicide application. Zebda, a Egyptian cultivar, is aesthetically unacceptable since it is green at maturity but do possess resistant to this disease. The cv. Kent and Haden are important in international trade but are highly susceptible to anthracnose when grown humid areas however they are produced in arid production areas (Ploetz 2007). Singh et al. (2014) have already released the draft genome of *Mangifera indica* with a small genome size of about 450 Mbp. Molecular genomics technologies are being used to assist our understanding of mango genetics and the diversity of the world mango gene pool. In recent years many studies using molecular marker to discover the diversity and phylogenetic relationships between mango varieties and related *Mangifera*species have been undertaken. The gene discovery for specific trait is yet to be taken in mango across the globe. In order to develop fruit fly tolerant varieties, *M.camptosperma*, EC95862 and VarateGiduga are being used as male parents at IIHR, Bengaluru.

7.7.2.6 Mutation Breeding

Mutation is an important tool in creating variability. Although mutations and chimeras occur frequently, they are generally marked by intra-somatic selections. In mango, mutants establishing as cultivars are rare. Two somatic mutants have been reported in mango. The variety 'David Haden' was reported to be sport of 'Haden'. It is larger than 'Haden' & matures early (Young and Ledin 1954). Another mutant 'Rosica' from seedless variety Peru was reported to be a mutant of 'Rosado de Ica'. It is high yielding and regular bearing (Medina 1977). A mericlonal chimera of Alphonso was observed by Roy (1950), it differed with respect to fruit shape and mutants of 'Puthi' was developed Roy and Viswesvariya (1951).

Induction of mutation was first tried using gamma irradiation by Siddique et al. (1966). The dormant buds of cv. Langra were irradiated and grafted on rootstocks. Those bud sticks which had undergone 3 kR of radiation bore fruits which were heavier, larger and had more creamish yellow pulp than control (Siddiqui 1985). Sharma and Majumder (1988b) observed that dosages above 5 kR are lethal for mango and the LD 50 doses lie between 2 and 4 kR. The chemical mutagens EMS and NMU were found to be optimum at 1.5% and 0.05%, respectively. The degree of mutations induced by physical and chemical mutagens was observed to be almost same, indicating the high sensitivity of only some loci. In spite of these few attempts, no worthwhile variety has been developed by mutation on mango.

7.8 Biotechnological Studies

The role of biotechnological methods in understanding the genetics and for making improvement is of recent origin. These come in handy in solving the mystery of the nomenclature. Monoembryonic and polyembryonic mango genotypes of India are reported to have different genetic bases (Ravishankar et al. 2004). Lopez-Valenzuela et al. (1997) used RAPD markers to differentiate mango cultivars by embryo type and geographical origin. The main components of mango genetic engineering are efficient somatic embryogenesis and induction of mutations in embryogenic cultures, selectable markers, and transformation with gene that mediates a horticultural trait (Litz 2004). Mango cultivars have been characterized using morphological traits which are highly influenced by environmental conditions and hence are not reliable (Tanksley et al. 1989), hence several genetic markers including isozymes, RAPD, AFLP, ISSR, SSRs, and VNTRs have been developed which can be used for more robust characterization of mango cultivars.

In mango, isozymes were used to detect possible genetic variation among the clones (Gan et al. 1981). Degani et al. (1992) used isozymes for characterization and parentage analysis in mango cultivars. They could distinguish 'Pico' and 'Carabao' by isozyme analysis. The isozyme banding pattern was used to correlate the origin of some of the cultivars such as 'Haden' from 'Mulgoba' and 'Zill' from 'Haden' (Campbell 1992) and 'Tahar' from 'Irwin' (Slor and Gazit 1982). The origin of some of the mango cultivars was refuted by isozyme analysis (Degani et al. 1990, 1992). Thus 'Edward' is commonly accepted to be hybrid of 'Haden' and 'Carabao' (Campbell 1992), the isozyme banding pattern showed that 'Carbao' cannot be the male parent. Isozyme analysis also revealed that 'Mulgoba' may not be a parent of 'Keitt' and 'Givataim' cannot be an offspring of the polyembryonic cultivar '13-1'. Several molecular markers such as RAPD (Schnell et al. 1995; Valdomiro and Paulo Sarmanho (2004)), mini- and microsatellite probes by Adato et al. (1995), and AFLP markers by Kasikush et al. (2001) have been used for identification of mango cultivars and generation of a preliminary genetic map. The major problem in tissue culture propagation of mango is phenol exudation and explant discoloration. Studies at Pune have revealed that plantlets were generated from monoembryonic varieties. Sahijram et al. (2005) were able to rescue hybrid embryos 6-8 weeks after pollination. This holds promise for increasing the number of progeny population, which is a limiting factor in mango hybridization. Ravishankar et al. (2015) found that the SSR marker characterization of Indian mango cultivars revealed two distinct groups of populations with geographical affiliation. Six SSR loci with low PI have been identified as universal markers for mango characterization.

7.8.1 Research Gap in Breeding

In a heterozygous crop like mango, it is all but natural that deriving genetical information is time consuming. There is a need for extensive information generation on the conservation, characterization, and evaluation of wild species in the light of biotic and abiotic stresses. The polyembryonic status of several species and their use as rootstock is still not clear. The large diversity within the *Mangifera indica* L needs to be evaluated for desirable traits and nomenclature ambiguity needs to be clarified. Rootstock breeding efforts need to be intensified. Promotion of heirloom varieties and on-farm conservation needs to be promoted by identifying custodians of genetic diversity.

7.9 Conclusion

Mango is one of the important fruit crops in tropical and subtropical countries owing to its nutritional benefits as well as industrial significance. Climate change has presented before mango growers huge challenges in the form of new pest and diseases, cultivation in problematic soils, and also changed consumer preferences in global market. The ever-reducing land and water resources warrant the development of technologies for increasing mango productivity per unit land as well as per unit resources used. This necessitates the development of dwarf varieties or dwarfing rootstock to be used for high density plantation as well as varieties with high fruit quality along with resistance to biotic and abiotic stresses for minimizing the use of injurious pesticides and to add to the quality of produce and consumer safety. Several studies have been conducted to understand the inheritance of traits, diversity of genotypes, identification of potential donors, molecular markers for diversity analysis, and their utilization in the improvement of mango for commercial cultivation and marketing.

References

Abirami K, Singh SK, Singh Room, Mohapatra T, Kumar AR (2008) Genetic diversity studies on polyembryonic and monoembryonic mango genotypes using molecular markers. Indian J Hort 65(3): 258–262

Adato A, Sharon D, Lavi U, Hillel I, Gazit S (1995) Application of DNA fingerprints for identification and genetic analysis of mango (*Mangifera indica*) genotypes. J Am Soc Hort Sci 120:259–264

Angeles DE (1991) *Mangiferaaltissima*. In: Verheij EWM, Coronel RE (eds) Edible fruits and nuts. Plant Resources of South East Asia 2. PUDOC. Wegeningen, pp 206–207

Aron YH, Czosnek GS, Degani H (1998) Polyembryony in mango (*Mangifera indica* L.) is controlled by a single dominant gene. Hort Sci 33(7):1241–1242

Arumuganathan K, Earle ED (1991) Nuclear DNA content of some important plant species. Plant Mol Biol Rep 9:208–218

Bally ISE, Hofman PJ, Irving DE, Coates LM, Dann EK (2009) The effects of nitrogen on postharvest disease in mango (Mangifera indica L. ‘Keitt’). Acta Hortic 820:365-370

Bhujanga Rao C, Rangacharlu VS (1958) Breeding new mango varieties in South India. Indian J Hort 15:173–183

Bhujanga Rao C, Swamy GS, Nagabhushanam M, Rama Rao BV (1963) Performance of some promising Andhra mango hybrids. Punjab Hort J 3:124–136

Bompard JM (1993) The genus Mangifera rediscovered: the potential contribution of wild species to mango cultivation. Acta Hort 341:69–77

Bompard JM, Schnell RJ (1997) Taxomomy and systematics. In: Litz RE (ed) The mango, botany, production and uses. CABI International, New York, pp 21–48

Brettell RIS, Johnson PR, Kulkarni VJ, Muller W, Bally ISE (2004) Inheritance of fruit characters in hybrid mangoes produced through controlled pollination. Acta Hort 645:319–326

Brooks RM, Olmo HP (1972) Register of new fruit and nut varieties. Mango. University of California Press. Berkeley, Los Angeles, London, pp 286–295

Burns W, Prayag SH (1921) The Book of Mango. Bulletin, Department of Agriculture, Bombay, p 103

Campbell RJ (ed) (1992) A guide to mango in Florida. Fairchild Tropical Garden, Miami, USA

Campbell RJ (2007) The Potential of New Mangifera Species in Florida. In: Proceedings of the Florida State Horticultural Society, Vol. 120, pp. 11-12

Chacko EK, Kohli RR, Randhava GS (1974) Investigations on the use of 2- chloroethylphosponic acid (Ethephon CEPA) for the control of biennial bearing in mango. Sci Hortic 2:389-398

Cordeiro MCR, Pinto ACQ, Ramos VHV, Faleiro FG, Fraga L (2006) Identification of plantlet genetic origin in polyembryonic mango (Mangifera indica, L.) cv. Rosinha seeds using RAPD markers. Rev Bras Frutic 28:454-457

Degani C, Batsri RE, Gazit S (1990) Enzyme polymorphism in mango. J Am Soc Hortic Sci 115:844–847

Degani C, Cohen M, Reuvani ORE, Gazit, S (1993) Frequency and characteristic of zygotic seedlings frompolyembryonic mango cultivars determined using isozymes asgenetic markers. Acta Hort 341:78–85

Degani C, Cohen M, El-Batsri R, Gazit S (1992) PGI isozymediversity and its genetic control in mango. Hort Sci 27:252–254

Deshmukh PA, Budrukkar NB (1985) Promising mango selection from Himatbag (Aurangabad). Abstracts: 2nd National Symposium on Mango, Banglore, p. 9

Dillon NL, Bally ISE, Wright CL, et al. (2013) Genetic diversity of the Australian National Mango Genebank. Sci Hortic 150:213–226

Dinesh MR, Vasugi CS (2002) Catalogue of mango germplasm. Indian Institute of Horticultural Research, Hesseraghatta Lake PO, Bangalore 560089:160

Dinesh MR (2003) Genetic studies in mango (*Mangifera indica* L.). J Appl Hort 5(1):27–28

Dinesh MR, Vasugi C, Ravishankar KV, Reddy YTN (2012) Mango Catalogue, ICAR-IIHR, p 468

Dinesh MR, Sankaran M, Ravishankar KV, Gowda S (2019) A unique mango (*Mangifera indica* L.) variety. J Hort Sci 14(2):76,

Dinesh MR, Sankaran M, Ravishankar KV, Sunil Gowda DC, Veena GL (2017) Mango Catalogue. ICAR-IIHR, Bengaluru, p 640p

Duval MF, Bunel J, Sitbon C, Risterucci AM (2005) Development of microsatellite markers for mango (*Mangifera indica* L.). Mol Ecol Notes 5:824–826

Eiadthong W, Yonemori K, Sugiura A, Utsunomiya N, Subhadrabandhu S (1999) Analysis of Phylogenetic Relationships in Mangifera by Restriction Site Analysis of an Amplified Region of cpDNA. Sci Hortic 80:145-155. http://dx.doi.org/10.1016/S0304-4238(98)00222-2

Fairchild D (1948) The mango relatives of Cochin China: those with fivestamen flowers. Proc Flo State Hort Soc 64:257

Fukino N, Yoshioka Y, Kubo N, Hirai M, Sugiyama M, Sakata Y, Matsumoto S (2008) Development of 101 novel SSR markers and construction of an SSRbased genetic linkage map in cucumber (*Cucumis sativus* L.). Breed Sci 58:475–483

Gan YY, Zaini S, Idris A (1981) Genetic variation in the grafted vegetatively propagated mango Mangifera indica. Pertanika 4:53–63

Gunjate RT, Burondkar MM (1993) Parthenocarpic mango developed through hybridization. Acta Hort 341:107–111

Gupta JH (1976) Reaction of mango varieties to powdery mildew (*OidiummangiferaeB*erthet) in Uttar Pardesh. Prog Hort 8:63–64

Hoda MN, Kumar R (1993) Improvement of mango. Proc. of the National Seminar on Irregular bearing in mango: problems and strategy, held from 12–13 July at Pusa. Pratan Kamal Printing Press, Muzaffarpur, Bihar, India, pp 34–35

Honsho C, Nishiyama K, Eiadthong W, Yonemori K (2005) Isolation and characterization of new microsatellites markers in mango (Mangifera indica). Mol Ecol Notes 5:152–154

Huang X, Wei X, Sang T, Zhao Q, Feng Q, Zhao Y, Li C, Zhu C, Lu T, Zhang Z, Li M, Fan D, Guo Y, Wang A, Wang L, Deng L, Li W, Lu Y, Weng Q, Liu K, Huang T, Zhou T, Jing Y, Li W, Lin Z, Buckler ES, Qian Q, Zhang Q-F, Li J, Han B (2010) Genome-wide association studies of 14 agronomic traits in rice landraces. Nat Genet 42:961-967

IP AUSTRALIA (2008) Mango (Mangifera indica), Variety: NMBP1243. Plant Varieties Journal, Cambridge, 21(3):84-85; 275-283

Iyer CPA (1991) Recent advances in varietal improvement in mango. Acta Hort 291:109–132

Iyer CPA, Subramanyam MD (1971) Possible role of embryo culture on mango breeding. Indian J Hortic 29:135–136

Iyer CPA, Subramanyan MD (1993) Improvement of mango. In: Chadha KL, Pareek OP (eds) Advances in horticulture, vol I. Malhotra Publishing, New Delhi

Iyer CPA, Subramanyam MD (1979) Improvement of mango by selection and hybridization. Annual Report, Indian Institute of Horticultural Research, Bengaluru, India, p 16

Iyer CPA, Subramanyam MD (1987) Improvement of mango by selection and hybridization. Annual Report of Indian Institute of Horticultural Research, Indian Institute of Horticultural Research, Bangalore, p 11

Iyer CPA, Degani C (1997) Classical breeding and genetics. In: Litz RE (ed) The mango, botany, production and uses. CAB International, New York, pp 49–68

Kasikush K, Jinggui F, Tomer E, Hillel J, Lavi U (2001) Cultivar identification and genetic map of mango (*Mangifera indica*). Euphytica 122:129–136

Kostermans AJGH, Bompard JM (1993) The Mangoes: their botany, nomenclature, horticulture and utilization. International Board for Plant Genetic Resources, Rome and the Linnean Society Publishers, London, 232 p

Kulkarni VJ (1986) Graft induced off-season flowering and fruiting in the mango (*Mangifera indica* L.). J Hort Sci. 61(1):141–145

Kurian RM, Iyer CPA (1992) Stem anatomical characters in relation to tree vigour in mango. Scientia Hort 50:245–248

Lavi U, Sharon D, Tomer E, Adato A, Gazit S (1993) Conventional and modern breeding of mango cultivars and root- stocks. Acta Hort 341:146–151

Lavi U, Tomer E, Gazit S (1989) Inheritance of agricultural important traits in mango. Euphytica 44:5–10

Lavi U, Tomer E, Hillel J (1998) Components of genetic correlations of mango traits. Scientia Hort 75:11–25

Sahijram L, Bollamma KT, Naren A, Soneji JR, Dinesh MR,Halesh GK (2005) In vitro hybrid embryo rescue in mango (Mangifera indica L.) breeding. Ind J Hort 62(3):235-237

Li H, Peng Z, Yang X, Wang W, Fu J, Wang J, Han Y, Chai Y, Guo T, Yang N, Liu J, Warburton ML, Cheng Y, Hao X, Zhang P, Zhao J, Liu Y, Wang G, Li J, Yan J (2013) Genome-wide association study dissects the genetic architecture of oil biosynthesis in maize kernels. Nat Genet 45:43-50

Litz RE (2004) Biotechnology and mango improvement. Acta Hortic 645:85–92

Liu ZW, Biyashev RM, SaghaiMaroof MA (1996) Development of simple sequence repeat DNA markers and their integration into a barley linkage map. Theor Appl Genet 96:869–876

Lopez-Valenzuela JA, Martinez O, Parades-Lopez O (1997) Geographic differentiation and embryo type identification in *Mangifera indica* L. cultivars using RAPD markers. Hort Sci 32:1105–1108

Majumder PK, Sharma DK (1990) Mango. In: Bose-and TK, Mitra SK (eds) Fruits: tropical and subtropical. Naya Prakashan, Kolkata, pp 1–62

Majumder PK, Sharma DK, Singh RN (1981) Breeding for dwarfness in mango (*Mangifera indica* L.). National Symposium on Tropical and Subtropical Fruit Crops, Bengaluru, pp 3

Majumder PK, Singh RN, Sharma DK, Mukherjee SK (1972) Preliminary studies on inheritance in *Mangifera indica* L. Acta Hort 24:101–106

Mallik PC (1951) Morphology and biology of the mango flower. Proc Indian Sci Congr 155

Marais Z (1992) Mango evaluation for breeding. Central Salt and Freshwater Research Institute Information Bulletin, pp 234–237

Maries C (1902)-Indian mangoes. J R Hort Soc 26:755–770;/901–2

Medina JP (1977) 'Rosica'-a new mango variety selectedin Ica, Peru. Fruit Var J 31:88–89

Milbourne D, Meyer RC, Collins AJ, Ramsay LD, Gebhardt C, Waugh R (1998) Isolation, characterisation and mapping of simple sequence repeat loci in potato. Mol Gen Genet 259:233-245

Mukherjee SK (1948) The varieties of the mango (Mangifera indica L.) and their classification. Bull Bot Soc Bengal 2:101–133

Mukherjee SK (1950) Mango: its allopolyploid nature. Nature 166:196–197

Mukherjee SK (1953) The Mango: Its botany, cultivation, uses and future improvements, especially as observed in India. Ecol Bot 7:130

Mukherjee SK (1963) Cytology and breeding of mango. Punjab Hort J 3:107-115

Mukherjee SK, Singh RN, Majumder PK, Sharma DK (1968) Present position regarding breeding of Mango (*Mangifera indica* L.) in India. Euphytica 17:462–467

Naik KC (1948) Improvement of the mango (*Mangifera indica* L.) by selection and hybridization. Indian J Agric Sci 18:35–41

Naik KC, Gangolly SR (1950) Classification and Nomenclature of South Indian Mangoes. The Madras Department of Agriculture, Printing Press, Madras, p 311

Olano CT, Schnell RJ, Quintanilla WE, Campbell RJ (2005) Pedigree analysis of Florida mango cultivars. Proc Fl State Hort Soc 118:192–197

Prakash Om, Srivastava KC (1987) Mango diseases and their management. WorldReview.Today and Tomorrow Printers, New Delhi

Oppenheimer C (1956). Study tour report on subtropical fruit growing and research in India and Ceylon. Special Bulletin 3

Oppenheimer CH (1967) Nimrod a new mango variety-selected in Israel. Proc Florida State Hort Soc 80:358–359

Pandey SN (1984) International Check List of Mango Cultivars. Division of Fruits and Horticultural Technology, Indian Agricultural Research Institute, New Delhi

Pandey SN (1998) Mango cultivars. In: Prakash S (ed) Mango Cultivation. Ram . International Book Distributing Co., Charbagh, Lucknow, pp 39–99

Pandey SN, Dinesh MR (2010) Mango. Published by Indian Council of Agricultural Research, New Delhi, p 153

Pasam R, Sharma R, Malosetti M, van Eeuwijk F, Haseneyer G, Kilian B, Graner A (2012) Genomewide association studies for agronomical traits in a world wide spring barley collection. BMC Plant Biol 12:16

Pinto ACQ, Byrne DH (1993) Mango hybridization studies in tropialSavannah('Cerrados') of Brazil Central Region. Acta Hort 341:98–106

Ploetz (2007) Anthracnose of mango: Management of the most important pre and post harvest disease. Report. University of Florida, TREC Homestead Department of Plant Pathology USA. 1-11

Popenoe W (1932) Manual of tropical and subtropical fruits. (facsimile of the 1920 ed. Publ. 1974) Hafner Press, New York

Prabhuram R (1998) breeding studies with special reference to red peel colour in Mango (Mangifera indica L.). Ph.D Thesis, New Delhi, pp 86

Rajan S, Yadava LP, Ram Kumar Sexena SK (2009) Genetic divergence in mango varieties and possible use in breeding. Indian J Hort 66(1):7–12

Rajput MS, Chadha KL, Negi SS (1996) Dashehari-51, a regular bearing and high yielding clone of mango cv. Dashehari. (Abstract). In: Proceedings of the 5th international mango symposium, Tel Aviv, Israel, pp. 52

Ram N, Kamalwanshi RS, Sachin IP (1987) Studies on mango malformation. Indian J Mycol Plant Pathol 17:29–33

Ravishankar KV, Mani BHR, Anand L, Dinesh MR (2011) Development of new microsatellite markers from mango (*Mangifera indica*) and crossspeciesamplification. Am J Bot:96–99

Ravishankar KV, Chandrashekara P, Sreedhara SA, Dinesh MR, Anand L, Prasad GVSS (2004) Diverse genetic bases of Indian polyembryonic and monoembryonic mango (*Mangi-feraindica*L) cultivars. Curr Sci 87(7):870–871

Ravishankar KV, Padmakar B, Bajpai Anju, Srivastava N, Mani BH, Vasugi C, Rajan S, Dinesh MR (2015) Genetic diversity and population structure analysis of mango (Mangifera indica) cultivars assessed by microsatellite markers. Trees 2015(29):775–783

Roy,B. 1950. A mango chimera. Current Science.19, 93

Roy B, Visweswaraiya SS (1951) Cytogenetics of mango and banana. Report, Maharashtra Association for the Cultivation of Sciences, Pune

Sachan SCP, Katrodia JS, Chundawat BS, Patel MN (1988) New mango hybrids from Gujarat. Acta Hort 231:103–105

Salvi MJ, Gunjate RT (1988) Mango breeding work in Konkan region of Maharashtra State. Acta Hort 231:100–102

Sankaran M, Dinesh MR (2018) Conservation and utilization of genetic resources of mango. In Book: genetic resources of fruit crops published by Jaya publishing house, New Delhi, page No. 16–37

Sankaran M, Dinesh MR, Gowda DCS, Venugopalan R (2020) Genetic Analysis in mango (Mangifera indica L.) based on fruit characteristics of 400 genotypes. J Hort Sci 15(2):161-172

Sant ram (1996) High density orcharding. Mango Res Bull No 122

Schnell RJ, Knight RJ (1992) Frequency of zygotic seedlings fromfive polyembryonic mango rootstocks. Hort Sci 27:174–176

Schnell RJ, Olano CT, Quintanilla WE, Meerow AW (2005) Isolation and characterization of 15 microsatellite loci from mango (Mangifera indica L.) and cross-species amplification in closely related taxa. Mol Ecol Notes 5:625-627

Schnell RJ, Ronning CM, Knight RJ (1995) Identification of cultivars and validation of genetic relationships in Mangifera indica L. using RAPD markers. Theor Appl Genet 90:269–274

Sen PK, Mallik PC, Ganguly BD (1946) Hybridization of mango. Indian Hort. 4:4

Sharma DK (1987) Mango breeding. Acta Hort 196:61–67

Sharma DK, Majumder PK (1988a) Further studies on inheritance inmango. Acta Hort 231:106–111

Sharma DK, Majumder PK (1988b) Induction of variability in mangothrough physical and chemical mutagens. Acta Hort 231:112–116

Sharma DK, Majumder PK, Singh RN (1972) Inheritance Pattern inMango (*Mangifera indica* L.). In: Proceedings of the symposium on recent advances in horticulture. UP Institute of Agri. Sci. Kanpur, pp 66–8

Sharma DK, Singh RN (1970) Self incompatibility in mango(*Mangifera indica* L.). Hort Res 10:108–115

Shukla AK, Vashishtha BB (2004) Mango: Fruit breeding approachesand achievements. International Book Distributing Co., Charbagh, Lucknow, U.P., India, pp 95–116

Siddique SH (1985) Induced somatic mutation in mango (*Mangifera indica* L.) cv, Langra. Pak J Bot 17:75–79

Siddiqui SH, Mujeeb KA, Wasti SM (1966) Evolution of newvarieties of mango (*Mangifera indica* L.) through induced somaticmutations by ionizing radiations. In: Proceedings of the 1st agricultural symposium atomic energy commission, Dacca, pp 34–37

Singh H (1996) Mango. ICAR, New Delhi, India

Singh H, Chadha KL (1981) Improvement of Dashehari by clonalselection. In: National symposium on trop. and sub-tropical on fruit crops. Hort Soc India, Bangalore, p. 5 (Abstr.)

Singh NK, Mahato AK, Sharma, N, Gaikwad, K, Srivastava M (2014) A draft genome of the king of fruit, mango (Mangifera indica L.). Plant and Animal Genome XXII Conference, San Diego, CA, USA

Singh RN, Majumder, PK, Sharma, DK, Mukherjee SK (1972) Some promising mango hybrids. Acta Hort 24:117-119

Singh J, Singh RR, Dubey PS, Singh UK (2004) Studies on genetic variability and character association for yield and quality traits in mango chance seedlings. J Appl Biol 14:37–39

Singh LB (1960) The Mango-Botany, cultivation and utilization. LeonardHill, London

Singh RN, Gorakh S, Rao OP, Mishra JS (1985) Improvement of Banarsi Langra through clonal selection. Prog. Hort. 17:273–277

Singh RN, Sharma DK, Majumder PK (1977) Improvement of mango through hybridization. In: Fruit Breeding in India. G.S. Nijar (Ed).Oxford and

Singh RN, Sharma DK, Majumder PK (1980) An efficient technique of mango hybridization. Scientia Hort 12:299–301

Singh SH (1963) Mango hybridization in Uttar Pradesh. Punjab Hort J 3:116–123

Slor E, Gazit S (1982) Tahar, a new mango cultivar(in Hebrew). Alon ha Notea 36:807

Sturrock TT (1968) Genetics of mango polyembryony. Fla State Hort Soc Proc 82:318-321

Surapaneni M, Vemireddy L, Begum H, Purushotham Reddy B, Neetasri C, Nagaraju J, Anwar SY, Siddiq EA (2013) Population structure and genetic analysis of different utility types of mango (*Mangifera indica* L.) germplasm of Andhra Pradesh state of India using microsatellite markers. Plant Syst Evol 299:1215–1229

Tanksley SN, Young A, Paterson B, Bonierbale M (1989) RFLP mapping in plant breeding; new tools for an old science. Bio/Technol 7:257–264

Tomer E, Lavi U, Degani C, Gazit S (1993) Noami: a new mango cultivar. Hort Sci 28:755–756

Truscott M (1992) Biochemical screening of polyploid mango seedlings. CSFRI Inf Bull 237:17–18

Ukoskit K (2007) Development of Microsatellite Markers in Mango (Mangifera indica L.) using 5' anchored PCR. Thammasat Int J Sci Tech, New York 12(3):1-7

Valdomiro, Paulo Sarmanho (2004) Genetic Variability in Mango Genotypes Detected by RAPD Markers. Acta Hortic 303-310

Viruel MA, Escribano P, Barbieri M, Ferri M, Hormaza JI (2005) Fingerprinting, embryo type and geographic differentiation in mango (Mangifera indica L., Anacardiaceae) with microsatellites. Mol Breed 15:383–393

Wagle PV (1929) A preliminary study of the pollination of the Alphonso Mango. Agric J India 24:259–263

Watt G (1891) Dictionary of Economic Products of India. 5

Whiley AW, Mayers PE, Saranah J, Bartley JP (1993) Breeding mangoes for Australian conditions. Acta Hortic 341:136–145

Wood house EJ (1909) Mangoes in Bhagalpur. A preliminary note. Q J Dept Agric Bengal 2:168–187

Yadav IS (1996) Occurrence of races of wild species of Mangifera (unpublished)

Yadav IS, Rajan S (1993) Genetic resources of mango. Advances in horticulture, vol 1. Fruit, part 1. In: Chadha KL, Pareek OP (eds) Malhotra Publishing House, New Delhi, pp 77–93

Yee W (1958) The Mango in Hawaii. Hawaii Agric Exp, Series Circular, p 388

Young TW, Leiden RB (1954) Mango breeding. Proc Fl State Hort Soc 67:241–244

In Vitro Culture and Genetic Transformation in Mango

8

César Petri, Richard E. Litz,
Sanjay K. Singh, and José I. Hormaza

Abstract

Mango is one of the world's most important fruit crops and is widely grown in the tropics and subtropics. Despite its importance in local economies and as an increasingly important export crop, very little progress has been made in terms of classical breeding and genetics that would facilitate mango cultivar improvement. Consequently, almost all the important mango cultivars in its area of origin in India and Southeast Asia are clonal selections that are several hundred years old. Biotechnology has the potential to revolutionize mango improvement, germplasm storage and production. Genetic transformation of embryogenic cultures opens the possibility of improving existing cultivars by incorporating into their genomes genes that target specific traits. Embryogenic mango cultures have the potential to rationalize germplasm storage and the international exchange of disease-indexed clonal materials. Finally, the development of efficient in vitro vegetative reproduction systems, whether this is based upon production of embryogenic or shoot tip and nodal cultures, could enable the availability of clonal rootstocks of exotic monoembryonic selections.

C. Petri (✉) · J. I. Hormaza
Departamento de Fruticultura Subtropical y Mediterránea, Instituto de Hortofruticultura Subtropical y Mediterránea La Mayora (IHSM La Mayora—UMA-CSIC), Avenida Dr. Wienberg, s/n, 29750 Algarrobo-Costa, Málaga, Spain
e-mail: cesar.petri@csic.es

J. I. Hormaza
e-mail: ihormaza@eelm.csic.es

R. E. Litz
Department of Horticultural Sciences, Tropical Research and Education Center, University of Florida, 18905 SW 280 Street, Homestead, FL 33031-3314, USA
e-mail: relitz@ufl.edu

S. K. Singh
Division of Fruits and Horticultural Technology, Indian Agricultural Research Institute, New Delhi 110 012, India
e-mail: sanjaydr2@gmail.com

8.1 Introduction

Mango (*Mangifera indica* L.) is a member of the family Anacardiaceae in the order Sapindales. *Mangifera indica* is the most widely cultivated species within the genus although there are 69 known *Mangifera* species. *Mangifera indica* has its natural distribution in South and Southeast Asia. The mango ranks fifth in total world production among fruit crops, after banana, citrus, grapes and apples. Mango domestication most probably began ca. 4,000 years ago in different regions of Asia. There are two distinct ecogeographic races that can be distinguished on the basis of their mode of reproduction and their centers of diversity: a subtropical group with

C. Kole (ed.), *The Mango Genome*, Compendium of Plant Genomes,
https://doi.org/10.1007/978-3-030-47829-2_8

monoembryonic seeds (Indian type) and a tropical group with polyembryonic seeds (Southeast Asian type) (Mukherjee and Litz 2009).

Detailed botanical descriptions of mango and some of the other *Mangifera* species are available (i.e., Mukherjee and Litz 2009).The chromosome number is $2n = 2x = 40$. Mango is a woody perennial evergreen tree that can reach more than 40 m height and can survive for hundreds of years. Mango trees that have been domesticated by selection from open-pollinated seedling populations show variation in tree architecture (i.e., shape and size), besides fruit quality traits that are unique with every chance seedling. Leaves are simple and alternate, with a morphology that varies among genotypes. The juvenile period of seedling trees is ca. 7 years. The root system consists of a long, vigorous taproot and abundant surface feeder roots. Both hermaphrodite and staminate flowers are borne on terminal pyramidal panicles and are glabrous or pubescent; the inflorescence is rigid and erect, up to 30 cm long, and is widely branched, usually tertiary, although the final branch is always cymose. The inflorescence is densely flowered with hundreds of small flowers (5–10 mm diameter). The mango fruit is a large, fleshy drupe, containing an edible mesocarp of varying thickness. It is a climacteric fruit, and an increased ethylene production takes place during its maturation.

Mango is currently cultivated throughout the tropics and subtropics from the equator to ca. 37° latitude in southern Europe (Spain). World production of mangoes can only be estimated since FAOSTAT includes mango, mangosteen and guava as a single production item. India is the main producing country by far with approximately a third of the world production, followed by China, Thailand, Indonesia and Mexico (FAOSTAT 2019).

Monoembryonic mango cultivars are propagated by grafting onto seedling rootstocks. In many countries, uniform nucellar seedlings from polyembryonic mangoes are utilized as rootstocks. Both types of cultivars, the monoembryonic and polyembryonic, when propagated continuously may have several distinct forms called 'strains', which are usually named after the place where they originated, e.g., 'Carabao' variety, 'Batangas' strain, where Batangas is one of the provinces of the Philippines (Pateña and Barba 2011).

Conventional breeding in mango has had limited impact on cultivar and rootstock development due to the long juvenile phase of ca. 7 years, low fruit set, high fruit drop, single seed per fruit, high degree of cross-pollination, polyembryony and allopolyploidy. Furthermore, little is known about the inheritance of horticulturally important traits because of its highly heterozygous nature, the difficulty in making controlled pollinations, and the high cost and time involved in screening large number of hybrid progenies (Litz and Hormaza 2020). Most monoembryonic cultivars are vegetatively propagated seedlings resulting from natural cross-pollinations.

Currently, most significant problems in mango production and post-harvest are alternate bearing habit, susceptibility to anthracnose (caused by *Colletotrichum gloeosporioides*) and issues related with postharvest handling, e.g., storage and long distance transportation, due to its climacteric nature. Commonly, cultivar breeding goals in mango include tolerance of regular bearing, dwarfing, medium size fruit (250–300 g), anthracnose tolerance, and a stable pleasant flavor combined with good storage performance (Mukherjee and Litz 2009; Litz and Hormaza 2020). Desired traits for rootstocks include scion compatibility, polyembryony, dwarfing, and tolerance of calcareous soils, salinity and soil-borne pathogens (Chandra et al. 2011; Litz and Hormaza 2020).

Although in most woody plants are regeneration from cell and tissue cultures and genetic transformation of elite selections are difficult tasks (Song et al. 2019), regeneration of several mango cultivars by somatic embryogenesis from maternal tissue explants (nucellus) has been described. On the other hand, micropropagation by shoot tip and nodal culture has been elusive due to high endogenous levels of microbial contamination, excessive polyphenol exudation and explant necrosis.

Many of scion cultivar breeding objectives can be achieved using conventional breeding procedures involving either controlled or open-pollinations between superior genotypes followed by selection among the offspring. Most of the important mango cultivars are not amenable to hi-tech cultivation practices and do not meet the requirements of modern horticultural production systems, e.g., dwarfing, precocity and regularity in bearing with high yield, resistance to diseases, pests and physiological disorders and good storage quality.

Employing biotechnology procedures to correct genetic flaws of existing cultivars has great potential. Inclusion of in vitro culture techniques, in vitro mutagenesis and genetic transformation in mango breeding programs would expedite the development of improved existing cultivars, as is currently the case with some other woody perennial fruit crops (Song et al. 2019).

8.2 In Vitro Shoot Establishment and Micropropagation

Mango is highly heterozygous and seed propagation does not ensure true-to-type progeny. Among the different vegetative propagation methods, the most commonly utilized are grafting and budding. These methods are time consuming and labor intensive. The standardization of micropropagation techniques in mango would facilitate the production of large quantities of elite plant material for nurseries and growers. Polyembryonic mango genotypes, particularly those that are used as rootstocks, are usually propagated by seed; a limited number of sexual, non-clonal seedlings also develop in polyembryonic seeds and this can cause problems in uniformity of seedlings used for clonal rootstock and thus mature trees may differ from the clone. Micropropagation could resolve the problem of clonal rootstock multiplication.

Efforts to micropropagate mango have met with little success (Table 8.1). Major problems that have been encountered with in vitro shoot establishment are latent microbial contamination, excessive polyphenol exudation, plant growth medium browning and explant necrosis (Fig. 8.1). Secondary metabolism is stimulated and polyphenols are oxidized after wounding and/or explant disinfestation. This leads to massive necrosis and explants cannot survive long enough to respond to culture conditions. This has made it difficult to achieve any conclusions from those experiments that would result in optimized procedures and plant growth conditions. The in vitro phenol exudation has been positively correlated with explant necrosis (Krishna et al. 2008). Later, Krishna et al. (2010) suggested pre-treatment, i.e., etiolation of new shoot sprouts + spray of imidazole (200 ppm) + agitation in 0.2% PVP before explant excision improved culture establishment, reduction of phenolics and microbial infection in shoot cultures. Explants inoculated on the medium comprising B5 macro-salts excluding $(NH_4)_2SO_4$ and $(NH_4)_2NO_3$ plus MS micro-salts and organics supplemented with 0.5 mg l^{-1} NAA + 2.0 mg l^{-1} BA during March was most effective giving higher culture establishment.

Several studies have attempted to overcome these problems and some important factors influencing mango in vitro shoot establishment have been determined (Yang and Lüdders 1993; Thomas and Ravindra 1997; Sharma and Singh 2002; Chandra et al. 2004; Samaan et al. 2007; Krishna et al. 2008).

Shoot tip and nodal explants have been utilized (Table 8.1). Genotype significantly influenced establishment; however, disinfestation treatment, basal plant growth medium, growth regulators, etc. are also important. Different basal culture media have significant different effects on in vitro culture establishment. Thus, while some reports (Thomas and Ravindra 1997; Sharma and Singh 2002; Chandra et al. 2004) described less phenolic exudation and improved explant survival with MS medium (Murashige and Skoog 1962), other reports (Yang and Lüdders 1993; Samaan et al. 2007; Krishna et al. 2008) observed better results with woody plant medium (WPM) (Lloyd and McCown 1981), B5 medium (Gamborg et al. 1968) and G medium (Yang et al. 1984).

It is clear that the response to the combination of plant growth regulators and their concentrations is highly genotype dependent (Yang and

Table 8.1 *In vitro* shoot establishment and micropropagation of *Mangifera indica* L

Genotype	Explant	Medium[a] + growth regulators[a]	Results	Reference
'Turpentine', 'Gomera', '13-1', 'Sabre'	Shoot tips and stem nodes	G medium + BA + Zeatin + 2iP + IAA + IBA + Casein hydrolysate	Shoot establishment and elongation	Yang and Lüdders (1993)
'Alphonso', 'Totapuri',	Shoot tips	Half MS medium	Shoot establishment	Thomas and Ravindra (1997)
'Amrapali', 'Dashehari', 'Langra', 'Chausa'	Shoot tips	MS medium + IAA + kinetin + adenine + casein hydrolysate	Bud establishment and elongation	Chandra et al. (2004)
'Zebda', 'HindySinnara', 'Sukkary', '13-1'	Stem nodes	(1) WPM + Kinetin (2) B5 medium + IBA, BA, Zeatin, 2iP + IAA + casein hydrolysate	(1) Shoot establishment (2) Shoot multiplication and rooting	Samaan et al. (2007)
'Amrapali'	Shoot tips	Modified B5 macros + MS micros & vit. + casein hydrolysate	Shoot establishment	Krishna et al. (2008)
'Irwin'	Nodal	Full- or half-strength WPM or ½WPM	Culture establishment and shoot proliferation	Tetsumura et al. (2016)

2iP: N^6-(2-Isopentenyl) adenine; BA: Benzyladenine; G: G medium (Yang et al. 1984); IAA: Indol-acetic acid; IBA: Indol-butyric acid; MS: Murashige & Skoog (Murashige and Skoog 1962); WPM: Woody Plant Medium (Lloyd and McCown 1980)
[a]According to the authors, the most appropriate combination among the tested

Fig. 8.1 **a** Exudation from an in vitro mango (cv. 'Keitt') shoot tip after 4 days of culture. **b** Medium browning observed after few days of in vitro culture of a 'Gomera 1' shoot

Lüdders 1993; Chandra et al. 2004; Samaan et al. 2007). For example, 1.0 mg/L of 2iP combined with 0.5 mg/L indol-butyric acid (IBA) stimulated shooting (80%) from nodal explants of 'Sukkary'; however, its effect was poor for '13-1' and 'HindySinnara' (viz. 'Hindi be Sennara') with only 10 and 20% positive explant response, respectively (Samaan et al. 2007). The optimum plant growth medium formulation must therefore be standardized for each genotype.

Different disinfestation treatments have been applied to disinfect explants without causing extreme phenolic compound exudation and tissue necrosis. Different antiseptic solutions alone and in combination (NaOCl, $Ca(OCl)_2$, $HgCl_2$, H_2O_2, ethanol, etc.), antibiotics, antifungal agents and detergents have been assayed. Often, adsorbents (PVP, active charcoal) and/or antioxidant molecules (citric or ascorbic acid) have been used to counter the effect of oxidative stress during disinfestation (Yang and Lüdders 1993; Thomas and Ravindra 1997; Sharma and Singh 2002; Chandra et al. 2004; Samaan et al. 2007; Krishna et al. 2008). These treatments were often used sequentially and frequent subculture to fresh plant growth medium was reported. Success was genotype dependent in terms of explant survival and contamination rates (Yang and Lüdders 1993; Thomas and Ravindra 1997; Chandra et al. 2004; Samaanet al. 2007; Hare Krishna et al. 2008). Furthermore, pretreatments of the explants by etiolation and PVP application (Krishna et al. 2008) or by spraying the tree with a disinfectant solution followed by a plant growth regulator solution (Chandra et al. 2004) reduced explant necrosis and improved the recovery of aseptic cultures. Earlier, Chavan et al. (2000) suggested dipping of apical and nodal segments in 0.5% PVP + 10% sucrose for 30 s to reduce in vitro explant browning. Mango tissues darken very quickly in vitro as a result of action of the enzyme polyphenol oxidase (PPO) (Litz and Vijayakumar 1988). Sharma and Singh (2002) reported that etiolated apical mango shoots upon in vitro culture showed a marked decline in PPO activity and tissue phenol content suggesting a negative correlation between PPO activity and phenol content and explant survival. Krishna (2006) also observed a marked decline in total phenols, peroxidase and PPO activities in etiolated mango shoots in comparison to the non-etiolated control. The in vitro phenol exudation was much lower in etiolated shoots leading to higher explant survival and reduced necrosis.

It seems that the season of explant collection can be a key factor to reduce phenolic compound exudation, microbial contamination and improve explant survival. The most appropriate periods for collection in the Northern Hemisphere are from May to June (Yang and Lüdders 1993; Thomas and Ravindra 1997) or from March to April and September to October (Chandra et al. 2004).

The age of the shoots also affects the results. One to four month-old shoots, with green foliage and dark green stems, were found ideal for culture establishment with the best response and with less polyphenol oxidation and microbial contamination (Thomas and Ravindra 1997; Chandra et al. 2004). Younger shoots (1–2 weeks), with reddish brown tender leaves and stems, showed maximum explant browning and phenolic exudation, whereas older shoots, with dark green mature foliage and thick brown stems, showed maximum contamination (Thomas and Ravindra 1997; Chandra et al. 2004). Explants taken from young seedlings showed least phenolic exudation and contamination but poor growth responses (Chandra et al. 2004). Other factors that reduced microbial contamination and increased explant survival were collection of explants from greenhouse-grown plants or from mycorrhizal plants (Krishna et al. 2008).

Size of the shoot tip explants is important (Yang and Lüdders 1993; Thomas and Ravindra 1997; Chandra et al. 2004; Hare Krishna et al. 2008), as it is related to endogenous microbial load and the presence of secondary metabolites that cause browning and necrosis. This seems to be related to the number of cut petioles that are in contact with the plant growth medium (Thomas and Ravindra 1997). The most appropriate shoot tip length ranges from 5 to 7 cm. Smaller explants showed less microbial contamination and browning but poor culture establishment (Yang and Lüdders 1993; Thomas and Ravindra 1997; Chandra et al. 2004).

Yang and Lüdders (1993) reported the best results on mango in vitro shoot establishment, although their study involved young seedling shoot tips and nodal explants, not elite material.

Explants were collected in May and June from 2-year-old mango seedlings growing in a greenhouse. Plant materials were surface disinfested by successive immersion in 2% NaOCl plus 0.1% Tween® 20 for 15 min, and after 5–10 h, re-surface disinfested for 10 min, and rinsed three times with sterile distilled water. Subsequently, explants were cultured on plant growth medium consisting of G salts and vitamins (Yang et al. 1984), 30 g/L sucrose, 300 mg/L casein hydrolysate, 300 mg/L glutamine, 100 mg/L myoinositol, 1.5 g/L gelrite and 1.5 g/L Difco Bacto-agar. Medium was supplemented with 1.0 mg/L benzyladenine (BA), 1.0 mg/L zeatin, 2.0 mg/L N^6-(2-isopentenyl)adenine (2iP) 1.0 mg/L indole acetic acid (IAA), and 0.5 mg/L IBA. Later, Krishna et al. (2010) suggested that regardless of genotype and duration of pre-treatments, etiolation of new shoot sprouts + spray of imidazole (200 ppm) + agitation in 0.2% PVP during March before culture improved culture establishment and reduced microbial infection of shoot cultures. The highest explant survival and least medium browning were recorded when 4.0 cm long explants were used. Likewise, explants excised from older shoots, i.e., from 3- to 4-month-old current season shoots, registered higher survival. Explants inoculated on the medium comprising B5 macro-salts excluding $(NH_4)_2SO_4$ and $(NH_4)_2NO_3$ plus MS micro-salts and organics supplemented with 0.5 mg/L NAA + 2.0 mg/L BA during March was most effective giving higher culture establishment in 'Amrapali' (85.59%) with 4 cm shoot tips, while it was minimum in 'Pusa Arunima' (58.10%) with 1 cm explants. Explants were subcultured onto fresh medium for several days (as needed) to salvage the tissues from oxidation products that accumulated in the plant growth medium. They were then subcultured at 3–4 week intervals. With this procedure, authors described nearly 100% explant survival for both sorts of explants for all genotypes tested (Yang and Lüdders 1993). Shoot tip explants responded at a frequency of ca. 64, 43 and 42% for 'Turpentine', 'Gomera', and 'Sabre', respectively. Other studies also reported, in the best of the cases, 50% of explant survival and 25% of explant response (Thomas and Ravindra 1997; Chandra et al. 2004; Samaan et al. 2007; Krishna et al. 2008); however, the procedures only succeeded to delay microbial contamination and, eventually, high contamination levels caused death (Chandra et al. 2004; Samaan et al. 2007).

With respect to in vitro shoot multiplication, very little information is available. Samaan et al. (2007) designed a proliferation medium for shoots of the rootstock 'Hindy Sinnara' (viz. 'Hindi be-Sennara'). Defoliated in vitro established shoots of 1–3 cm length were cultured on B5 (Gamborg et al. 1968) basal medium supplemented with IBA (0.5 mg/L), BA, zeatin and IAA (each at 1.0 mg/L) and 2iP (2.0 mg/L), casein hydrolysate and glutamine (300 mg/L), 3% sucrose and 2.5 g/L phytagel. They described the production of 3.7 lateral shoots per initial shoot (Samaan et al. 2007).

For successful in vitro shoot establishment, it is necessary to deal with the (i) growth response, (ii) necrosis response, and (iii) microbial contamination. Strong disinfestation treatments reduce contamination but increase necrosis; the use of young shoots not only reduces contamination but also reduces the explant response and stimulates exudation and necrosis. According to Krishna et al. (2008), factors responsible for successful establishment involve ontogenetic age and pre-treatments to mother plants and explants and future research must be focused on the integration of micropropagation technology with microbial biotechnology. Explant disinfestation must not only remove microbes from surfaces, but bacteria inhabiting inner tissues and organs. In vitro conditions are designed for plant growth and development, but these conditions are also often ideal for bacterial and/or fungus multiplication (Orlikowska et al. 2017).

8.3 In Vitro Regeneration via Organogenesis and Somatic Embryogenesis

Plant regeneration, both in nature and in vitro, is de novo organogenesis, in which plant cuttings or explants first turn out in ectopic apical meristems and subsequently develop shoots and/or roots. Plants can also regenerate through somatic embryogenesis in vitro, whereby isolated protoplasts or cells first develop zygotic embryo-like structures that afterwards generate whole plants. De novo regeneration can occur directly from parental tissues or indirectly through the formation of calluses. For more details about cellular origins and molecular mechanisms involved in plant regeneration, see Ikeuchi et al. (2016).

8.3.1 Shoot and Root Organogenesis

Singh et al. (1997) reported induction of callus on different explants such as epicotyl segment, leaf petiole and shoot tip excised from in vitro germinated embryos. Maximum callus was recorded on epicotyl segments, while direct root organogenesis was noted in epicotyl and shoot tip cultures with low level (0.5 mg/L) of 2,4-D alone. Raghuvanshi and Srivastava (1995) described indirect shoot regeneration from fully expanded, young 'Amrapali' leaves (Table 8.2). The pretreatment of explants in an agitated liquid plant growth medium was essential to overcome the problem of phenol exudation and improved explant survival. Before cultivation of leaf pieces on agar-gelled MS medium, explants were cultured in 100 ml liquid MS medium supplemented with 0.05% PVP in 250 ml flasks on an orbital shaker at 75 rpm for 24 h with the liquid medium refreshed at 2-h intervals. For the shoot induction medium, multiple shoots were produced with several combinations of plant growth regulators. The optimum medium contained 1.1 μM IAA and 17.8 μM BA, with >4 shoots per explant (Raghuvanshi and Srivastava 1995).

Root formation has been obtained from in vitro shoots. Ara et al. (1998) developed an efficient two-step protocol for microshoots obtained from nucellar somatic embryos ('Amrapali'). Maximum rooting success (89.71%) was achieved when explants were first cultured in liquid rooting induction medium supplemented with 5 mg/L IBA for 24 h. Subsequently, micro-shoots were transferred to auxin-free semisolid rooting medium. A dark incubation period of 2 weeks, before culture under a 16:8-h light:dark photoperiod (60 μmol $m^{-2}s^{-1}$), stimulated rooting compared with shoots under the photoperiod regime (Ara et al. 1998). Other studies also reported successful rooting with plant growth medium supplemented with IBA but with lower efficiencies; 20% and 0.4% for 'Amrapali' and 'Hindy Sinnara' (viz. 'Hindi be Sennara') shoots, respectively (Raghuvanshi and Srivastava 1995; Samaan et al. 2007). In these cases, cultures were directly incubated at 25 °C with a 16 h photoperiod provided by cool white fluorescent lights (50–70 μmol $m^{-2}s^{-1}$).

8.3.2 Somatic Embryogenesis

Contrary to organogenesis, somatic embryogenesis has been successfully achieved with different cultivars (Table 8.2). The main source of explants used for initiation of the culture involves nucellar tissue (maternal) excised from immature fruits. Nucellus derived plants are generally free from viruses and other endophytic disease causing organisms, due to the absence of vascular connection between the surrounding maternal tissue and the nucellus. Therefore, efficient recovery of somatic embryos particularly in monoembryonic mango cultivars would eliminate systemic pathogens (Litz 1984). Embryogenic cultures have also been reported from immature cotyledons of zygotic embryos (Xiao et al. 2004). Mango somatic embryogenesis from nucellar explants was first reported by Litz and collaborators (Litz et al. 1982, 1984; Litz and Schaffer 1987). Two publications by DeWald and collaborators in 1989 have formed the basis for most subsequent work (DeWald et al. 1989a, b). The procedure broadly comprises four stages (Fig. 8.2): (1) induction, (2) maintenance (proliferation), (3) maturation, and (4) germination of somatic embryos.

Table 8.2 Regeneration of *Mangifera indica* L. by organogenesis and somatic embryogenesis

Genotype	Explant	Medium + growth regulators	Results	Reference
'Turpentine N2-17-2', 'Heart', 'Sabre', 'Ono', 'Chino'	Ovules	Modified MS + CW	Somatic embryogenesis	Litz et al. (1982)
'Chino', 'Ono'	Nucellus & globular adventitious proembryos	(1) Modified MS + CW or 2, 4-D (2) Modified MS (3) Modified MS + BA	(1) PEM, proliferation (2) Somatic embryogenesis (3) SE germination	Litz et al. 1984
'Parris', 'Peach', 'Irwin', 'Keitt', 'Tommy Atkins'	Nucellus	Modified MS + 2, 4-D	Somatic embryogenesis	Litz and Schaffer (1987)
'Parris', 'James Saigon', 'Tommy Atkins', 'Tutehau'	Nucellus	(1) Modified B5 macros + MS micros & vit. + 2, 4-D (2) Modified B5 macros + MS micros & vit. + 2, 4-D + kinetin (3) B5 macros + MS micros & vit + CW (4) B5 macros + MS micros & vit + CW (5) Half B5 + CW + casein	(1) PEM (2) Proliferation (3) Somatic embryogenesis (4) SE maturation (5) Germination	DeWald et al. (1989a, b); Litz and Yurgalevitch (1997)
'Alphonso', 'Mundan', 'Baneshan'	Nucellus	(1) MS + 2, 4-D & GA3 (2) Half MS + ABA + CW (3) MS + BA	(1) PEM, direct somatic embryogenesis (2) SE maturation (3) SE germination	Jana et al. (1994)
'Hindi'	Nucellus	(1) B5 macros + MS micros & vit. + 2, 4-D (2) B5 macros + MS micros & vit. + BA (3) B5 macros + MS micros & vit. + AC + CW + ABA	(1) PEM (2) Somatic embryogenesis (3) SE maturation	Monsalud et al. (1995)
'Amrapali'	Young fully expanded leaves	(1) MS + BA + IAA or NAA (2) MS + IBA	(1) Bud regeneration & elongation (2) Root formation	Raghuvanshiand Srivastava (1995)
'Carabao'	Zygotic embryos	B5 macros + MS micros & vit. + CW	Zygotic embryo maturation and germination	Pliego-Alfaro et al. (1996a)
'Hindi'	Nucellus	(1) B5 macros + MS micros & vit. + 2, 4-D (2) B5 macros + MS micros & vit. (3) B5 macros + MS micros & vit. + ABA +mannitol	(1) PEM (2) Somatic embryogenesis (3) S. E. maturation	Pliego-Alfaro et al. (1996b)

(continued)

Table 8.2 (continued)

Genotype	Explant	Medium + growth regulators	Results	Reference
'Carabao'	Nucellus	(1) B5 macros + MS micros & vit. + 2, 4-D (2) B5 macros + MS micros & vit. (3) B5 macros + MS micros & vit. + Kinetin + BA	(1) PEM, proliferation (2) Somatic embryogenesis (3) S. E. maturation	Lad et al. (1997)
'Hindi', 'Nam Doc Mai', 'Lippens', 'Tommy Atkins'	Nucellus	Modified B5 macros + MS micros & vit. + 2, 4-D + nurse culture	PEM	Litz et al. (1998)
'Amrapali', 'Chausa'	Nucellus	(1) Modified MS + 2, 4-D (2) B5 macros + MS micros (3) Half MS + GA_3	(1) PEM, proliferation (2) Somatic embryogenesis & S.E: maturation (3) Germination	Ara et al. (2000b)
'Arka Anmol'	Nucellus	(1) RO + AC + 2, 4-D + GA_3 (2) Modified B5/RO + AC + 2, 4-D + GA3 + CW + Casein (3) Modified B5/RO + ABA (4) Modified B5 + TDZ	(1) PEM, Somatic embryogenesis (2) S. E. development (3) S.E. maturation (4) Germination	Thomas (1999)
'Amrapali'	Protoplasts from nucellus	(1) B5 macros + MS micros & vit + 2, 4-D (2) B5 macros + MS micros & vit + Kinetin (3) B5 macros + MS micros & vit + GA3 (4) B5 macros + MS micros & vit	(1) Microcalli formation (2) Somatic embryogenesis, proliferation (3) Germination (4) Shoot and root elongation	Ara et al. (2000a)
'Carabao'	Nucellus	(1) BP medium + 2, 4-D + CW (2) BP medium + CW + AC	(1) PEM, Somatic embryogenesis, maturation, germination (2) Plantlet conversion	Pateña et al. (2002)
'Ataulfo'	Nucellus	1)Modified B5 macros + MS micros & vit + 2, 4-D (2) Modified B5 macros + MS micros & vit + BA (3) Modified B5 macros + MS micros & vit + CW	(1) PEM, proliferation (2) Somatic embryogenesis (3) SE maturation, germination	Rivera-Dominguez et al. (2004)
'Zi-Hua'	Immature cotyledons	(1) MS + 2,4-D + GA_3 + CW + AC (2) B5 macros + MS micros & vit + IBA + AC (3) B5 macros + MS micros & vit + Kinetin	(1) PEM (2) Embryogenesis, maturation (3) Germination	Xiao et al. (2004)

(continued)

Table 8.2 (continued)

Genotype	Explant	Medium + growth regulators	Results	Reference
'Amrapali'	Nucellus	(1) Modified MS (2) Modified MS + 2,4-D or NAA (3) B5 macros + MS micros & vit (4) Modified B5 macros + MS micros & vit + GA_3	(1) PEM (2) PEM maintenance (3) Somatic embryogenesis (4) Maturation, germination	Ara et al. (2004)
'Chaunsa', 'Anwar Rataul'	Nucellus	MS + 2,4-D	PEM	Usman et al. (2005)
'Kurakkan'	Nucellus	(1) MS + 2,4-D + malt extract + spermidine (2) MS +2,4-D (3) MS + ABA + IAA + PEG (4) MS + NAA +kinetin + glutamine	(1) PEM (2) PEM maintenance (3) Somatic embryogenesis (4) Maturation, germination	Mishra et al. (2010)
'Carabao'	Nucellus	(1) B5 macros + MS micros & vit + 2,4-D + CW (2) B5 macros + MS micros & vit + AC + kinetin	(1) PEM, proliferation, germination (2) Plantlet conversion	Pateña and Barba (2011)
'Zebda', 'Sedeek', 'Hindi'	Nucellus	(1) B5 + 2,4-D (2) B5 (3) B5 macros + MS micros	(1) PEM, Proliferation (2) Maturation (3) Germination	Nower (2013)
'Baramasi'	Nucellus	(1) B5 macros + MS micros & vit + 2,4-D + BA + Glu + ME (2) B5 macros + MS micros & BA (3) B5 macros + MS micros & vit + Kinetin + GA_3	(1) PEM (2) Proliferation, Maturation (3) Germination	Al-Busaidi et al. (2016)
'Alphonso', 'Carabao', 'Turpentine'	Nucellus	(1) B5 macros + MS micros & vit + 2,4-D + BA (2) B5 macros + MS micros & vit (3) B5 macros + MS micros & vit + IAA + ABA (4) B5 macros + MS micros & vit	(1) PEM (2) Proliferation (3) Maturation (4) Germination	Shukla et al. (2016)
'Olour', 'Carabao'	Nucellus	(1) RO + MS micros & vit + AC + 2,4-D + GA_3	(1) PEM, Proliferation	Sajana et al. (2019)

2,4-D: 2,4- dichlorophenoxy acetic acid; ABA: Abcisic acid; AC: activated charcoal; B5: (Gamborg et al. 1968); BA: Benzyladenine; BP = Barba & Pateña formulation (Pateña et al. 2002); CW: Coconut water; GA_3: Giberellic acid; IAA: Indol-acetic acid; IBA: Indol-butyric acid; MS: (Murashige and Skoog 1962); NAA: Naphtalen-acetic acid; PEM: Pro-Embryogenic Masses; RO: Rugini Olive medium (Rugini 1984); SE: Somatic embryos; TDZ: Thidiazuron; WPM: Woody Plant Medium (Lloyd and McCown 1980)

Fig. 8.2 **a** Embryogenic 'Tommy Atkins' mango suspension. **b** Globular stage 'Hindi be Sennara' somatic embryos. **c** Heart stage somatic embryos of 'Keitt'. **d** Mature 'Nam Doc Mai' somatic embryos beggining to germinate. **e** Plant recovery from 'Keitt' somatic embryo

8.3.2.1 Explant Collection and Induction of Cultures

Immature fruits from 30 to 40 days after pollination are used as the source of nucellar explants. Embryogenic competence has been related to the developmental stage of the nucellus at the time of explanting and is influenced by the physiological condition of the tree (Litz and Hormaza 2020). The fruits are surface-sterilized for 10 min with ethanol 70% (v/v), 30 min with 20–30% (v/v) domestic bleach containing Tween 20, and then rinsed twice in sterile distilled water. The ovules are removed from fruits that have been bisected along the longitudinal axis. The embryo mass is discarded from each bisected ovule. Ovule halves are positioned so that the nucellar tissue is in contact with the induction medium (DeWald et al. 1989a). Alternatively, the nucellus can be removed from the bisected ovules and placed on the induction medium (Litz and Hormaza 2020).

DeWald et al. (1989a) optimized a number of parameters for induction of somatic embryogenesis from nucellar explants. They determined that growth of the cultures in suspension was more efficient than on solid medium. However, subculturing onto semisolid medium with gellan gum was crucial to obtain morphological normal somatic embryos. They also determined that the optimum induction growth medium consisted of modified B5 basal medium (Gamborg et al. 1968) (with 25 mM KNO_3 as the only source of inorganic nitrogen) and 5–6% sucrose to augment the recovery of normally differentiated heart-shaped embryos.

The plant growth regulator 2,4-dichlorophenoxyacetic acid (2,4-D) is essential for induction of embryogenic competence from the nucellus. Lad et al. (1997) determined that a minimum of 7-day pulse with 4.5 μM 2,4-D was necessary for induction from 'Carabao' nucellar explants and embryogenic competence was optimum after a minimum of 28-day exposure to 2,4-D. Most reports indicate that 2,4-D is essential for induction (DeWald et al. 1989a, b; Pliego-Alfaro et al. 1996b; Lad et al. 1997; Ara et al. 2000b; Singh et al. 2002; Sulekha and Rajmohan 2004), although other studies used 2,4-D together with other growth regulators, such as gibberellic acid, NAA or kinetin (Jana et al. 1994; Thomas 1999; Ara et al. 2004; Shukla et al. 2016). When immature cotyledons were used as explants, indole butyric acid (IBA) successfully induced the generation of embryogenic cultures (Xiao et al. 2004). The induction of embryogenic competence has been shown to be highly genotype dependent but unrelated to polyembryony or monoembryony (Litz et al. 1998; Ara et al. 2000b; Wei et al. 2013, Shukla et al. 2016). An interesting study with 'Tommy Atkins' (monoembryonic) and 'Tutehau' (polyembryonic) revealed more ethylene production in the monoembryonic cultures than in those of the polyembryonic genotype (Litz and Yurgalevitch 1997). Induction of embryogenic competence is more sensitive to the activity of spermidine synthase than to the activity of SAM decarboxylase, which confirmed earlier studies involving the effects of spermidine on induction of somatic embryogenesis in mango (Litz and Schaffer 1987).

Currently, a 'standard' procedure with nucellar explants utilizes a basal medium consisting of B5 (Gamborg et al. 1968) major salts without $(NH_4)_2SO_4$, MS minor salts and vitamins (Murashige and Skoog 1962), 60.0 g/L sucrose, 400.0 mg/L glutamine, 4.5 μM 2,4-D and 2.0 g/L gellan gum (Litz and Hormaza 2020). The cultures are incubated in darkness at 25 ± 1 °C. Medium must be refreshed frequently (sometimes as frequently as 1 or 2-day intervals) to reduce phenolic exudation and explant necrosis.

8.3.2.2 Maintenance

Three to nine weeks after nucellus explanting, embryogenic cultures usually appear and are creamy-white and friable, and consist of proembryonic masses (PEMs) (DeWald et al. 1989a). PEM proliferation is normally performed by inoculating ca. 200 to 300 mg of PEMs and culturing it in liquid maintenance medium of the same formulation as the induction medium (Litz et al. 1984; DeWald et al. 1989a; Pliego-Alfaro et al. 1996b; Lad et al. 1997; Ara et al. 2004;

Rivera-Domínguez et al. 2004). The PEMs are subcultured into liquid medium in Erlenmeyer flasks and maintained on a rotary shaker at about 100 rpm at 22–25 °C in darkness (Fig. 8.2A). Subcultures to fresh medium are usually done at 5–7 day intervals. The establishment of rapidly proliferating suspension cultures is highly cultivar dependent (Litz and Hormaza 2020).

The PEMs develop globular somatic embryos (Fig. 8.2B) in the presence of a primary induction agent, generally 2,4-D. In suspension culture and semisolid medium, the PEMs increase in size and their protoderm dedifferentiates becoming embryogenic. After that, in the presence of 2,4-D, secondary globular somatic embryos develop from embryogenic cells in the protoderm (Litz and Hormaza 2020). Although culture into liquid medium reduces tissue browning, some authors maintain the cultures on semisolid medium. Shifting the base medium from B5 medium (Gamborg et al. 1968) to Barba and Pateña (BP) medium (Pateña et al. 2002) has been reported to control browning.

8.3.2.3 Somatic Embryo Production and Maturation

Somatic embryo development is stimulated by transferring globular PEMs from the maintenance medium onto semisolid or into liquid medium without growth regulators. DeWald et al. (1989a) found subculture onto semisolid medium essential for high-frequency recovery of morphologically normal somatic embryos. They also reported that sucrose concentration (5–6% w/v) and filter sterilized coconut water (CW) at 20% (v/v) enhanced embryo development (DeWald et al. 1989a, b). Some reports indicate that the addition of cytokinins in the proliferation medium is beneficial. Addition of kinetin (4.6 μM) improved the subsequent production of 'Amrapali' somatic embryos that could mature normally (Ara et al. 2000a). BA (2.2 μM) allowed the generation of large numbers of somatic embryos from nucellar explants of 'Alphonso', 'Carabao' and 'Turpentine' and accelerated the development from PEMs and globular somatic embryos into heart shaped and early cotyledonary stage somatic embryos (Shukla et al. 2016).

Approximately 3 weeks are required for heart stage somatic embryos to develop from PEMs. Subsequently, somatic embryos are transferred to semisolid maturation medium for 2–5 more months (Fig. 8.2C). Maturation is an important step to establish bipolarity, minimize fasciation and synchronize the development of somatic embryos.

Mango somatic embryos that develop on semisolid medium usually grow more vigorously and they appear morphologically normal. Somatic embryos that develop in suspension culture can exhibit developmental abnormalities, i.e., polycotyledony, loss of bipolarity, fasciation and hyperhydricity. Typically, somatic embryos become hyperhydric during the early cotyledonary stage of development. Following subculture onto semisolid maturation medium these hyperhydric somatic embryos rapidly increase in size; however, they are unable to develop to maturity and they never germinate (Monsalud et al. 1995).

Zygotic embryos of mango require 3–4 months to reach maturity. Because of the long maturation period for somatic embryos, manipulation of the culture medium is necessary to inhibit precocious germination and abnormal embryo development (Fig. 8.2D). The addition of abscisic acid (ABA) to the maturation medium, high osmolarity or low temperatures reduced precocious germination and permitted the production of good quality somatic embryos (Monsalud et al. 1995; Pliego-Alfaro et al. 1996a, b). The increase of sucrose concentration up to 6% for 'Parris' improved the maturation of normal somatic embryos. Similar results were observed with ABA combined with coconut water with lower sucrose concentration (3%), but the effect of ABA was masked when 6% sucrose was used (DeWald et al. 1989b).

Standard maturation medium consists of modified B5 major salts, MS minor salts and vitamins, 2.74 mM glutamine, 0.025% (w/v) casein hydrolysate, 20% (v/v) coconut water,

40 g/L sucrose and 2 g/L gellan gum (DeWald et al. 1989b). The sucrose concentration is gradually reduced to 10 g/L during sequential subcultures to fresh plant growth medium. Cultures are maintained in darkness at 25 ± 1 °C (Litz and Hormaza 2020).

8.3.2.4 Germination and Conversion into Plantlets

Mango somatic embryos are able to germinate after a 3- to 5-month maturation period, at which time they are transferred to light conditions (16-h photoperiod and 60–70 μmol m^{-2} s^{-1} light intensity) and sucrose concentration is reduced to 10 g/L. During germination the hypocotyl elongates, followed by growth of the single tap root from the radicular end. The shoot apex emerges approximately 2 weeks after germination.

Survival rate and conversion to plantlet is variable among genotypes and is affected by different factors. Germination rate was significantly increased in liquid medium reaching up to 71.56 and 51.56% for 'Chausa' and 'Amrapali' somatic embryos, respectively, compared with 12.67 and 6.82% in semisolid medium (Ara et al. 2000b). Furthermore, conversion into plantlets was only achieved when embryos were cultured in liquid medium (Ara et al. 2000b). However, germination and conversion of 'Parris' and 'Carabao' somatic embryos have been reported in semisolid medium (DeWald et al. 1989b; Pateña and Barba 2011) (Fig. 8.2E).

The germination medium has been modified by the addition of gibberellic acid (GA_3) or cytokinins such as kinetin, BA or thidiazuron (TDZ) (Lad et al. 1997; Thomas 1999; Ara et al. 2000a, b, 2004; Xiao et al. 2004). Nevertheless, germination and plantlet development have been achieved without any plant growth regulators but with the addition of complex organic addenda such as coconut water (CW), malt extract or casein hydrolysate (DeWald et al. 1989b; Pliego-Alfaro et al. 1996a; Rivera-Domínguez et al. 2004; Shukla et al. 2016). Modified B5 medium supplemented with CW and casein hydrolysate significantly increased 'Parris' embryo germination rate (63%) compared with full-strength or half-strength B5 medium without organic addenda (35 and 27%, respectively) (DeWald et al. 1989b).

Plantlet conversion rate from germinated embryos has averaged 30–40% (Ara et al. 2000a, b, 2004; Rivera-Domínguez et al. 2004; Xiao et al. 2004; Pateña and Barba 2011). In vitro plants can be directly potted in soil or *ex vitro* grafted. Successful acclimatization of directly potted plants has been reported for several cultivars such as 'Alphonso' (67% survival rate), 'Carabao' (27% survival rate), 'Turpentine' (50% survival rate) and 'Parris' (survival rate not indicated) (DeWald et al. 1989b; Shukla et al. 2016). As the root system develops slowly, periodic applications of macro- and microelements as a foliar spray have been necessary for successful acclimatization of 'Parris' plantlets (DeWald et al. 1989b). For 'Carabao', *ex vitro* grafting performed significantly better than direct transfer of plantlets to soil, with percent survival ranging among trials from 14.3% to 50.0% compared to 0%–16.7% for the potted plants. Plants grafted *ex vitro* grew and produced new leaves and shoots within 14–27 days (Pateña and Barba 2011).

8.4 Encapsulation and Cryopreservation

Cryopreservation is a method for long-term storage of plant cells or tissues in liquid nitrogen (−196 °C), keeping them viable, with no metabolic deterioration and cell division, free of germs, and revivable upon thawing. A typical cryopreservation cycle consists of stages of selection of tissue, preparation of material, freezing, storage in liquid nitrogen, thawing, washing, reculturing, and, finally, regeneration of plantlets (Reed 2008). Another important method for preservation is the encapsulation-dehydration method. Explants such as shoot tips or somatic embryos are encapsulated in alginate beads to form artificial seeds. After revival, these can be used for direct germination. The use of somatic embryos for synthetic seed production has been considered. Cotyledonary-stage somatic embryos originating from nucellar explants of 'Amrapali' were encapsulated individually in 2% alginate gel (Ara et al. 1999). The encapsulated somatic embryos germinated successfully (73.61 ± 7.08% germination rate) on 0.6% agar-gelled medium containing B5 major salts, MS minor salts and organics, 3% sucrose and 2.9 mM GA_3. From these germinated somatic embryos, about 45% developed into plantlets and were successfully established in soil (Ara et al. 1999). This report demonstrated the possibility of plant regeneration from synthetic seeds. Wu et al. (2007) found embryogenic mango cultures ('Zihua') to be more responsive to regeneration after cryopreservation.

8.5 Embryo Culture

High fruit drop and embryo abortion in hand-pollinated panicles in mango are the major limiting factors, which lead to meager seedling (F_1) recovery for evaluation. Hence, *in ovulo*, embryo culture was proposed to improve breeding efficiency, which can be achieved by pollinating more flowers per panicle (8–12) followed by in vitro embryo culture. Immature fruitlets (45 days) of different cross combinations ('Amrapali' × 'Sensation', 'Amrapali' × 'Tommy Atkins') were cultured on solidified MS medium supplemented with 1.0 mg/L BA, 0.5 mg/L IAA, 400 mg/L glutamine, 200 mg/L casein hydrolysate, 100 mg/L activated charcoal, and 45 g/L sucrose. The hybridization efficiency was enhanced with the higher recovery of hybrid embryos per panicle (Fig. 8.3). The cultured embryos could be effectively matured, germinated and hardened in glass jars filled with sterile peat: perlite (4:1) moistened with ¼ MS salts solution (pH 5.8). This technique was also effective in rescuing dropped developing fertilized fruitlets due to adverse weather conditions. The technique increased the breeding efficiency in mango where

Fig. 8.3 **a** In vitro embryo culture initiation. **b** Initiation of germination of matured embryo. **c** Formation of a rooted hybrid seedling. **d** 'Amrapali' x 'Sensation' hybrid seedling in primary in vitro hardening

the final in vivo fruit recovery is not more than 1 percent, i.e., one or two fruits per panicle until maturity. This procedure may also aid in recovering hybrids in cross combinations where embryo degenerates due to pre- or post-zygotic barriers in conventional breeding.

Later, Sahijram et al. (2005) standardized the method for raising hybrids in the cross 'Alphonso' × 'Kerala Dwarf' in 3 months. They further proposed *ex vitro* shoot-tip grafting for rescuing in vitro raised hybrid plants on in vivo raised rootstocks under shade net. Graft union was successful in 67% of the hybrids in 4–6 weeks. Perez-Hernandez and Grajal-Martın (2011) suggested the role of coconut water in embryo culture in 'Lippens' and 'Keitt'. Complete plantlets could be obtained at a frequency above 83% for both cultivars. Recently, Singh (2016) proposed bio-hardening of in vitro raised mango plantlets. Hybrid plantlets (OP: 'Amrapali', 'PusaArunima', 'Hybrid 8-11', HP: 'Amrapali' × 'Sensation' and 'Amrapali' × 'Tommy Atkins') were successfully biohardened using arbuscular mycorrhizal fungi and *Aspergilus niger* during primary hardening.

8.6 Genetic Transformation

Genetic transformation of mango, with selectable and reporter marker genes, has been demonstrated using *Agrobacterium tumefaciens* (Mathews et al. 1992, 1993; Cruz-Hernandez and Litz 1997; Samanta et al. 2007) as well as by biolistics (Cruz-Hernandez et al. 2000) (Table 8.3). Most of these publications showed uniformly transformed PEM and/or somatic embryos but failed on the recovery of a complete plantlet following transformation. Mathews et al. (1993) reported the recovery of transgenic 'Keitt' plants.

Genetic transformation has targeted PEMs, using the *npt*II marker gene (resistance to aminoglycoside antibiotics; kanamycin) to select for transformed tissues. The sensitivity of non-transformed tissues to kanamycin is dependent on the stage of the embryogenic culture at the time of treatment and the form of exposure (Mathews and Litz 1990). Growth of non-transformed PEMs in liquid selection medium ceased at 12.5 mg/L kanamycin, whereas the growth of PEMs and mature somatic embryos on semisolid medium was inhibited by 200 mg/L kanamycin (Mathews and Litz 1990).

8.6.1 Transformation Procedures

Agrobacterium tumefaciens-mediated transformation procedures are briefly described according to Mathews et al. (1992). Embryogenic cultures are established according to the procedure described by DeWald et al. (1989a, b).

For bacterial culture preparation, a single colony from semisolid LB medium of a disarmed *A. tumefaciens* strain harboring the desired binary plasmid is inoculated into 5 ml of liquid LB with the appropriate antibiotics for plasmid and bacterium strain selection. After ca. 16 h at 28 °C, the culture is transferred to 50 ml liquid LB supplemented with 30 μM acetosyringone, and incubated at 200 rpm for 16 h at 28 °C.

Actively growing PEMs are infected with *A. tumefaciens*. For co-cultivation, 3.0 g of PEMs of < 1000 μm diameter maintained in liquid medium are lightly wounded with a sterile artist's paint brush and cultured in 50 ml of liquid maintenance medium to which 0.05 ml of a log-phase of the acetosyringone-activated *A. tumefaciens* culture is added. Flasks are maintained on a rotary shaker at 120 rpm. PEMs are transferred to fresh maintenance medium every 24 h for 3 days.

After 3 days of co-cultivation, infected PEMs are plated on semisolid maintenance medium containing 200 mg/L kanamycin (for transgenic cell selection) and 500 mg/L cefotaxime (for *A. tumefaciens* elimination). Later, PEMs are subcultured onto the same medium every 3 weeks for 10 months and are thereafter maintained on selection medium (200 mg/L kanamycin) without cefotaxime. Following co-cultivation, PEMs turn black within 5–7 days after planting on selection medium containing kanamycin. *Agrobacterium*-inoculated cultures show vigorous growth on selection medium over a period of 150 days whereas non-transformed control

Table 8.3 Genetic transformation in mango

Cultivar	Explant	Technique	Gene(s)	Reference
'Hindi'	PEM from nucellus	*Agrobacterium tumefaciens* (strain A280)	*npt*II, *gus*	Mathews et al. (1992)
'Keitt'	Segments of somatic embryos derived from zygotic embryos	*A. tumefaciens* (strain C58C1)	*npt*II	Mathews et al. (1993)
'Hindi'	PEM from nucellus	*A. tumefaciens* (strain LBA4404)	*npt*II, *gus*, antisense *ACC synthase*	Cruz-Hernandez and Litz (1997)
			*npt*II, *gus*, antisense*ACC oxidase*	
			nptII, gus, antisense *alternative oxidase*	
'Carabao'	PEM from nucellus	Biolistic	*npt*II, *gus* (transient transf.) *npt*II, *gfp* (stable transf.)	Cruz-Hernandez et al. (2000)
'Kesington Pride'				
'Vellaikolumban'	PEM from nucellus	*A. tumefaciens* (strain LBA4404)	*npt*II, *gus*	Samanta et al. (2007)
'Kent', 'Haden', 'Madame Francis'	Somatic embryos	*Agrobacterium rhizogenes*	*npt*II, *gus*	Chavarri et al. (2010)

PEM Pro-embryogenic masses

PEMs do not show signs of cell proliferation. Such areas of fresh PEM proliferation in the transformed cultures are very distinct and easily separable from the dark tissue. These fresh PEMs are subcultured onto fresh selection medium and continue to proliferate. Cultures generally consist of both transformed and non-transformed PEMs (Mathews et al. 1992). Therefore, a stepwise selection strategy has been used to eliminate chimeras: (1) initial screening at 200 mg/L kanamycin to enable growth of transformed cells; (2) further screening at 400 mg/L kanamycin to reduce chimeras; and (3) recovery of pure transformed clones of PEMs in liquid selection medium with 100 mg/L kanamycin.

Pro-embryos from the liquid selection medium are transferred to a liquid embryogenesis medium (maintenance medium without 2,4-D, supplemented with 0.22 μM BA) containing 100 mg/L kanamycin for 30–50 days for somatic embryo development. All cultures in liquid selection medium are transferred at 3–5 day intervals to fresh medium.

8.6.2 Applications of Genetic Transformation

Mango genetic transformation could assist breeding programs to address some of the major breeding objectives, such as resistance to anthracnose and delay or control of ripening.

Anthracnose, caused by the fungus *Colletotrichum gloeosporioides* Penz., is the main disease affecting mango fruit production and postharvest. Most commercial mango cultivars are highly susceptible and conventional breeding to introduce greater resistance into new cultivars has been unsuccessful (Jayasankar and Litz 1998). Alternative approaches could involve the use of in vitro selection and genetic transformation to enhance resistance to anthracnose in existing cultivars. Jayasankar and Litz (1998) induced somaclonal variation and selected resistant embryogenic cultures from 'Carabao' and 'Hindi' through recurrent selection with increasing concentrations of *C. gloeosporioides* culture filtrate and purified phytotoxin. Resistance to the pathogen was related to the excretion of pathogenesis-related proteins such as chitinases and glucanases (Jayasankar and Litz 1998). Some DNA markers (RAPDs) were associated with the resistance trait, with eight primers in 'Carabao' and five primers in 'Hindi' (Jayasankar et al. 1998). To the best our knowledge, quantitative trait loci (QTLs) or genes associated with anthracnose resistance have not been described in mango. As soon as some candidate genes are identified, strategies involving overexpression and/or silencing of those genes/QTLs could be applied through genetic transformation techniques.

Cruz-Hernandez and Litz (1997) transformed PEMs from 'Hindi' nucellar tissues with mango antisense sequences of genes involved in fruit ripening, i.e., *ACC synthase* and *ACC oxidase* (ethylene synthesis), and *alternative oxidase* (respiratory metabolism). The antisense DNA strategy causes silencing of the target sequence. Antisense mRNA anneals to the endogenous target mRNA. Consequently, dsRNA is formed and post-transcription algene silencing specific to the target sequence is triggered.

Stable integration of the transgenes into mango PEMs and torpedo stage somatic embryos was confirmed by PCRs and Southern blot analyses. Although transgenic plantlets were not obtained, they observed a faster proliferation of the antisense *ACC synthase* lines in the maintenance medium that could be explained by the relationship between the ethylene synthesis reduction and the acquisition of embryogenic competence (Cruz-Hernandez and Litz 1997). A similar approach has been applied in plum, which was transformed with an antisense construct of a peach *ACC oxidase* gene (Gonzalez Padilla et al. 2003). In some of the transgenic lines that were produced, ethylene production and softening were delayed (Callahan and Scorza 2007).

8.7 Conclusions and Future Work

Although mango is one of the five most important fruit crops worldwide in terms of production, classical genetic and biotechnological advances in this species lag well behind those in other woody perennial crops. New and increasingly cheaper sequencing technologies will result in a qualitative change in breeding programs in mango, especially since genome data are increasingly been released in this crop. Biotechnology approaches can address several important issues that limit mango production in different regions of the world. One example includes the optimization of in vitro propagation procedures that could reduce the dependence on polyembryonic seedlings as rootstocks, facilitating the development of clonal rootstocks with selected interesting traits such as tolerance to salinity, drought or soil fungus diseases. Similarly, cryopreservation of embryogenic cultures for use as a backup for mango germplasm collections would also facilitate the international exchange of rapidly deteriorating genetic resources. New and promising genomic edition technologies such as CRISPR-Cas9 to address different major breeding objectives, e.g., fruit ripening, control of physiological disorders, tree size or resistance to pests and diseases, might also be explored in mango.

References

Al-Busaidi KT, Shukla M, Al-Burashdi, AH, Al-Blushi, GS, Al-Jabri, MH, Al-Kalbani, BS, Al-Hasani, HD (2016) *In vitro* regeneration of mango (*Mangifera indica* L.) cv. Baramasi through nucellar embryogenesis. J Hort For 8(5):37–43

Ara H, Jaiswal U, Jaiswal VS (1998) Rooting of microshoots of *Mangifera indica* L. cv, Amrapali. Curr Sci 74:240–242

Ara H, Jaiswal U, Jaiswal VS (1999) Germination and plantlet regeneration from encapsulated somatic embryos of mango (*Mangifera indica* L.). Plant Cell Rep 19:166–170. https://doi.org/10.1007/s002990050728

Ara H, Jaiswal U, Jaiswal VS (2000a) Plant regeneration from protoplasts of mango (*Mangifera indica* L.) through somatic embryogenesis. Plant Cell Rep 19:622–627. https://doi.org/10.1007/s002990050783

Ara H, Jaiswal U, Jaiswal VS (2000b) Somatic embryogenesis and plantlet regeneration in Amrapali and Chausa cultivars of mango (*Mangifera indica* L.). Curr Sci 78:164–169

Ara H, Jaiswal U, Jaiswal VS (2004) An improved method of proliferation of proembryogenic calli of *Mangifera indica* L. var. Amrapali for scale-up of somatic embryo production. Indian J Biotechnol 3:229–234

Callahan A, Scorza R (2007) Effects of a peach antisense ACC oxidase gene on plum fruit quality. Acta Hort 738:567–573

Chandra R, Padaria JC, Shubhra S (2004) Factors influencing in vitro establishment of mango shoot buds. Indian J Plant Physiol 9:136–144

Chandra R, Pati R, Mishra M (2011) Mango. In: Singh HP, Parthasarathy VA, Babu KN (eds) Advances in horticulture biotechnology—Regeneration systems—Fruit crops, plantation crops and spices, 1st edn. Westville Publishing House, New Delhi, India, pp 73–86

Chavan SS, Ranade SS, Deore AC, Deshpande RS, Dhonukshe BL (2000) Cloning of Alphonso mango through vegetative explants. Ann Plant Physiol 14 (2):178–181

Marleny Chavarri, Vegas LA, ZambranoYN Gutierrez Z (2010) Insertion of *Agrobacterium rhizogenesrol*B gene in mango. Interciencia 35(7):521–525

Cruz-Hernandez A, Litz RE (1997) Transformation of mango somatic embryos. Acta Hort 455:292–298

Cruz-Hernandez A, Town L, Cavallaro A, Botella JR (2000) Transient and stable transformation in Mango by particle bombardment. Acta Hort 509:237–242

DeWald SG, Litz RE, Moore GA (1989a) Optimizing somatic embryo production in mango. J Am Soc Hort Sci 114:712–716

DeWald SG, Litz RE, Moore GA (1989b) Maturation and germination of mango somatic embryos. J Am Soc Hort Sci 114:837–841

FAOSTAT (2019) FAOSTAT. http://www.fao.org/faostat/en/

Gamborg OL, Miller RA, Ojima K (1968) Nutrient requirements of suspension cultures of soybean root cells. Exp Cell Res 50:151–158

Gonzalez Padilla IM, Webb K, Scorza R (2003) Early antibiotic selection and efficient rooting and acclimatization improve the production of transgenic plum plants (*Prunus domestica* L.). Plant Cell Rep 22:38–45. https://doi.org/10.1007/s00299-003-0648-z

Ikeuchi M, Ogawa Y, Iwase A, Sugimoto K (2016) Plant regeneration: cellular origins and molecular mechanisms. Development 143:1442–1451. https://doi.org/10.1242/dev.134668

Jana MM, Nadgauda RS, Rajmohan K, Mascarenhas AF (1994) Rapid somatic embryogenesis from the nucelli of monoembryonic mango varieties. Vitro Cell Dev Biol - Plant 30:55–57. https://doi.org/10.1007/BF02632120

Jayasankar S, Litz RE (1998) Characterization of embryogenic mango cultures selected for resistance to *Colletotrichum gloeosporioides* culture filtrate and phytotoxin. Theor Appl Genet 96:823–831. https://doi.org/10.1007/s001220050808

Jayasankar S, Litz RE, Schnell RJ, Hernandez AC (1998) Embryogenic mango cultures selected for resistance to *Colletotrichum gloeosporioides* culture filtrate show variation in random amplified polymorphic DNA (RAPD) markers. Vitro Cell Dev Biol - Plant 34:112–116. https://doi.org/10.1007/BF02822774

Krishna Hare 2006. Studies on in vitro propagation of mango (*Mangifera indica* L.). Ph D thesis, Post Graduate School, IARI, New Delhi

Krishna H, Sairam RK, Singh SK et al (2008) Mango explant browning: Effect of ontogenic age, mycorrhization and pre-treatments. Sci Hort 118:132–138. https://doi.org/10.1016/j.scienta.2008.05.040

Krishna H, Singh SK, Sairam RK, Verma SK (2010) Effect of different factors on in vitro shoot tip culture establishment in mango. Indian J Hort 67(3):293–300

Lad BL, Jayasankar S, Pliego-Alfaro F, Moon PA, Litz RE (1997) Temporal effect of 2,4-D on induction of embryogenic nucellar cultures and somatic embryo development of "Carabao" mango. Vitro Cell Dev Biol - Plant 33:253–257. https://doi.org/10.1007/s11627-997-0045-3

Litz RE, Schaffer B (1987) Polyamines in adventitious and somatic embryogenesis in mango (*Mangifera indica* L.). J Plant Physiol 128:251–258. https://doi.org/10.1016/S0176-1617(87)80239-0

Litz RE, Yurgalevitch C (1997) Effects of 1-aminocyclopropane-1-carboxylic acid, aminoethoxyvinylglycine, methylglyoxal bis-(guanylhydrazone) and dicyclohexylammonium sulfate on induction of embryogenic competence of mango nucellar explants. Plant Cell Tiss Org Cult 51:171–176. https://doi.org/10.1023/A:1005978807215

Litz RE, Hormaza JI (2020) *Mangifera indica* Mango. In: Litz RE, Pliego-Alfaro F, Hormaza JI (eds) Biotechnology of fruit and nut crops, 2nd edn. CAB International, Wallingford, Oxfordshire, UK, pp 27–43

Litz RE, Hendrix RC, Moon PA, Chavez VM (1998) Induction of embryogenic mango cultures as affected by genotype, explanting, 2,4-D and embryogenic nurse culture. Plant Cell Tiss Org Cult 53:13–18. https://doi.org/10.1023/B:TICU.0000009707.63691.82

Litz RE, Knight RL, Gazit S (1982) Somatic embryos from cultured ovules of polyembryonic *Mangifera indica* L. Plant Cell Rep 1:264–266

Litz RE, Knight RJ Jr, Gazit S (1984) *In vitro* somatic embryogenesis from *Mangifera indica* L. callus. Sci Hort 22:233–240

Litz RE, Vijayakumar N (1988) In vitro somatic embryogenesis from the nucellus of mango. Acta Hort 231:473–475

Lloyd G, McCown B (1981) Commercially feasible micropropagation of mountain laurel, *Kalmia latifolia* by use of shoot tip culture. Proc Int Plant Propagators' Soc 30:421–427

Mathews H, Litz RE (1990) Kanamycin sensitivity of mango somatic embryos. HortSci 25:965–966. https://doi.org/10.21273/hortsci.25.8.965

Mathews H, Litz RE, Wilde HD et al (1992) Stable integration and expression of b-glucuronidase and *npt*II genes in mango somatic embryos. Vitro Cell Dev Biol - Plant 28:172–178

Mathews H, Litz RE, Wilde HD, Wetzstein HY (1993) Genetic transformation of mango. Acta Hort 341:93–97

Mishra M, Yukti Shree, Rajesh Pati, Seal S, Shukla M, Kamle M, Chandra R (2010) Micropropagation of *Mangifera indica* L. cv. Kurakkan through somatic embryogenesis. Indian J Genet Plant Breed 70(1):85–90

Monsalud MJ, Mathews H, Litz RE, Gray DJ (1995) Control of hyperhydricity of mango somatic embryos. Plant Cell Tiss Org Cult 42:195–206. https://doi.org/10.1007/BF00034238

Mukherjee SK, Litz RE (2009) Introduction: botany and importance. In: Litz RE (ed) The mango: botany, production and uses, 2nd edn. CAB International, Wallingford, Oxfordshire, UK, pp 1–18

Murashige T, Skoog F (1962) A revised medium for rapid growth and bio assays with tobacco tissue cultures. Physiol Plant 15:473–497. https://doi.org/10.1111/j.1399-3054.1962.tb08052.x

Nower AA (2013) *In vitro* production of somatic embryos from nucellus of mango (*Mangifera indica* L.). Life Sci J 10(2):1164–1174

Orlikowska T, Nowak K, Reed B (2017) Bacteria in the plant tissue culture environment. Plant Cell Tiss Org Cult 128:487–508. https://doi.org/10.1007/s11240-016-1144-9

Pateña LF, Barba RC (2011) The development of techniques for tissue culture of mango (*Mangifera indica* L.) var. Carabao and successful transfer of *ex vitro*-grafted plants to soil and the field. Vitro Cell Dev Biol - Plant 47:629–636. https://doi.org/10.1007/s11627-011-9412-1

Pateña LF, Carlos-Refuerzo LR, Barba RC (2002) Somatic embryogenesis and plantlet regeneration of mango (*Mangifera indica* L.). Vitro Cell Dev Biol - Plant 38:173–177

Perez-Hernandez JB, Grajal-Martın MJ (2011) *In vitro* culture of immature zygotic mango embryos and plantlet development. Hortsci 46(11):1528–1532

Pliego-Alfaro F, Litz RE, Moon PA, Gray DJ (1996a) Effect of abscisic acid, osmolarity and temperature on *in vitro* development of recalcitrant mango nucellar embryos. Plant Cell Tiss Org Cult 44:53–61. https://doi.org/10.1007/BF00045913

Pliego-Alfaro F, Monsalud MJR, Litz RE, Gray DJ, Moon PA (1996b) Effect of abscisic acid, osmolarity and partial desiccation on the development of recalcitrant mango somatic embryos. Plant Cell Tiss Org Cult 44:63–70. https://doi.org/10.1007/BF00045914

Raghuvanshi SS, Srivastava A (1995) Plant regeneration of *Mangifera indica* using liquid shaker culture to reduce phenolic exudation. Plant Cell Tiss Org Cult 41:83–85. https://doi.org/10.1007/BF00124092

Reed BM (2008) Plant cryopreservation: a practical guide. Springer, Berlin

Rivera-Domínguez M, Manzanilla-Ramírez MA, Robles-González M, Gómez-Lim MA (2004) Induction of somatic embryogenesis and plant regeneration of "Ataulfo" mango (*Mangifera indica*). Plant Cell Tiss Org Cult 79:101–104. https://doi.org/10.1023/B:TICU.0000049442.54898.6f

Rugini E (1984) *In vitro* propagation of some olive cultivars. Sci Hort 24:123

Sahijram L, Bollamma KT, Naren A, Soneji JR, Dinesh MR, Halesh GK (2005) *In vitro* hybrid embryo rescue in mango (*Mangifera indica* L.) breeding. Indian J Hort 62(3):235–237

Sahijram L, Soneji JR, Bollamma KT, Sreevalli Y, Naren A, Dinesh MR, Reddy, YTN (2009) *Ex vitro* shoot-tip grafting for rescuing hybrid *vitro* plants of mango (*Mangifera indica* L.) developed through embryo culture. Acta Hort 820:320–336

Sajana S, Thomas P, Nandeesha P, Kurian Reju M (2019) Factors affecting induction of callus from nucellus tyissue of polyembryonic mango cv. Vellaikolumban and cv. Olour. Int J Curr Microbiol App Sci 8 (10):686–692

Samaan M, Wafaa H, Rawash M (2007) Factors affecting in vitro establishment and multiplication of some mango (*Mangifera indica* L.) rootstocks. J Biol Chem Environ Sci 2:79–93. https://doi.org/10.13140/RG.2.2.27066.70080

Samanta S, Ravindra MB, Dinesh MR et al (2007) In vitro regeneration and transient gene expression in mango cv. "Vellaikolumban". J Hort Sci Biotechnol 82:275–282. https://doi.org/10.1080/14620316.2007.11512229

Sharma RR, Singh SK (2002) Etiolation reduces phenolic content and polyphenoloxidase activity at the pre-culture and *in vitro* exudation of phenols from mango explants. Trop Agric 79(2):94–9

Shukla M, Al-Busaidi KT, Al-Blushi GS et al (2016) Nucellar embryogenesis and plantlet regeneration in monoembryonic and polyembryonic mango (*Mangifera indica* L.) cultivars. Afr J Biotechnol 15:2814–2823. https://doi.org/10.5897/ajb2016.15713

Singh SK, Sharma HC, Singh SP, Kishore PBK(1997) Callus induction and root initiation from explants of zygotic embryos in mango hybrids. Plant tissue culture and biotechnology: emerging trends. Proceeding of symposium, Hyderabad, pp 129–133

Singh A (2016) Bio-hardening and hybridity analysis of in vitro embryo cultured mango (*Mangifera indica* L.) plantlets, Ph D thesis, Post Graduate School, IARI, New Delhi

Singh, SK, Sharma HC, Singh SP (2002) In vitro polyembryony in monoembryonic mango cultivars (*Mangifera indica* L.). In: Kapoor AC, editor.

Sustainability of Hill Agriculture: Emerging Trends and Possible Solutions, pp 295-259

Singh SP, Sharma HC, Goswami AM, Singh SK (1999) Role of *in vitro* embryo rescue in *Mangifera* sp. breeding. In: National seminar on new horizons in production and post-harvest management of tropical and subtropical fruits, 8–9 Dec, The Horticultural Society of India, New Delhi, Book of Abstracts, pp 36

Song G, Prieto H, Orbovic V (2019) *Agrobacterium*-mediated transformation of tree fruit crops: methods, progress, and challenges. Front Plant Sci. https://doi.org/10.3389/fpls.2019.00226

Sulekha GR,Rajmohan K(2004) Relative response of varieties and explants in the induction of somatic embryogenesis in mango (*Mangifera indica* L.). South Indian Hort 52(1–6):5–12

Tetsumura T, Sakota T, Nagano H, Izaka S, Nguyen TMH, Tamura S, Ishimura S, Honsho C (2016) Plant regeneration from nodal explants of 'Irwin' mango seedlings. Acta Hort 1113:127–134. https://doi.org/10.17660/ActaHortic.2016.1113.18

Thomas P (1999) Somatic embryogenesis and plantlet regeneration from nucellar tissue of monoembryonic mango. J Hort Sci Biotechnol 74:135–139. https://doi.org/10.1080/14620316.1999.11511086

Thomas P, Ravindra MB (1997) Shoot tip culture in mango: influence of medium, genotype, explant factors, season and decontamination treatments on phenolic exudation, explant survival and axenic culture establishment. J Hort Sci 72:713–722. https://doi.org/10.1080/14620316.1997.11515563

Usman M, Jaskani M, Rehman MY, Fatima B (2005) In vitro regeneration of monoemrbyonic mango cultivars. Int J Biol Biotechnol 2(1):23–27

Wei J, Chen Y, Zhang Y, Gao A, Liu D (2013) Induction of somatic embryogenesis of three different mango (*Mangifera indica* L.) genotypes. Acta Hort 992:283–288

Wu YJ, Huang XL, Chen QZ, Li XJ, Engelmann F (2007) Induction and cryopreservation of embryogenic cultures from nucelli and immature cotyledon cuts of mango (*Mangifera indica* L. var. *zihua*). Plant Cell Rep 26:161–168

Xiao JN, Huang XL, Wu YJ, Li X-J, Zhou M-D, Engelmann F (2004) Direct somatic embryogenesis induced from cotyledons of mango immature zygotic embryos. Vitro Cell Dev Biol - Plant 40:196–199. https://doi.org/10.1079/IVP2003513

Yang Z, Lüdders P (1993) Effect of growth regulators and media on in vitro shoot tip culture of different cultivars of mango (*Mangifera indica* L.). Acta Hort 341:240–247. https://doi.org/10.5829/idosi.wasj.2014.31.05.14332

Yang ZH, Hu NY, Lu GM (1984) *In vitro* shoot tips culture of peach. Acta Univ Sept Occ Agri 1:13–19

Molecular Mapping and Breeding in Mango

9

Pumipat Tongyoo, Janejira Duangjit, Nimisha Sharma, and Julapark Chunwongse

Abstract

The Mango (*Mangifera indica* L.) is a member of family Anacardiaceae. It is one of the most important tropical fruit crops in the world. Most of the cultivars grown commercially are selections rather than introducing by breeding programs. Although, traditional breeding techniques cannot fulfill the demand of improved cultivars to promote superior quality traits due to many constraints like long juvenile phase, high heterozygosity, alternate bearing, polyembryony, heavy fruitlet drop, etc. Further, there is lack of understanding in heritability of the most important horticulture traits in mango. Therefore, by linking the important traits with molecular characteristics will help to improve the efficiency of mango breeding programs. The genetic markers and location in the genome that associated with important traits are commonly identified by utilizing genetic map. Consequence in an improvement of strong association between traits and molecular marker in marker-assisted selection. With the help of genetic map, mango breeders can easily identify the potential parents and progeny at early stage. Therefore, in this chapter, we have summarized the linkage and genetic map generated by mango researchers using various molecular markers, represented all 20 linkage groups that could be utilized as valuable tool and database by mango breeders for trait-specific breeding.

P. Tongyoo (✉)
Center for Agricultural Biotechnology, Kasetsart University, Kamphaeng Saen Campus, Nakhon Pathom 73140, Thailand
e-mail: pumipat.tong@ku.th

J. Duangjit
Department of Horticulture, Faculty of Agriculture, Kasetsart University, Bangkhen Campus, Bangkok 10900, Thailand
e-mail: janejira.d@ku.th

N. Sharma
Division of Fruits and Horticultural Technology, Indian Agricultural Research Institute, New Delhi 110012, India
e-mail: nims17sharma@gmail.com

J. Chunwongse
Department of Horticulture, Faculty of Agriculture at KamphaengSaen, Kasetsart University, Kamphaeng Saen Campus, Nakhon Pathom 73140, Thailand
e-mail: julapark.c@ku.th

9.1 Introduction

The improvement in mango breeding program is still challenging. There is huge varietal wealth accessible; however, targeted breeding of mango is cumbersome. There are several constraints generally faced by breeders like long generation interval, inbreeding depression, poor fruit set, unpredictable outcome due to high level of heterozygosity, polyembryony in some popular

C. Kole (ed.), *The Mango Genome*, Compendium of Plant Genomes,
https://doi.org/10.1007/978-3-030-47829-2_9

commercial cultivars, early post-zygotic auto-incompatibility and extensive land area required (Miller and Gross 2011). It remains an expensive and time-consuming process up to date. The undesirable mango characters are incorporated by many environmental and physiological effects, which are commonly controlled by multiple genes. To conquer these problems, molecular biology has vast potential and further it reduced the uncertainty in trait-specific breeding of mango. Recently, modern biotechnology, especially genomics, reviewed as a potential tool in the efficiency improvement in plant breeding program by selecting traits of interest at the seedling stage using marker-assisted selection (MAS). Systematic hybridization program with the help of advanced molecular approaches is in progress in many countries. In molecular breeding program, the segregating markers in a new generation are required which normally time consuming for developing these segregating population in fruit crops. Most economic traits in crop plants show quantitative variation and are thought to be controlled by multiple genes, most of which have minor effects and only a few genes have major effects (Falconer 1989). Establishing efficient selection based on genetic marker information from quantitative trait locus (QTL) mapping would be able to increase gain per breeding cycle because of increased confidence and certainty in the genetic selection. Due to the progress of DNA sequencing technology in the past few decades, SNPs provide the greatest marker density, best potential for automation, lowest cost per data point and as unambiguous markers. The technology is also appropriate in data sharing because of there is no platform-dependent information. Consequence in the reliable and producible analysis results without any uncertainty. It allows us to investigate a complete genome sequence of organisms. Because most trees' genome is large, highly repetitive, and complex structure, genome sequencing became an essential tool for the genetic improvement of trees through a construction of physical map (Jiang et al. 2014; Poursarebani et al. 2014; Pan et al. 2016). It will facilitate DNA marker traits linked identification. In the meanwhile, chromosomal rearrangements (CRs) are considered as another genetic diversity factor during an evolutionary (Ayala and Coluzzi 2005), which are also important to consider in breeding programs. It is obvious that molecular tools can facilitate development of new strategies for more efficient breeding in mango. Access to molecular markers linked to desirable alleles underlying phenotypic traits greatly enhances the breeding efficiency because of the possibility of genetic screening at a very young age. In addition, the new plant breeding technologies (NPBTs) that facilitated cisgenesis and genome editing approaches can enhance either pathogen resistance or qualitative traits improvement by promoting or detracting at certain genes immediately which a minimal modification in the target genome (Dalla Costa et al. 2017). Therefore, this will become a powerful approach in crop improvement research under 'green' biotechnologies for long life crops. Thus, this chapter aims to describe how to implement molecular genetics in association to mango breeding.

9.2 Molecular Mapping and Breeding in Mango

The traditional way of performing plant breeding is selection based on phenotypes. This method, however, is a time-consuming especially in fruit trees. The step that takes the longest time is juvenile stage to become a fruit-bearing tree. In mango, it takes around 3–10 years for this juvenile process (Iyer and Degani 1997; Bally et al. 2009). Even though, the struggles can be tackling by grafting the desirable varieties onto the mature mango tree or by applying chemicals, the outcome can be deviated due to evolution (Bally et al. 2009). Together with the fact that mango also has a long generation time, high natural fruit shedding, high heterozygosity, cultivar incompatibility and polyembryony, breeding mango traditionally has been slow and challenging, and often times takes more than 20 years.

Selection at an early stage is needed to reduce the total time of the mango breeding, additionally, this will also reduce amount of work, space, and other resources. Early selection can be done based only in some desired traits such as disease resistance. However, other traits can be done using marker-assisted selection or MAS, which is a technique using molecular markers that linked to a gene or quantitative trait loci (QTLs) of interest. Nevertheless, molecular map construction needs to be made prior to the QTL mapping. The development of dense molecular maps has sped up recently due to the advent of molecular markers.

Bi-parental crosses are required to construct a genetic linkage map, each parent must have distinguishing features on either the morphological or molecular mapping level (e.g. difference in fruit color, difference in disease susceptibility). It is preferred to use nearly homozygous lines (such as inbred lines) for genetic mapping as used in self-pollinated species when possible. However, it is more complicated to develop in bred lines from cross-pollinated species like mango. Moreover, there are not possible to get inbred lines because of inbreeding depression. Consequence to the used of half-sib or full-sib progenies that developed from specific crosses can be applied for genetic mapping of polymorphic molecular markers.

In most cases, F1 was utilized as shown in Table 9.1. Generally, a full sib hybrid population derived from differing phenotype parents are commonly used in breeding program rather than half-sib population from open-pollinated parents. Furthermore, a sufficient number of individuals are also required for developing an accurate genetic map (Staub et al. 1996). Most genetic maps constructed in mango used mapping population in a range of 31–173 individuals (Table 9.1). The three largest populations in Australia were constructed from their primary commercial mango cultivar 'Kensington Pride' by crossing to 'Irwin', 'Tommy-Atkins', and 'Creeper'. While the two F_2 populations derived from self-pollination of 'Tommy-Atkins' and 'Haden' were also constructed in Florida (Arias et al. 2012).

9.2.1 Mango Molecular Mapping

Several genetic linkage maps were constructed based on different molecular technologies (Table 9.1). For example, a partial genetic linkage map from a cross between 'Alphonso' and 'Palmer' was constructed based on 31 F_1 progenies (Chunwongse et al. 2015). In this study, 29 linkage groups covering a genetic distance in total of 529.9 centiMorgan (cM). The genetic map is shown in Fig. 9.1. This map consisted of 67 restriction fragment length polymorphism (RFLP) markers and 9 microsatellite markers. The same group also used RFLP and amplified fragment length polymorphism (AFLP) markers to study the genome of 'Alphonso' and 'Palmer' (Phumichai 2001). A two-way pseudo-testcross mapping strategy was used in this study. Genetic maps consisting of 74 linkage groups covering 1,559.8 cM, and 71 linkage groups covering 1,910.3 cM were constructed for 'Alphonso' and 'Palmer', respectively. Another genetic linkage map based on a cross between 'Keitt' and 'Tommy-Atkins' was also constructed by Kashkush et al. (2001). The linkage map displays 13 linkage groups that cover 161.5 cM by utilizing 34 AFLP markers. There are some studies that successfully constructed several linkage groups similar with the expected number of chromosomes of mango haploid genomes (20). A consensus map of mangoes from seven mapping populations was developed using 726 single nucleotide polymorphism (SNP) markers. The map was constructed by JoinMap and OneMap software. The details of each linkage group are summarized in Table 9.2.

A mango haploid genome size is estimated at 439 Mb, which consisted of 20 chromosome each haploid genome. The average chromosome size is estimated at 22 Mb. The total genetic distance in haploid genome is 2,890 cM, which leads to the estimation for a cM at 150 Kb. A specific locus-amplified fragment (SLAF) library construction was used to generate 20 linkage groups with total length 3,148.28 cM

Table 9.1 Details of genetic linkage map studies in mango (*Mangifera indica* L.) based on different crosses and molecular markers

Mapping population	Population type	Population size	Number of markers	Type of markers	Total length	Average marker spacing (cM)	References
'Alphonso' and 'Palmer'	F1	31	197	RFLP[(1)]	1437.7	10.4	Phumichai (2001)
			650	AFLP[(2)]			
'Keitt' and 'Tommy-Atkins'	F1	29	34	AFLP[(2)]	na[(4)]	4.75	Kashkush et al. (2001)
'Keitt' and 'Tommy-Atkins'	F1	60	81	AFLP[(2)]	354.1	4.37	Fang et al. (2003)
'Alphonso' and 'Palmer'	F1	31	9	Microsatellite	529.9	6.97	Chunwongse et al. (2015)
			67	RFLP[(1)]			
'Jin-Hwang' and 'Irwin'	F1	173	6,594	SLAF[(3)]	3148.28	0.48	Luo et al. (2016)
'Tommy-Atkins' and'Tommy-Atkins'	Self pollinated	60	726	SNPs[(5)]	2890.6	4.13	Kuhn et al. (2017)
'Haden' and 'Tommy-Atkins'	F1	225					
'Haden'and 'Haden'	Self Pollinated	40					
'NMBP1243' and 'Kensington Pride'	F1	100					
'Creeper' and 'Kensington Pride'	F1	70					
'Irwin' and 'Kensington Pride'	F1	180					
Tommy-Atkins'and 'Kensington Pride'	F1	100					

(1) restriction fragment length polymorphism
(2) amplified fragment length polymorphism
(3) specific-locus amplified fragment
(4) not applicable
(5) single nucleotide polymorphism

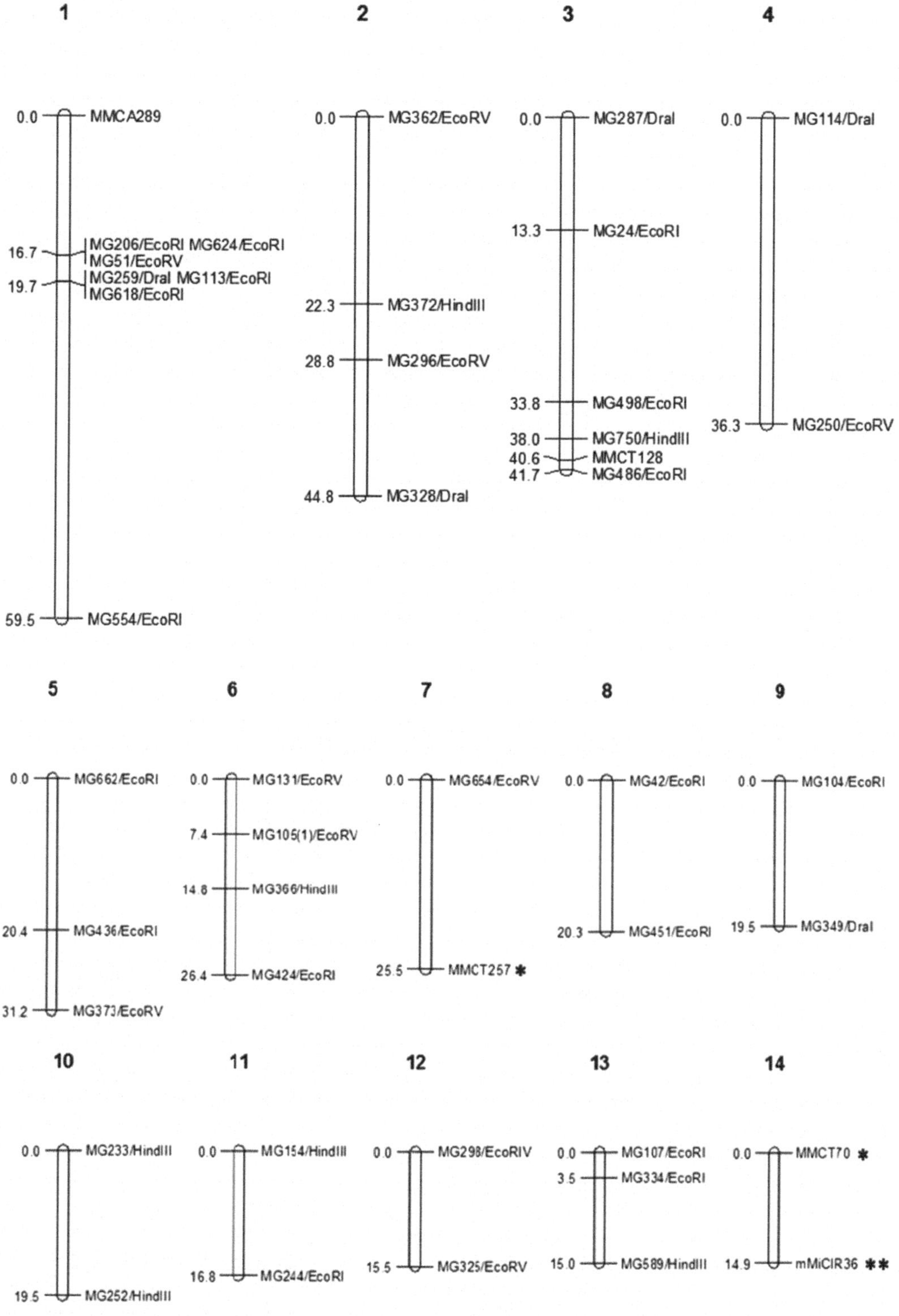

Fig. 9.1 The 29 genetic map of mango from SSR markers (Chunwongse et al. 2015)

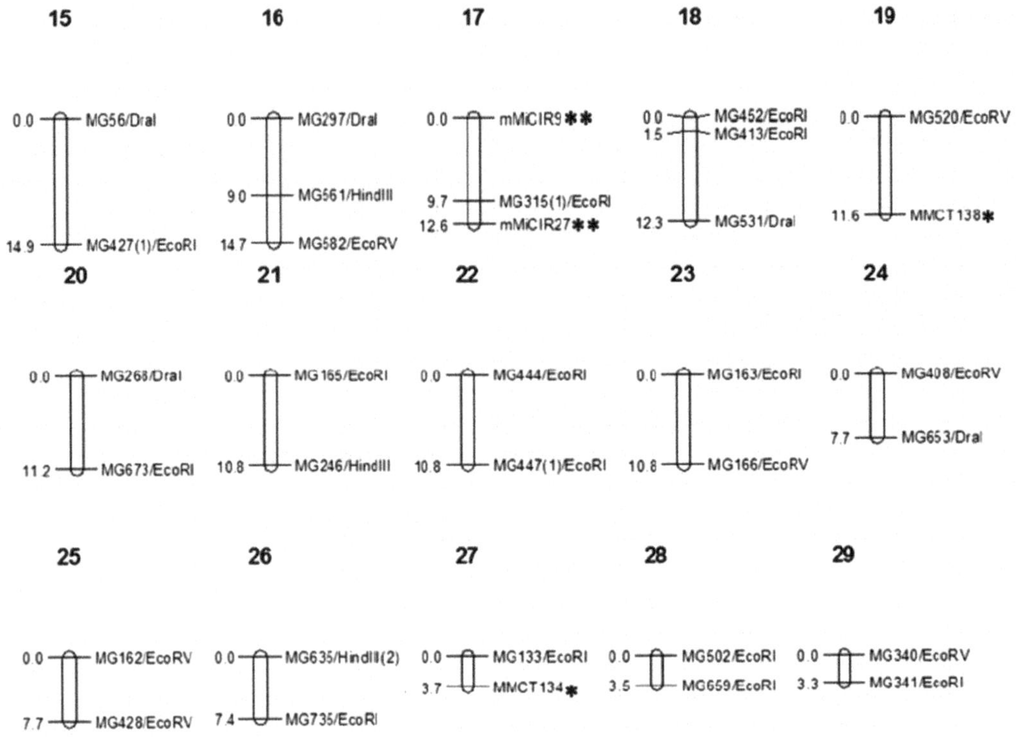

Fig. 9.1 (continued)

Table 9.2 Genetic map summary of mango developed by Kuhn et al. (2017) and Luo et al. (2016)

Linkage group	Consensus map Kuhn et al. (2017)		Integrated map (Luo et al. 2016)	
	Length (cM)	Average distance (cM)	Length (cM)	Average distance (cM)
1	111.2	4.1	168.96	0.47
2	135.6	4.5	180.35	0.64
3	79.4	3.2	198.74	0.57
4	223.2	6.4	188.39	0.35
5	126.3	4.2	144.75	0.58
6	80.4	3.4	146.29	0.47
7	151.1	5.4	125.31	0.54
8	247.8	6	167.78	0.8
9	143.1	4.2	182.83	0.54
10	186.5	4.5	119.61	0.43
11	77.2	3.1	178.46	0.38
12	148.8	4.4	136.15	0.55
13	154.9	3.7	160.1	0.41
14	114.9	4.4	116.84	0.45
15	166.2	3.8	140.72	0.54
16	228	3.3	206.09	0.35

(continued)

Table 9.2 (continued)

Linkage group	Consensus map Kuhn et al. (2017)		Integrated map (Luo et al. 2016)	
	Length (cM)	Average distance (cM)	Length (cM)	Average distance (cM)
17	156.7	2.8	132.75	0.59
18	76.5	3.8	140.96	0.64
19	126.7	3.8	193.2	0.4
20	156.1	3.7	122.95	0.42
Total	2,890.6	4.135	2,144.23	0.48

using HighMap strategy (Liu et al. 2014). The genetic map utilized 6,594 SLAFs from 173 F_1 populations constructed by a cross between 'Jin-Hwang' and 'Irwin' (Luo et al. 2016).

By comparing the average distance between these two genetic maps, it shows the potential for constructing a high-density genetic map that allows to implement a fine QTL mapping and MAS in mango. The consensus genetic map of mango developed by Kuhn et al. (2017). High density map of mango is presented in Fig. 9.2. In addition, the significant association between traits and SNPs markers for several traits was also explored in this study. Mapped markers can be facilitated as anchors for further genetic linkage mapping and genome sequencing analysis of mango breeding programs in the future.

Moreover, molecular markers from the genetic analysis can be used in the estimation of outcrossing rates in mango. Santos and Neto (2011) demonstrated it by using AFLP and microsatellite markers in a cross between 'Haden' and 'Tommy Atkins'. These rates were calculated using to tackle an outcrossing perennial crops issue. A multi-locus outcrossing rate (tm) was estimated by counting number of markers. The outcrossing rates ranged from 0.85 to 0.87 estimating by 7 AFLP markers and ranged from 0.83 to 0.91 by 3 microsatellite loci. For preservation of these genetic variability of mango species, it must be concerned that they are predominantly open-pollinated. Any biometrical models applied to mangoes to estimate genetic parameters should consider the deviation from random outcrossing.

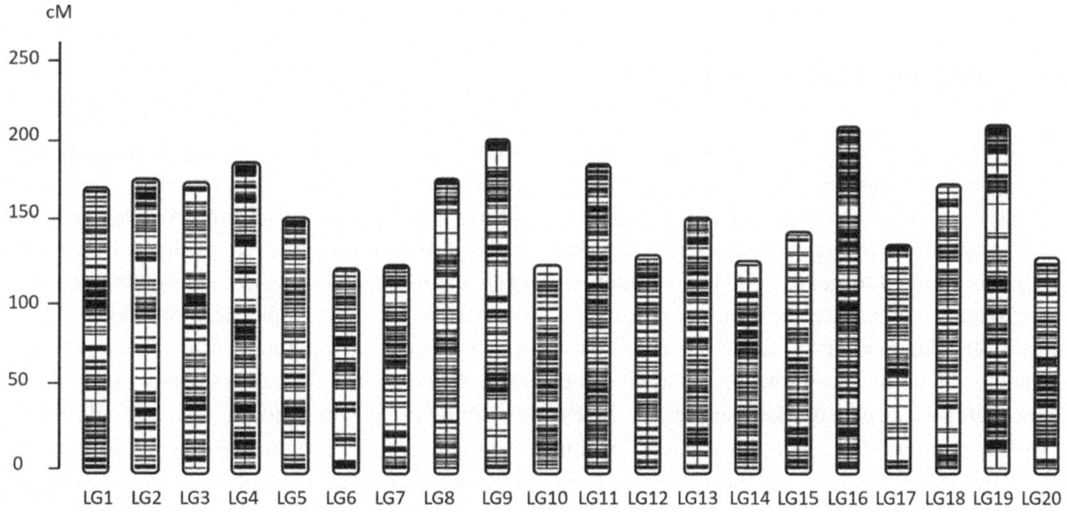

Fig. 9.2 High density genetic map of mango (Modified from Luo et al. 2016)

9.2.2 Genes and QTL Mapped

In the past decade, the molecular marker and linkage map have not been applied directly to mango breeding programs, especially in complex characteristics such as quality traits. By the number of mango germplasm in the Indian and IndoChinese, this opens a possibility of utilizing GWAS as well as linkage disequilibrium strategies on mango. Moreover, it will lead to an genetic related traits utilization analysis as a next-generation breeding program (Barabaschi et al. 2016; Kang et al. 2016; Can et al. 2019).

The mango genetic maps were produced from different populations as basis of QTL analyses. But one major drawback of this approach is the population size limitation which normally not large enough which leads to the limitation of recombination rates that can be occurred. Consequence in a low efficiency QTL detection. There are successful reports between traits and SNP markers association in branching habit and some fruit traits which are beak shape, blush intensity, pulp color, and ground skin color. One of the important mongo characteristics which is embryo type was also reported by using crosses between a monoembryonic maternal parent and a polyembryonic paternal parent. The significant association was shown in a single locus on a linkage group 8 (LG8) of a consensus map (Kuhn et al. 2017).

9.2.3 MAS Experiments in Mango

Even though a genetic map is not necessary to identification of markers and traits of interest associated in MAS approach, the QTL analysis can help to increase the chance to identify the genetic area linked to genes or QTLs of interest. Unfortunately, there are not many reports in using molecular marker for selecting a progeny in mango breeding program up to date. The major useful molecular markers especially SNPs up to date can facilitate germplasm identification and a hybrid clarification as well as a genetic diversity analysis in various cultivars.

The successfully applied of three pairs of microsatellite markers to identify 82 seedlings (14.69%) of 558 seedlings derived from 364 seeds of cultivar 'Kaew007' as a hybrid were reported. Many mango cultivars also gave a very low success of hybrid seedling. By using 14 random amplified polymorphic DNA (RAPD) primers, only 3% and 5% were recognized as a hybrid seedling in cultivars 'Manila' and 'Ataulfo', respectively. We found the height of a mango seedling 'Kaew007' was not associated with a zygotic seedling (unpublished data). The evaluation of polyembryony in the mango cultivars 'Manila' and 'Ataulfo' was also performed using 14 RAPD primers in 100 seeds of each cultivar. As a result, more than 80% of 'Manila' and 'Ataulfo' seeds give two to four embryos (Ochoa et al. 2012). Polyembryony in mango was proposed as controlling by only single dominant gene (Aron et al. 1998). A significant association in two of the three mapping populations was also reported as a single locus for polyembryony trait (Kuhn et al. 2017).

9.2.4 Association Mapping Studies in Mango

In case of mango, association mapping (AM) is particularly suited to overcome some constraints like long generation time, labored trait evaluation, long maturation process as well as the polyploidy issue which cannot be surpassed during pedigree-based mapping.

Generally, mango requires 5–7 years to become juvenility which is considering as economic input before phenotypic evaluation (Davenport 2007). A full sib crosses from two outbred is usually used in most of fruit crops QTL mapping analysis. As a result, there is a low resolution of QTLs detection due to a single meiotic event of maximum four alleles segregation. Whereas AM utilized a large alleles number distributed over the genomes to be calculated the association with curtain phenotypes to identify the significant loci. Hence, AM needs a large set of molecular markers to detect any linked to the

Table 9.3 Association mapping studies in mango (*Mangifera indica* L.)

Study	Markers	Conclusion	References
Resistance gene analogs (RGAs) in mango	EST-SSRs	134 unique EST-SSR loci for resistance breeding	Lantican et al. (2020)
Association of polyembryony in germplasm collection	SNPs	Marker Mi_0173 associate with polyembryony	Kuhn et al. (2019)
Association analysis for pomological traits in mango	SSRs	A set of five genic-SSR markers associated with six critical pomological traits like regular bearing, fruit quality, yield related etc.	Lal et al. (2017)
Association analysis for cultivar identification	SNPs (50 K SNP Chip)	Population structure of Indian mango varieties were determined	Singh et al. (2016)

recombination events within genomes (Khan and Korban 2012). The utilization of SSRs is wildly implemented in many crops such as pear (Oraguzie et al. 2010), apple (Cevik et al. 2010), peach (Font i Forcada et al. 2013) as well as in mango (Lal et al. 2017). Detail of association studies in mango presented in Table 9.3. Furthermore, this high density genetic marker using in association analysis might reveal new candidate genes associated with interested as either major or minor QTLs from the association mapping(Oraguzie et al. 2007).

9.2.5 Genome-Wide Association Studies (GWAS)

The genome-wide association studies (GWAS) overcome several limitations of traditional gene mapping. It not only provides higher resolution at gene level, but the existing well studied populations were also can be used in the association analysis as a result in higher number of individuals in the analysis. Next-generation sequencing technologies provide huge data at genomic and transcriptomic level. This information provides a useful dataset of SNP which beneficially in the identification of small haplotype blocks that significantly associated with QTLs. The GWAS-identified SNPs can be used in combinations with SNP biomarkers in MAS breeding program. A web genomic resource MiSNPDb is available at webtom.cabgrid.res.in/mangosnps. These web genomic resources could be utilized to generate high density linkage map, identification of QTL, differentiation of closely related varieties, as well as a development of genomic selection SNP chip in the future (Iquebal et al. 2017). Presently in October 2020, a total of 129 GWAS tools and resources are freely available at bioinformaticshome.com/tools/gwas/gwas.html. Some of the bioinformatics software implemented verities statistical models to facilitate the association analysis in GWAS such as PLINK (Chang et al. 2015), METAL (Willer et al. 2010), GAPIT (Lipka et al. 2012), and TASSEL (Bradbury et al. 2007). Additionally, there are improvements of analytical method, for instance, the development of multi-locus and multi-trait, combining of linkage association mapping, etc. (Gupta et al. 2019; Zhang et al. 2019). The GWAS model numbers are increasing to overcome both computational efficiency and detection power. A few hundred, thousand to multiple millions of sample size are required (Zhang et al. 2018). Efficiency of software mainly depends on the availability of molecular markers and population size, used in the analysis. Increasing a sample size in GWAS is a key feature to explain the heritability better, and it has become a recommendation for great success in GWAS research.

9.3 Concluding Remark and Future Prospects

There are a very limited number of mango breeding programs worldwide that implementing molecular genetic information as well as utilizing genomic resources. Therefore, in this chapter, we have summarized the high density markers like SNPs generated using genomic and transcriptomic data, SNPs database, consensus genetic map, association analysis, and linkage studies, which leads to the identification of regions in the genome that associated with economically important traits that beneficial for MAS in a specific hybridization to challenge the drawbacks in mango breeding programs. A set of genetic markers including with precise analysis methods will become the valuable materials for mango breeders. There will be a tool to help them in the identification of potential parents from germplasm resource as well as validating of hybrid progenies. The combination of tradition and molecular breeding approach will be a significant step toward mango breeding program efficiently. The global collaboration is also an important aspect in mango and other perennial crops. For example, the FruitBreedomics program (Laurens et al. 2018) was purposed as a gap filling between genomics and breeding in fruit crops, prioritized in apple and peach genome sequence initiative. It has been made available for the public as a resource of molecular and bioinformatics tools. It also provides pre-breeding materials for initiating various breeding programs in fruit crops. This kind of community of fruit trees will contribute to mango improvement inescapably.

References

Arias RS, Borrone JW, Tondo CL, Kuhn DN, Irish BM et al (2012) Genomics of tropical fruit tree crops. In: Schnell RJ, Priyadarshan PM (eds) Genomics of tree crops. Springer, New York, pp 209–239

Aron Y, Czosnek H, Gazit S, Degani C (1998) Polyembryony in mango (*Mangifera indica*L.) is controlled by a single dominant gene. HortScience 33: 1241–1242

Ayala FJ, Coluzzi M (2005)Chromosome speciation: humans, Drosophila, and mosquitoes. Proc Natl Acad Sci USA 102 (Suppl 1): 6535–6542

Bally IS, Lu P, Johnson PR (2009) Mango breeding. In: Jain SM, Priyadarshan PM (eds) Breeding plantation tree crops: tropical species. Springer, New York, pp 51–82

Barabaschi D, Tondelli A, Desiderio F, Volante A, Vaccino P et al (2016) Next generation breeding. Plant Sci 242:3–13

Boonsong T, Cenchamas S, Vichit C,Thiamchai T (1984) Studies on the ontogeny of polyembryony in mango (*Mangifera indica* L.) cv. Okrong Thong. In: Proceedings of the 22nd national conference Bangkok, Thailand, 30 January–3 February 1984, pp 137–142

Bradbury PJ, Zhang Z, Kroon DE, Ramdoss Y, Casstevens TM, Buckler ES (2007) TASSEL: software for association mapping of complex traits in diverse samples. Bioinformatics 23(19):2633–2635

Can H, Kal U, Ozyigit II, Paksoy M, Turkmen O (2019) Construction, characteristics and high throughput molecular screening methodologies in some special breeding populations: a horticultural perspective. J Genet 98:86

Cevik V, Ryder CD, Popovich A, Manning K, King GJ, Seymour GB (2010) A Fruitfull-like gene is associated with genetic variation for fruit flesh firmness in apple (*Malus domestica*Borkh.). Tree Genet Genomes 6:271–279

Chang CC, Chow CC, Tellier LC, VattikutiS S, Purcell SM, Lee JJ (2015) Second-generation PLINK: rising to the challenge of larger and richer datasets. Gigascience 4:7

Chunwongse C, Phumichai C, Tongyoo P, Juejun N, Chunwongse J (2015) Development of di-nucleotide microsatellite markers and construction of genetic linkage map in mango (*Mangifera indica* L.). Songklanakarin J Sci Technol 37:119–127

Dalla Costa L, Malnoy M, Gribaudo I (2017) Breeding next generation tree fruits: technical and legal challenges. Hort Res 4:17067

Davenport TL (2007) Reproductive physiology of mango. Braz J Plant Physiol 19(4):363–376

Falconer DS (1989) Introduction to quantitative genetics. Longman Wiley, Burnt Mill, Harlow, Essex, England, New York

Fang J, Liu D, Ma Z (2003) Constructing mango (*Mangifera indica* L.) genetic map using markers for double heterozygous loci. Mol Plant Breed 1(3):313–319

Font i Forcada C, Guajardo V, Chin Wo SR, Moreno MÁ (2019) Association mapping analysis for fruit quality traits in *Prunuspersica* using SNP markers. Front Plant Sci 9:2005

Font i Forcada C, Oraguzie N, Igartua E, Moreno MA, Gogorcena Y (2013) Population structure and marker-trait associations for pomological traits in peach and nectarine cultivars. Tree Genet Genomes 9:331–349

Gupta PK, Kulwal PL, Jaiswal V (2019) Association mapping in plants in the post-GWAS genomics era. Adv Genet 104:75–154

Iquebal MA, Jaiswal S, Mahato AK, Jayaswal PK, Angadi UB et al (2017) MiSNPDb: a web-based genomic resources of tropical ecology fruit mango (*Mangifera indica* L.) for phylogeography and varietal differentiation. Sci Rep 7:14968

Iyer CPA, Degani C (1997) Classical breeding and genetics. In: Litz RE (ed) The mango, botany, production and uses. CAB International, Wallingford, UK, pp 49–68

Jiang Y, Xu P, Liu Z (2014) Generation of physical map contig-specific sequences. Front Genet 5:243

Kang YJ, Lee T, Lee J, Shim S, Jeong H et al (2016) Translational genomics for plant breeding with the genome sequence explosion. Plant Biotechnol J 14:1057–1069

Kashkush K, Jinggui F, Tomer E et al (2001) Cultivar identification and genetic map of mango (*Mangifera indica*). Euphytica 122:129–136

Khan MA, Korban SS (2012) Association mapping in forest trees and fruit crops. J Exp Bot 63(11):4045–4060

Kuhn DN, Bally ISE, Dillon NL, Innes D, Groh AM et al (2017) Genetic map of mango: a tool for mango breeding. Front Plant Sci 8:577

Kuhn DN, Dillon N, Bally I, Groh A, Rahaman J et al (2019) Estimation of genetic diversity and relatedness in a mango germplasm collection using SNP markers and a simplified visual analysis method. Sci Hort 252:156–168

Lal S, Singh AK, Singh SK, Srivastava M, Singh BP et al (2017) Association analysis for pomological traits in mango (*Mangifera indica* L.) by genic-SSR markers. Trees Struct Funct 31:1391

Lantican DV, Cortaga CQ, Manohar ANC, dela Cueva FM, Sison M LJ (2020) Transcriptome-wide analysis of expressed resistance gene analogs (RGAs) in mango. bioRxiv. https://doi.org/10.1101/2020.02.08.939736

Lipka AE, Tian F, Wang Q, PeifferJ, Li M et al (2012) GAPIT: genome association and prediction integrated tool. Bioinformatics 28:2397–2399

Liu D, Ma C, Hong W, Huang L, Liu M et al (2014) Construction and analysis of high-density linkage map using high-throughput sequencing data. PLoS ONE 9: e98855

Luo C, Shu B, Yao Q, Wu H, Xu W, Wang S (2016) Construction of a high-density genetic map based on large-scale marker development in mango using specific-locus amplified fragment sequencing (SLAF-seq). Front Plant Sci 7:1310

Miller AJ, Gross BL (2011) From forest to field: perennial fruit crop domestication. Am J Bot 98:1389–1414

Ochoa ECM, Andrade-Rodriguez M, Rodriguez MR, Monter AV (2012) Identification of zygotic and nucellar seedlings inpolyembryonic mango cultivars. Pesq AgropeqBras 47:1629–36

Oraguzie NC, Wilcox PL, Rikkerink EHA, De Silva HN (2007) Linkage disequilibrium. In: Oraguzie NC, Rikkerink EHA, Gardiner SE, De Silva HN (eds) Association mapping in plants. Springer, New York, pp 11–39

Oraguzie NC, Whitworth CJ, Brewer L, Hall A, Volz RK et al (2010) Relationships of PpACS1 and PpACS2 genotypes, internal ethylene concentration and fruit softening in European (*Pyrus communis*) and Japanese (*Pyrus pyrifolia*) pears during cold air storage. Plant Breed 129:219–226

Pan Y, Wang X, Liu L, Wang H, Luo M (2016) Whole genome mapping with feature sets from high-throughput sequencing data. PLoS ONE 11:e0161583

Phumichai C (2001) DNA marker studies in mango cultivars Alphonso, Palmer and their segregation population. Master of Science Thesis, Kasetsart university, Thailand

Poursarebani N, Nussbaumer T, Simkova H, Safar J, Witsenboer H et al (2014) Whole-genome profiling and shotgun sequencing delivers an anchored, gene-decorated, physical map assembly of bread wheat chromosome 6A. Plant J 79:334–347

Santos CAF, Neto FPL (2011) Outcrossing rate between 'Haden' and 'Tommy Atkins' mangoes estimated using microsatellite and AFLP markers. Pesq Agropec Bras 46:899–904

Schnell RJ, Olano CT, Quintanilla WE, Meerow AW (2005) Isolation and characterization of 15 microsatellite loci from mango (*Mangifera indica* L.) and cross-species amplification in closely related taxa. Mol Ecol 5:625–627

Singh NK, Mahato AK, Jayaswal PK, Singh A, Singh S et al (2016) Origin, diversity and genome sequence of mango (*Mangifera indica* L.). Indian J Hort Sci 51 (2.2):355–368

Staub JE, Serquen FC, Gupta M (1996) Genetic markers, map construction, and their application in plant breeding. HortScience 31(5):729–741

Willer CJ, Li Y, Abecasis GR (2010) METAL: fast and efficient meta-analysis of genomewide association scans. Bioinformatics 26:2190–2191

Zhang YM, Jia Z, Dunwell JM (2019) Editorial: The applications of new multi-Locus GWAS methodologies in the genetic dissection of complex traits. Front Plant Sci 10:100

Zhang Y, Qi G, Park JH, Chatterjee N (2018) Estimation of complex effect-size distributions using summary-level statistics from genome-wide association studies across 32 complex traits. Nat Genet 50:1318–1326

The Genome Sequence and Transcriptome Studies in Mango (*Mangifera indica* L.)

10

Nagendra K. Singh, Ajay K. Mahato, and Pawan K. Jayaswal

Abstract

Mango (*Mangifera indica* L.) is one of the most important fruits of the world both in terms of value and volume. Wild *Mangifera* species are distributed throughout South and South-East Asia, but mango cultivars have originated in India which produces more than fifty percent of the world's mango with more than thousand varieties. The genome size of mango estimated by flow cytometry is 402 ± 10 Mbp ($2n = 40$). Here, we present a brief account of the global efforts for sequencing of mango genome and transcriptome studies for the analysis of trait-related differential gene expression. High heterozygosity of about 2.5% made it difficult to assemble the genome of mango using Illumina short sequence reads of 100–150 bp because it did not permit the assembly of its maternal and paternal genomes into an integrated mosaic genome assembly. The problem was overcome by using PacBio SMRT long sequence reads for genome assembly using long overlaps of 500 bp and high mismatch of 15% to take care of the heterozygosity, resulting in the first draft genome assembly of 323 Mb of mango variety 'Amrapali' (NCBI Accession no. LMWC00000000 v.1). The draft genome assembly was updated to a reference quality assembly of 403 Mbp with 4312 scaffolds anchored to 20 chromosome pseudomolecules (LMWC01000000 v.2). The latest v.3 Amrapali assembly, assisted by BioNano optical fingerprinting and Hi-C conformation capture sequencing, has 2314 scaffolds with a high N50 value of 11.78 Mpb. Mapping sequence reads from 18 different transcriptome studies showed an average 96% coverage of the gene space, and BUSCO analysis of 1440 genes showed 93.4% coverage of the conserved eukaryote orthologs. A total of 46,395 protein-coding genes have been predicted showing maximum homology with *Citrus sinensis*. The Amrapali genome has 45% repeats and large segmental duplications, indicating at least one recent (15.87–31.74 Mya) and one ancient (253.96–269.84 Mya) whole genome duplication. Mosaic genome assemblies of mango variety 'Tomy Atkins' from USA and 'Kensington Pride' from Australia have also been presented at the annual Plant and Animal Genome (PAG) conferences. Recently, two more mosaic genome assemblies of mango varieties 'Hong Jian Ha' and 'Alphonso' have been reported using PacBio SMRT sequencing with Falcon-unzip and 'Canu' software to separate the primary contigs (P-contigs) and haplotigs (H-contigs). However, there is still a need to develop clean

N. K. Singh (✉) · A. K. Mahato · P. K. Jayaswal
ICAR-National Institute for Plant Biotechnology, Pusa Campus, New Delhi 110012, India
e-mail: nksingh4@gmail.com

C. Kole (ed.), *The Mango Genome*, Compendium of Plant Genomes,
https://doi.org/10.1007/978-3-030-47829-2_10

phase-separated reference assemblies of the maternal and paternal genomes of highly heterozygous mango using the recent trio-binning approach. The reference genome will help accelerate breeding of dwarf, stress tolerant, and high-quality mango varieties.

10.1 Introduction

Mango (*Mangifera indica* L.) is known as the 'king of fruits' for its rich taste, flavor, color, production volume, and diverse end usage. It belongs to the plant family Anacardiaceae and has a small genome size of 439 Mb ($2n = 40$) according to Arumuganathan and Earle (1991). However, based on flow cytometry a significantly smaller genome size of 402 ± 10 Mbp was estimated for the mango variety 'Amrapali' (Singh et al. 2018). Ancient Sanskrit literature indicates the origin of cultivated mango in India, although wild mango species are distributed throughout the South and South-East Asia. Recovery of Paleocene mango leaf fossils near Damalgiri in the West Garo Hills of Meghalaya in India points to the origin of genus *Mangifera* in the peninsular India before the joining of Indian and Asian continental plates (Mukherjee 1951; Mehrotra et al. 1998; Singh et al. 2016). India produces more than fifty percent of the world's mango and grows more than thousand varieties some of which are commercially viable, e.g., Alphonso, Banganapalli, Bombay Green, Chausa, Dashehari, Fazli, Gulabkhas, Kesar, Kishen Bhog, Langra, Neelam, Totapuri among others. India produced 20.8 million tons of mango from 2.3 Mha in 2018–19, of which 46,510 metric tons of fresh mango worth USD 60.26 million was exported according to the Agricultural and Processed Food Products Export Development Authority (APEDA), India (http://apeda.gov.in/apedawebsite/SubHead_Products/Mango.htm).

Mango is consumed largely as fresh fruit but now increasingly in processed forms like pulp, pickles, chutney, juice, etc. Biochemical studies have shown that mango fruit is rich in nutrients like vitamins A, C, E, and K; carotenoids (α, β); carbohydrate; fat; dietary fiber; protein; and minerals (Sodium, Potassium, Calcium, Copper, Iron, Magnesium, Manganese, and Zinc) important for human health (Sogi et al. 2013; Maldonado-Celis et al. 2019; https://fdc.nal.usda.gov/fdc-app.html#/food-details/169910/nutrients). Mango byproducts specifically seeds and peels generated from the fruit processing industries are a cheap source of food for the tribal population and also used in the cosmetic and pharmaceutical industries for valuable phenolic compounds (Schieber et al. 2003; Coelho et al. 2019). Recent studies have emphasized the utilization of mango kernel fat as a cocoa butter substitute in confectionery products, particularly chocolate, as a rich source of stearic, oleic, palmitic, arachidic, behenic, lignoceric, and linolenic acids (Jahurul 2014). Despite its huge economic significance genomic resources for mango have been limited for long and the genetics of useful horticultural and pomological traits are poorly understood. Here, we present a brief account of the global efforts to generate mango genome sequence and its use in the analysis of genetic diversity, population structure, and association studies. A major milestone was reached with the publication of the draft genome of Indian mango cultivar 'Amrapali' which is a hybrid of popular North Indian variety 'Dashehari' and a regular bearing South Indian variety Neelam (Mukherjee et al. 1968; Singh et al. 1972). The project was spearheaded by ICAR-National Institute for Plant Biotechnology, New Delhi (formerly National Research Centre on Plant Biotechnology) in collaboration with four other ICAR Institutes, namely Indian Agricultural Research Institute (IARI), New Delhi; Central Institute for Subtropical Horticulture (CISH), Lucknow; Indian Institute of Horticultural Research (IIHR), Bengaluru and Indian Agricultural Statistics Research Institute (IASRI), New Delhi. The genome assembly has been continuously improved and presented in the annual Plant and Animal Genome (PAG) Conference at San Diego (Singh et al. 2014, 2016b, 2017, 2018, 2019) and XIV NAAS Agricultural Science Congress, New Delhi (Jayaswal et al. 2019). Apart from ICAR, four other groups in three different countries have also initiated genome sequencing of different mango varieties, namely 'Kensington Pride' in Australia (Dillon et al. 2016), 'Tomy Atkins' in USA (Baruch et al. 2017; Kuhn et al. 2018), and 'Hong Xiang Ya' (Li et al. 2020) and Alphonso (Wang et al. 2020) in China.

10.2 Mango Karyotype, Molecular Markers, Genetic Diversity, Population Structure, and High-Density Linkage Maps

Cytological studies on mango and its related species *M. sylvatica, M. caloneura, M. zeylanica, M. caesia, M. foetida,* and *M. odorata* revealed that all of them have a diploid chromosome number of $2n = 20$ (Mukherjee 1950, 1957; Roy and Visweswariya 1951). Successful interspecific crosses between *M. zeylanica* and *M. odorata* (Mukherjee 1963) showed no sterility problem, and there was uniformity in chromosome karyotypes of these species. First, random amplified polymorphic DNA (RAPD) markers were used for the analysis of intra- and inter-varietal diversity in mango (Bally et al. 1996; Ravishankar et al. 2000), Subsequently, simple sequence repeat (SSR) markers were developed by sequencing of plasmid clones from microsatellite-enriched genomic libraries to study genetic diversity in a limited number (15–36) of mango varieties (Duval et al. 2005; Honsho et al. 2005; Ravishankar et al. 2011). Later on, transcriptome sequence-based 100 genic-SSR markers were used to analyze 60 diverse mango varieties (Mahato et al. 2006; Lal et al. 2019). Phenotyping of 17 critical pomological traits in 60 mango varieties was mapped by a genome-wide association study using generalized linear model (GLM) and mixed linear model (MLM) approaches. Population structure analysis using the genotyping data from 87 SSR markers revealed three distinct sub-populations corresponding to the geographical origin of these varieties. Association analysis using GLM identified 23 genic-SSR markers associated with 13 traits. Two alleles of the MSSR190 locus were significantly associated with the fruit diameter, pulp to stone ratio and fruit weight, and one allele of the MSSR146 locus was associated with fruit length and width using both GLM and MLM approaches with a high proportion of phenotypic variance explained (Lal et al. 2017). Further, the first draft of Amrapali genome was used to identify 122,332 SSR loci and develop 8,451 Type 1 and 835 highly variable SSR (HSSR) markers for high level of polymorphism (Singh et al. 2016b).

Transcriptome sequence has been used to identify large number of single nucleotide polymorphism (SNP) markers in mango (Sherman et al. 2015; Mahato et al. 2016). Further, sequencing of double-digested restriction-site-associated genomic DNA (ddRAD) of 84 diverse mango cultivars identified 1.67 million high-quality SNPs, two major sub-populations, and a SNP database has been created (Singh et al. 2016b; Iquebal et al. 2017). High-density linkage maps of mango have been developed using SNP markers genotyped using Fluidigm EP-1 platform or specific-locus amplified fragment (SLAF) sequencing, which helped in genetic anchoring of the mango genome sequence and quantitative trait loci (QTL) mapping of important traits (Luo et al 2016; Kuhn et al. 2017). Recently, SNPs genotyped by ddRAD sequencing were used to study diversity and population structure of 106 mango cultivars (Warschefski et al. 2019). There are two distinct approaches, namely linkage mapping and association mapping to identify the markers linked to quantitative traits. Linkage mapping explores trait-linked markers within a structured bi-parental population, whereas association mapping looks for marker-trait associations in a set of random genotypes without structured population. The genetic mapping and molecular characterization of complex trait-linked markers have facilitated genomics-assisted breeding for the development of high-yielding, disease resistant, and abiotic stress-tolerant varieties in rice (Singh et al. 2016a; Bhandari et al. 2019).

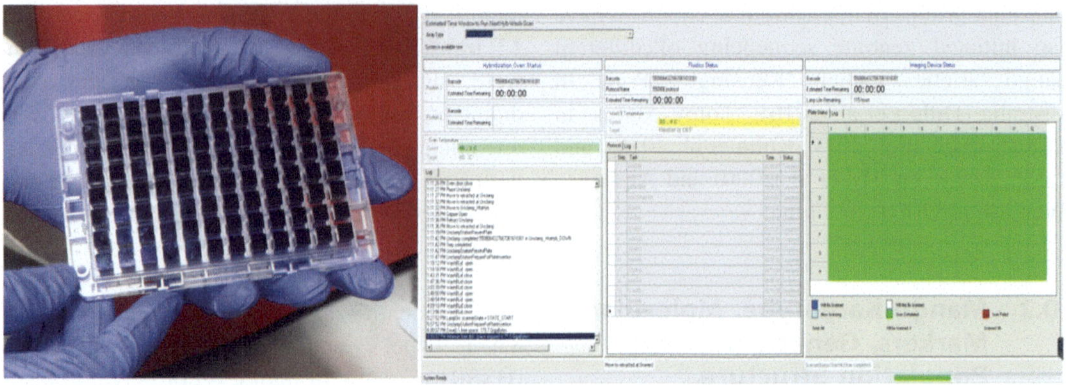

Fig. 10.1 A snapshot of 100% sample success rate of the mango 80 K genic-SNP chip array designed for Affymetrix GeneTitan genotyping platform

Singh et al. (2019) reported 10.5 million SNPs/Indels by re-sequencing a set of 24 diverse mango varieties and developed an 80 K genic-SNP genotyping chip for Affymetrix GeneTitan platform (Fig. 10.1). The chip was used for analyzing genetic diversity, population structure, and genome-wide association studies (GWAS) of fruit quality traits including fruit size, acidity, and pulp percentage in a set of 368 mango cultivars (NK Singh et al. unpublished). The population structure of cultivars was delineated using a subset of 60 genome-wide unlinked SNPs with 100% call rates revealing two sub-populations that corresponded to their geographical origin. Three different SNPs located on chromosome 5, 12, and 19 were significantly associated with mango pulp percentage, and one SNP on chromosome 10 was associated with fruit acidity with very high LOD score of 22. Haplotype-based phylogenetic analysis revealed the evolutionary relationship and mutational changes among the different groups of mango cultivars. Unweighted Pair Group Method with Arithmetic Mean (UPGMA) phylogenetic tree of 384 mango varieties based on the 60 unlinked SNPs also clustered them into two major groups, and revealed their evolutionary relationships (NK Singh et al. unpublished).

10.3 Sequencing of the Mango Genome

10.3.1 Genome Sequence of cv. Amrapali

In January 2014, NK Singh presented the first draft genome assembly of the mango variety 'Amrapali' using Illumina MiSeq overlapping paired-end reads at San Diego PAG conference (Singh et al. 2014). However, the assembly size (492 Mbp in 211,141 contigs) was larger than the estimated size of mango genome, i.e., 439 Mbp (Arumuganathan and Earle 1991), indicating redundancy in the assembled contigs. High heterozygosity (about one SNP/Indel every 40 bp) in the mango genome made it nearly impossible to produce a reference-quality assembly using short sequence reads (100–150 bp) of Illumina, because it did not permit integrated assembly of the maternal and paternal genomes into an integrated mosaic assembly, particularly in the genomic regions of high heterozygosity. The problem was overcome using 3rd generation single molecule real-time (SMRT) sequencing on PacBio RXII platform with long read lengths of 3.5–10 kbp, and the first de novo draft genome assembly of the mango variety

'Amrapali' was reported and sequence deposited in the NCBI public repository with Bio-Project no. PRJNA300605, WGS Accession no. LMWC00000000v.1 (Singh et al 2016b). PacBio sequence data of total 70 × genome coverage was generated using P4C2 and P5C3 chemistries and genome assembly was performed with an experimental version of the PacBio 'Falcon' assembler using custom assembly parameters of 500 bp sequence overlap and 15% mismatch to take care of the heterozygosity. This produced the first maternal-paternal mosaic genome assembly of 323 Mpb in 9,550 contigs with N50 value of 98.3 Kbp, the largest contig size of 1.09 Mbp and genome coverage of 73.2%. Total of 43,247 protein-coding genes were predicted in the size range of 150–12,102 bp with an average size of 894 bp, and functional annotation of 33,365 (77.14%) of these genes. They also identified 122,332 genomic-SSR of which 8,451 were type 1 SSR and 835 were highly variable SSR (HSSR) markers. Subsequently, a 403 Mbp reference genome assembly of 'Amrapali'was produced in 4,312 scaffolds using additional sequence data of 20 × genome coverage with PacBio P6C4 chemistry (Table 10.1, NCBI Accession no. LMWC01000000, v.2, Singh et al. 2017, 2018). The present BioNano optical fingerprint integrated assembly (NCBI Accession no. LMWC02000000, v. 3.0) has 2,314 scaffolds with genome coverage of 408 Mbp and a high N50 of 11.78 Mpb as compared to the version 2.0 assembly with an N50 of 282 kbp, providing almost complete coverage of the mango genome size estimated by flow cytometry, i.e., 402 ± 10 Mbp. About seventy-five percent of the scaffolds was anchored to the 20 chromosomes of mango with 6,297 (97.5%) of the SLAF markers in the high-density linkage map (Luo et al. 2016). Mapping of 18 different transcriptome sequence read archive (SRA) reads on the assembly showed 96% coverage of the gene space, while BUSCO analysis of 1440 conserved eukaryotic genes showed 93.4% coverage of genes (Singh et al. 2019).

10.3.2 Genome Sequence of cv. Tommy Atkins

A draft genome assembly of mango variety 'Tommy Atkins' was presented in the PAG conference at San Diego (Baruch et al. 2017). The genome assembly was based purely on Illumina's short-read paired-end and mate-pair sequence data, with an assembly size of 760 Mbp, clearly having phase redundancy due to high heterozygosity. It was also obvious from 85% of the assembly being anchored to 50 linkage groups using genetic maps from two different populations. An updated version of the Tommy Atkins genome assembly was presented in the subsequent PAG conference (Kuhn et al. 2018). For the updated assembly, they used 100 × genome coverage

Table 10.1 Comparison of the main features of the three published maternal-paternal mosaic assemblies of the highly heterozygous mango genome

Details	Amrapali[a]	Hong Xiang Ya[b]	Alphonso[c]
Sequencing platform	PacBio + Illumina + BioNano	PacBio + Illumina	PacBio + Illumina + Hi-C
Genome size (Mbp)	402±10	389	360
Assembly size (Mbp)	403	371.6	392.9
Sequencing depth (X)	101.98	613.79	240.0
No of sequence contigs	4312	420	672
Repeat sequence (%)	45.02	47.94	40.5
Scaffold N50 (Kbp)	11,780	4820	17,600
No. of predicted Genes	46,395	34,529	41,251

[a]Singh et al. (2018)
[b]Li et al. (2020)
[c]Wang et al. (2020)

Illumina sequence data integrated with 10 × Genomics data, assembled using 'DenovoMagic' genome assembler of NRGene. The resulting assembly was of 490 Mpb in 571 super-scaffolds. The assembly was improved by correcting and ordering of the scaffolds in pseudomolecules and incorporation of floating scaffolds into pseudomolecules using more than 5000 SNP markers from the published genetic maps (Luo et al. 2016; Kuhn et al. 2017). The final assembly is in 20 pseudomolecules with a size range of 11–22 Mb corresponding to the 20 haploid mango chromosomes, and 27,000 annotated protein-coding genes in the genome assembly.

10.3.3 Genome Sequence of cv. Kensington Pride

NL Dillon presented the initial draft genome assembly of Australia's most significant mango variety 'Kensington Pride' in the PAG conference at San Diego (Dillon et al. 2016). Under the mango genomics initiative of Queensland Department of Agriculture, Forestry and Fisheries (DAFF), the Horticulture Innovate Australia (HIA) had led an international collaboration between DAFF (Australia), Beijing Genomics Institute (BGI, China), and International Crops Research Institute for Semi-Arid Tropics (ICRISAT, India) to sequence the 'Kensington Pride' genome. The 407 Mpb assembly was completely based on the Illumina short reads sequencing. The heterozygosity issue was not addressed in this assembly, therefore redundancy of the contigs cannot be ruled out.

10.3.4 Genome Sequence of cv. Hong Xiang Ya

Recently, a genome assembly of mango variety 'Hong Xiang Ya' has been reported by Li et al. (2020) in the bioRxv preprint repository. They estimated a genome size of 389 Mbp using K-mer frequency distribution approach and generated sequence data with 225.38 × genome coverage on Illumina and 388.41 × genome coverage on PacBio Sequel II platform. The genome assembly was performed using Falcon and Falcon-unzip software to generate primary contigs (P-contigs) and haplotigs (H-contigs) assembly of 371.62 Mb in 120 contigs and 168.19 Mb in 1,190 contigs, respectively. The Primary contigs assembly is having N50 value of 4.82 Mbp, the maximum contig length of 11.46 Mbp, H-contigs N50 value of 458.69 Kbp, and the largest contig length of 2.0 Mpb. The published genetic map of 6500 SLAF markers (Luo et al. 2016) was used to anchor 112 of the P-contigs (367.05 Mbp) to 20 pseudomolecules, covering 98.77% of the total assembly (Table 10.1). They identified 34,529 protein-coding genes with an average gene length of 1,295 bp, average CDS length of 1,143 bp, and 5.6 exons per gene. Overall, 90.71% of the gene models were functionally annotated with significant similarities in the public databases and 284,679 SSRs were predicted constituting 1.1% (4.066 Mb) of the genome. They have deposited the genome assembly and sequenced reads in the BIG Genome Sequence Archive database under Accession number PRJCA002248.

10.3.5 Genome Sequence of cv. Alphonso

Another genome assembly of mango variety 'Alphonso' was reported this year by Wang et al. (2020). Alphonso was selected after a preliminary screening of 22 mango varieties for having the lowest level of 1.5% heterozygosity. The estimated genome size of Alphonso based on k-mer size distribution was estimated as 360 Mbp. Sequence data with 216 × genome coverage was generated using PacBio Sequel II platform and assembly was performed using 'Canu' assembler and improved further using paired-end and mate-pair Illumina reads.

Chromosomal level scaffolding was achieved using 33.77 Gbp of Illumina Hi-C data, and the scaffolds were arranged in pseudomolecules with the help of 6549 SLAF markers of the genetic map (Luo et al. 2016). The final genome assembly size is 392.9 Mbp in 252 large scaffolds (20 chromosome pseudomolecules, 2 organelles, and 230 floating scaffolds), with scaffold N50 of 17.6 Mbp and contig N50 of 3.5 Mbp, 90.1% of the assembly was anchored to 20 linkage groups with the pseudomolecule size rage of 12.2–29.4 Mbp. The Core Eukaryotic Genes Mapping Approach (CEGMA) analysis showed 95.1% and the Benchmarking Universal Single-Copy Orthologs (BUSCO) analysis showed 95.9% genome completeness based on the single-copy orthologous genes, thus confirming the high-quality of assembled genome (Table 10.1). Total of 41,251 protein-coding genes were predicted with an average gene length of 3503 bp, of which 37,424 (90.7%) were placed on the 20 pseudomolecules with a gene density of 105 genes per Mbp and 39,473 (95.68%) of the genes got functionally annotated. Further, Wang et al. (2020) have changed the chromosome numbering of mango in their paper based on descending pseudomolecule size.

10.4 Gene Content of the Mango Genome

We predicted total of 46,395 protein-coding genes in the version 2 assembly of 'Amrapali' genome, of which the highest number of 11,349 genes showed homology with sweet orange (*Citrus sinensis* L.) followed by cacao (*Theobroma cacao* L.) genome (Fig. 10.2). These genes were grouped into 21 broad functional categories, namely physiological functions (5363 genes), transportation (4804 genes), protein metabolism (4610 genes), transcription factors (3603 genes), phosphatases and kinases (2477 genes), resistance-like and defense response genes (2267 genes), growth and development (2247 genes), cell signaling (2178 genes), carbohydrate metabolism (1451 genes), nucleic acid metabolism (1975 genes) unknown function (1187 genes), lipid metabolism (1136 genes), membrane proteins (1030 genes), structural proteins (998 genes), secondary metabolites (904 genes), uncharacterized (644 genes), miscellaneous (584 genes), abiotic stress (567 genes), storage proteins (290 genes), acid metabolism (197 genes), and the largest category of 7884 hypothetical genes with no transcript support. Further, the 3603 transcription factor genes, which play very important roles in regulating the gene expression were divided into 36 categories with MYB factors (264 genes) other Zf factors (258 gens), NAC factors (213 genes), BHLH factors (183 gene), and ERF factor (162 genes) forming the largest five categories. These genes provide important reference points for further polymorphism survey and marker-trait association studies for plant breeding applications using bi-parental or association mapping panels.

10.5 Repeat Elements in the Mango Genome

The major proportion of higher eukaryotic genomes consists of different types of repetitive elements (RE) like transposable elements (TE) and simple sequence repeats like mini and microsatellites, which play important role in the genome evolution through cut-paste and copy-paste mechanisms. Identification and annotation of REs in the large eukaryotic genomes is a challenging task that requires both de novo and homology-based approaches (Lerat 2010). De novo analysis of REs in the mango genome using Repeat Modeler revealed that *Mangifera indica* does contain a large proportion of repetitive DNA. Singh et al. (2018) reported 45.02% interspersed REs (187.75 Mbp) masked in the total 403 Mbp 'Amrapali' reference assembly v.2 (Table 10.1). Proportion of different categories of sequences in the total REs was long terminal repeat (LTR) (36.72%), DNA transposons (22.28%), unclassified (33.024%), simple repeats (3.034%), low-complexity repeats (0.608%), and non-LTR (4.34%) (Fig. 10.3). Proportion of different categories of REs in the total 323 Mbp of the Amrapali draft genome (Table 10.2) was

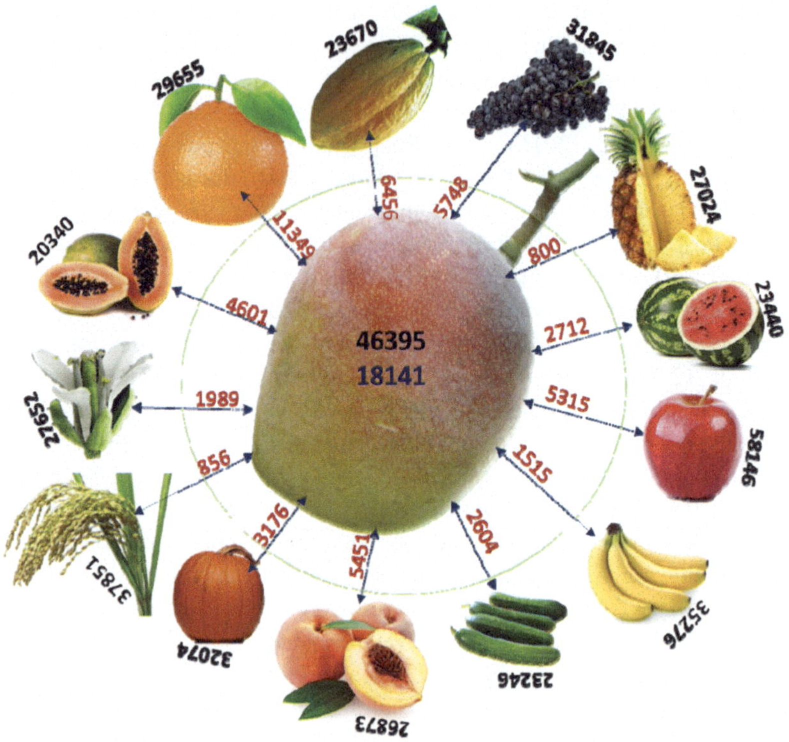

Fig. 10.2 Two-way Venn diagram of mango (*Mangifera indica* L. cv. Amrapai) genes conserved in 13 other plant species (shown in red fonts). Total number of genes in the respective species are shown in black fonts. Of the total 46,395 predicted genes, 18,141 were unique to mango

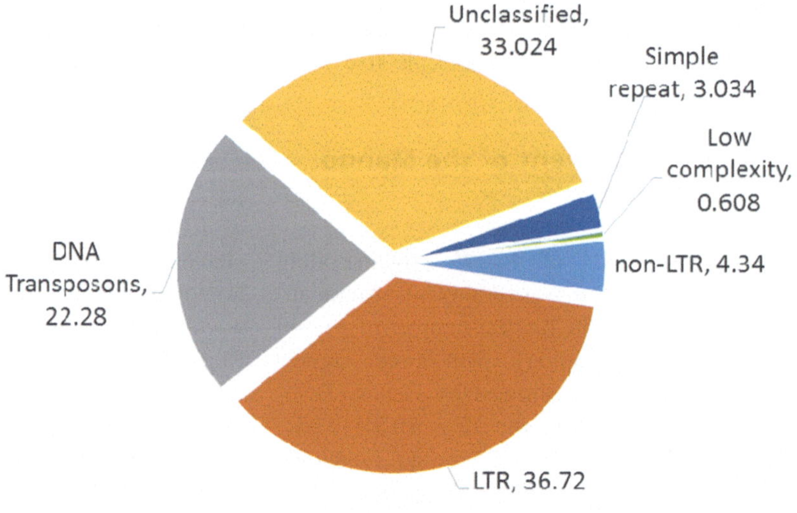

Fig. 10.3 Proportion of different types of repeat elements in the Amrapali mango genome ver. 2 assembly

short interspersed nuclear elements (SINEs) (0.04%), long interspersed nuclear elements (LINEs) (0.40%), LTRs (9.04%), DNA transposons (3.58%), unclassified (31.22%), simple repeats (1.09%), and low-complexity repeats (0.27%). The largest proportion (31.22%) of REs in the mango genome were unique to mango. Li et al. (2020) reported 46.18% TEs in the genome of mango variety Hong Xiang Ya, covering 171.58 Mbp out of the total 371.62 Mbp genome assembly. However, Wang et al. (2020) reported only 40.54% REs in the Alphonso genome

Table 10.2 Comparison of major repeat elements reported in the genomes of three different mango cultivars

TE Category	cv. Amarpali[a]		cv. Hong Xiang Ya[b]		cv. Alphanso[c]	
	Masked length (bp)	Percent in genome	Masked length (bp)	Percent in genome	Masked length (bp)	Percent in genome
SINEs	123,043	0.04	..	..	7713	0.00001
LINEs	1,282,443	0.40	2,297,579	0.62	3860235	0.98
LTR	29,209,092	9.04	85,763,725	23.08	59676610	15.19
DNA elements	11,569,382	3.58	16,187,808	4.36	21877011	5.57
Unclassified	100,892,375	31.22	67,335,389	18.12	73881229	18.80
Total interspersed	143,076,335	44.28	171,584,501	46.18	159302798	40.54
Simple repeats	3,530,710	1.09	5,673,978	1.53	..	..
Low complexity	886,364	0.27	906,641	0.24	..	..

[a]Singh et al. (2016b)
[b]Li et al. 2020
[c]Wang et al. (2020)

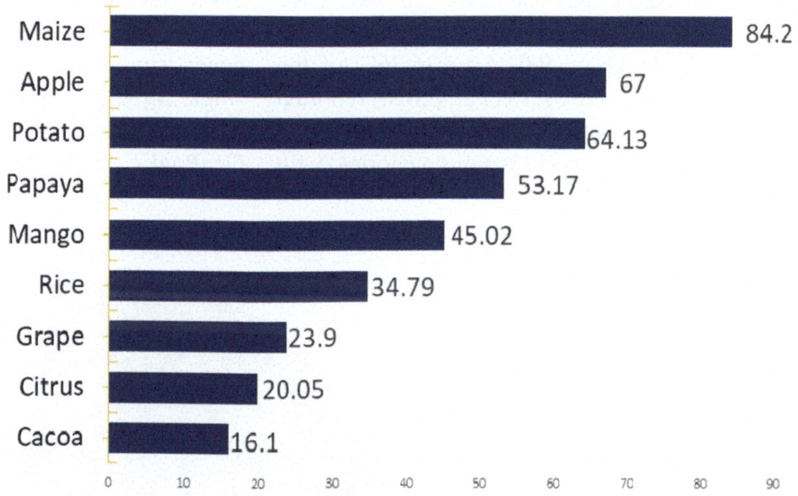

Fig. 10.4 Proportion of repeat elements in the Amrapali mango genome ver. 2 assembly in comparison to some other major fruits and crop plants

covering 159.30 Mbp of the total 392.9 Mbp assembly (Table 10.2). In all the three published mango genomes, LTRs are the most predominant elements over other class I and II RE. The reported REs in the mango genome is higher than 20.5% in sweet orange (Xu et al. 2011), 16.1% in Cacao (Argout et al. 2011), 23.90% in grape (Velasco et al. 2007), 25.03% in cucumber (Huang et al. 2009), but lower than 53.17% in papaya (Ming et al. 2008) and 67% in apple (Velasco et al. 2010) genome (Fig. 10.4).

10.6 Non-coding RNA in the Mango Genome

Most of the higher eukaryotes possess a small proportion of genes which are transcribed into non-coding RNA (ncRNA), which have very important biological functions. The computational discovery and characterization of novel ncRNA in plant species enriches our understanding of gene regulation in these plants.

Recent studies have described tissue-specific long non-coding RNA (lncRNA), which plays important role in regulating gene expression such as splicing, gene inactivation, and translation (Franco-Zorrilla et al. 2007; Ben et al. 2009; Liu et al. 2013). Apart from this, different types of canonical ncRNAs such as ribosomal RNA (rRNA) and transfer RNA (tRNA) are present in all species. Other important ncRNAs, such as miRNA, snRNA, and snoRNAs are also present in specific cellular locations. Singh et al. (2017, 2018) have reported 1,653 non-coding RNA elements in the Amrapali genome using Infernal and tRNAScan software. These belong to ten different categories such as tRNA (n = 601), sRNA (n = 12), snRNA (n = 529), rRNA (n = 150), ribozyme (n = 51), nodulin (n = 1), miRNA (n = 281), catalytic intron (n = 18), cis-regulatory elements (n = 7), and anti-sense RNA (n = 3). Among all identified categories of the ncRNA, rRNA, miRNA snRNA, and tRNA are the most dominant over other ncRNAs and individually make up 9.07% to 36.36% of the total ncRNA (Fig. 10.5). Li et al. (2020) reported in the Hong Xiang Ya genome total of 1125 ncRNAs belonging to six major categories, with tRNA having the highest copy number (598) followed by miRNA (n = 235), snRNA (n = 200), snoRNA(n = 47), and ribosomal RNA (n = 45). Wang et al. (2020) reported in the Alphonso genome total of 1893 ncRNA, including 599 tRNAs, 459 miRNAs, 560 snRNAs, and 275 rRNAs). The cultivar specific ncRNA are useful to understand the mechanism of interaction with the coding sequences and may be helpful in defining novel functional networks for the genetic improvement of mango.

10.7 Mango Chloroplast Genomes

The first chloroplast genome of mango was reported for variety 'Langra' using a combination of Sanger and 454-GS FLX sequencing (Azim et al. in 2014). It was assembled using CLC Genomics Workbench software (CLC bio, Denmark) into a cp genome of 151,173 bp in 31 contigs with 38.18% GC content. Total of 139 genes were predicted of which 119 genes were single copy, 20 duplicate genes, 29 tRNA genes, four rRNA, and 91 protein-coding genes (NCBI GenBank Accession number FJ212316).

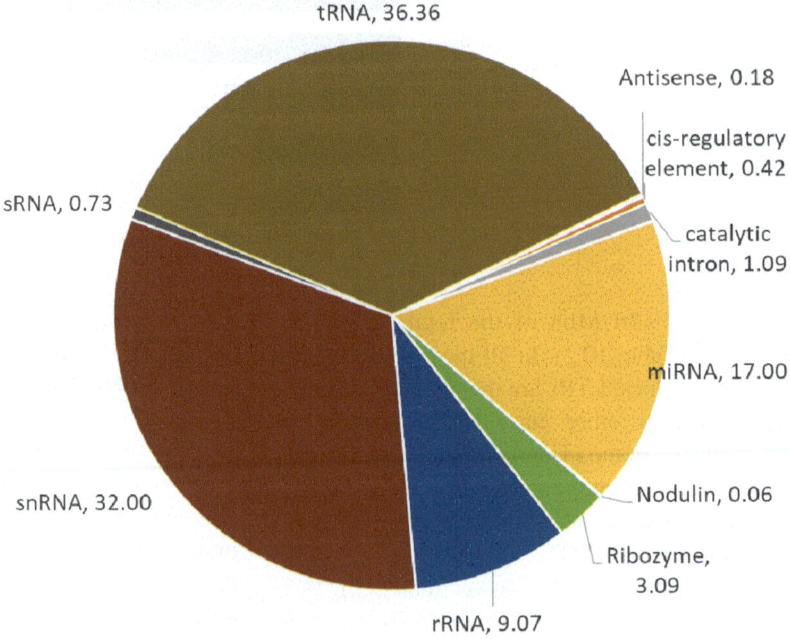

Fig. 10.5 Proportions of different kinds of non-coding RNA in the Amrapali mango genome ver. 2

Recently, the first full-length cp genome of a wild mango (*Mangifera sylvatica* Roxb.) has been reported by Zhang et al, (2020). The short read sequence data was generated using Illumina's Hiseq 2500 platform, and de novo assembly was performed using SOAPdenovo software. The full chloroplast genome assembly of 158,063 bp has a GC content of 37.89% and total of 112 genes were predicted and annotated of which 78 were protein-coding genes, 30 tRNA genes, and four rRNA genes. The sequence is available in the NCBI GenBank database under accession number MK790101.

10.8 Insights into Mango Genome Evolution by Whole Genome Duplication

Exponential increase in the whole genome and transcriptome sequence data provides a valuable resource to study the evolutionary consequences of gene duplications in different taxa, and unravel common principles underlying evolution by genome duplication followed by retention, diversification, or loss of genes. Singh et al. (2007) reported the duplication and retention of conserved single-copy genes between rice and wheat genomes, after an ancient whole genome duplication (WGD) about 70 million years ago (Patterson et al. 2004). Studies have shown that genome duplication through polyploidy has led to increase in the number of species in different angiosperm lineages like Poaceae, Brassicaceae, Solanaceae, and Leguminosae (Wood et al. 2009; Landis et al. 2018). The WGD is one of the most important factors leading to the evolution and diversification of plant species. In mango genome, two independent studies have reported whole genome duplication. Singh et al. (2018) reported at least three ancient whole genome duplication events (α, β and γ) in the Amrapali genome. They used SyMAP v4.2 software to identify the syntenic regions among the 20 chromosomes of mango and reported multiple recent and ancient WGD in the mango genome. The Synteny analysis identified total of 91 non-overlapping collinear blocks among which 18 blocks were large with duplicated fragment size of more than 1 Mbp. Three large blocks each were identified between chromosome pairs 1:20, 14:15, and 9:16 while chromosome pair 10:19 has two duplicated blocks. Further, seven different chromosome pairs were identified having one duplicated block each, namely chromosomes 11:16, 12:17, 8:13, 9:13, 4:17, 8:18, and 9:19. Among the eighteen duplicated segments identified there were at least one recent (15.87–31.74 Mya) and one ancient (253.96–269.84 Mya) duplication event (Fig. 10.6).

Later, Wang et al. (2020) have identified these duplications in the mango variety Alphonso and reported that mango diverged from litchi and sweet orange about 70 million years ago. In WGD analysis, Wang et al. reported 41 collinear blocks which were distributed across the 20 chromosomes of mango and also concluded that a WGD event may have occurred about 33 Mya. It is broadly accepted that macro-evolutionary changes have occurred due to whole genome duplications. Singh et al. (2018) also reported that a large number of these duplicated segments show collinearity with *Citrus sinensis* and *Theobroma cacao* genomes.

10.9 Transcriptome Studies for Identifying Differentially Expressed Genes for Mango Traits

During the last decades, RNA sequencing has been used extensively for understanding the developmental, cellular and physiological processes as well as domestication, disease resistance and alternate bearing in mango and other fruits. The main reason for this is the high throughput data generation and low cost of RNA sequencing (Martinelli et al. 2012; Wu et al. 2014; Dautt-Castro et al. 2015; Mahato et al. 2016; Srivastava et al. 2016; Yadav et al. 2020; Jayaswal et al. 2020). RNA-Seq offers important insights into gene expression under different conditions and allows discovery of new genes and pathways. RNA-Seq has demonstrated its potential in identifying tissue-specific long non-

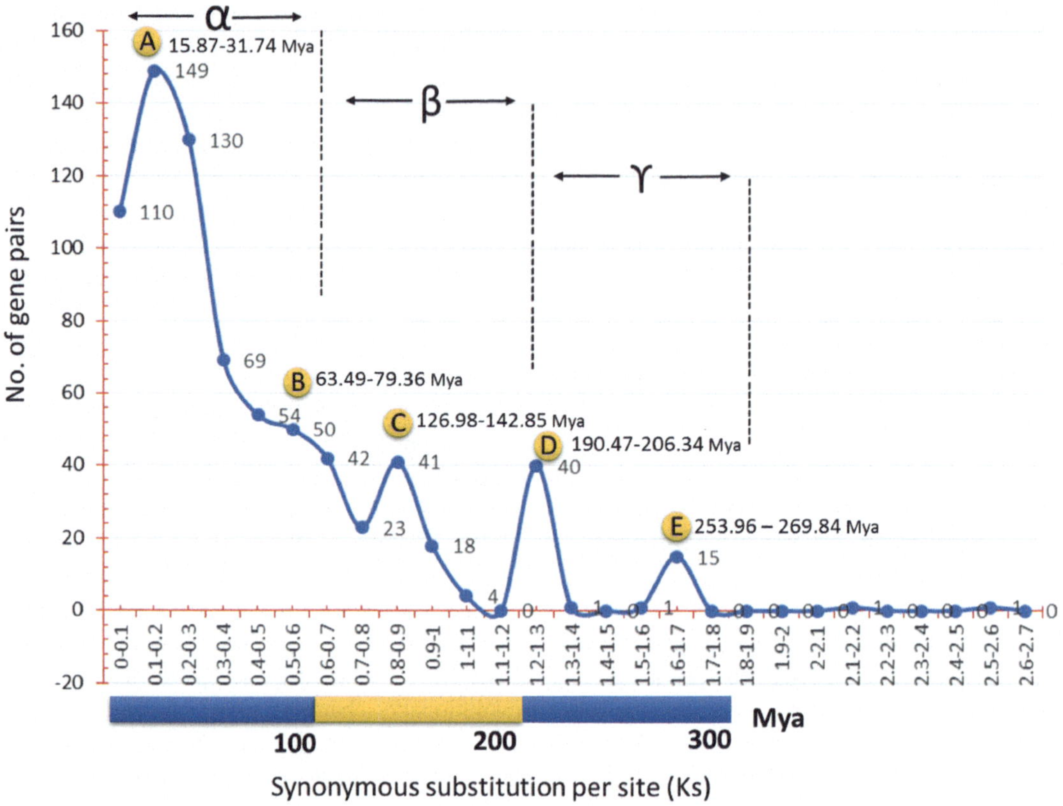

Fig. 10.6 The time range of whole genome and segmental duplications in the Amrapali mango genome based on synonymous substitution analyses of 18 large duplicated segment

coding RNA and how do they regulate the gene expression (Budak et al. 2020). Here, we have summarized some of these transcriptome-based differential gene expression studies on identification of candidate genes for important traits in mango.

10.9.1 Differential Gene Expression During Mango Malformation

Mango cultivation is widely affected by the fungal pathoen *Fusarium mangiferae,* which is associated with deformities in the vegetative and floral organs including shortened internodes and leaves in the mango plant (known as mango malformation), thus preventing normal fruiting in thousands of hectares (Chakrabarti 2011; Zhan et al. 2012). Flowers become either sterile or the buds abort shortly after fertilization, causing 80–100% fruit yield losses (Nor et al. 2013). Different measures are adopted to control the malformation in mango, such as fungicides, plantation of pathogen-free materials, and destruction of the infected branches of the tree but none of these are effective to control the disease completely. RNA-seq studies have been carried out to identify the differentially expressed genes (DEGs) between healthy and malformed tissues for understanding the molecular mechanism of mango-*F. mangiferae* interaction. Liu et al. 2016 analyzed transcriptome profiles of *F. mangiferae* (strain MG06) infected buds of mango cv. Keitt, at 45, 75, and 120 days after inoculation to generate 99 million sequence reads, and assembled these into 121,267 unigene contigs, of which 23,915 unigenes were assigned to 273 KEGG pathways. Variations were seen in the expression of genes with reference to the control sample and three other *F. mangiferae* infected samples, for example, 15,830 DEGs were identified in control vs 45 days, of which 6802 were up-regulated and 9028 were down-regulated. Of the 26,061 DEGs in control vs 75 days, 11,582 were up-regulated and 14,479 were down-regulated, and of the 20,146 DEGs in control vs. 120 days, 7222 were up-regulated and 12,924 were down-regulated. Apart from the DEG between control and infected samples, Liu et al. 2016 also studied plant defense response genes, e.g., (a) Pathogenesis-related (*PR*) genes (b) WRKY transcription factors (c) phenylpropanoid biosynthesis genes (d) signal transduction genes, and (e) protein kinase genes. The *PR* genes play important role in plant defenses against pathogen infection. Here, 46 *PR* genes of different categories (*PR1, PR2, PR4, PR10, PR5,* chitinases, peroxidases, proteinase inhibitors) were affected by infection with *F. mangiferae.* The WRKY transcription factors which regulate the plant immunity via interaction with different protein partners like resistance protein, histone deacetylases, calmodulin, MAP kinase kinases, MAP kinases, 14-3-3 proteins, and other WRKY transcription factors can act as activator or repressor in the plant signaling process (Pandey et al. 2009; Amorim et al. 2017). The study indicated that the WRKY genes were differentially expressed in the malformed tissues and may play crucial role in the mango plant defense. Phenylalanine ammonia-lyase (PAL) has a crucial role in secondary phenylpropanoid metabolism by thickening of cell walls and activating salicylic acid and jasmonic acid defense pathways (Kim and Hwang 2014). Liu et al. (2016) also showed that genes for O-hydroxycinnamoyl transferase, chalcone synthase, isoflavone 7-O-methyltransferase, and PALs were induced after Fusarium infection. Total of 93 DEGs were identified that were related to hormonal signaling during mango-Fusarium infection. Some of these are lipoxygenase isoform 3, ET-responsive transcription factors (ERFs), and 1-aminocyclopropane-1-carboxylate oxidases (ACOs) involved in the biosynthesis of JA and ET biosynthesis, respectively (Nascimento et al. 2018). There were total of 275 differentially expressed protein kinase genes between healthy and *F. mangiferae* infected mango buds. Some of these, e.g., calcium-dependent protein kinases, G-type lectin, S-receptor-like serine/threonine-protein kinases, and proline-rich receptor-like protein kinase genes were up-regulated, while leucine-rich repeat (LRR) receptor-like serine/threonine-protein kinase, and serine/threonine-protein kinase genes were down-regulated.

In another study on mango malformation in variety Amrapali, Yadav et al. (2020) sampled three infected tissues, namely single swollen malformed bud stage, multiple malformed bud stage, malformed panicle stage together with two healthy control samples, namely healthy bud stage (HB-1), and healthy panicle stage (HB-2), to generate total of 20.31, 20.77, 20.32, 27.92, and 18.59 million sequence reads, respectively, using Illumina NextSeq platform. They found higher expressions of seven flowering-related genes (*MYB30, TPL, bHLH, FTIP1, CDKC2, CPK33,* and *ATH1*) in both healthy buds and panicle as compared to the malformed buds and panicle. Among the other differentially expressed flowering genes in six possible combinations, the highly up-regulated genes included *UBP12, EFS, AGL8, AGL14, AGL20, AGL24, KIN10, MYB30, SUS2, FTIP1, CCT,* and *LDL2*, whereas highly down-regulated genes included *TIL1, TIC, DCL3, GA20OX3, CCT, AP1, AGL6, AGL8, MYB30, AGL8, GCT,* and *GA3OX1*. The data set provides information on transcripts putatively associated with embryonic flower, early flowering, flowering time control, and terminal flower in healthy and malformed tissues. Out of the observed DEGs, the transcript profiles of GA20OX3, AGL24, and LDL2, the key genes regulating floral transition and differentiation, were validated through qRT-PCR. The study provides a valuable resource for exploring the complex molecular mechanisms in flower development and malformation in mango.

10.9.2 Differentially Expressed Genes at Different Developmental Stages of Mango

In the early stages of mango fruit development, the tissue differentiation occurs for exocarp, endocarp, and mesocarp. The transcriptome-based analysis has enlarged our understanding of the gene expression during fruit development. Deshpande et al. (2017) reported transcriptome-based developmental studies on cv. Alphonso at eight different fruit development stages, including flowering, 30 days after pollination, 60 days after pollination, pulp 90 days after pollination, skin 90 days after pollination, 5 days after harvest: green stage, 10 days after harvest: mid-ripe stage, 15 days after harvest the ripe stage. For each tissue, more than 100 million reads were generated using Illumina sequencing. They reported total of 20,755 unique transcripts encoding 142 different biological pathways including ethylene, flavor-related pathway, starch, sucrose, amino acids, and fatty acids metabolism-related pathways. Specifically, the comparative analysis of differentially expressed transcripts showed total of 524 genes were down-regulated, including alpha-amylase and subtilisin inhibitor-like, carbonic anhydrase, chitinase and maternal effect embryo arrest 59, whereas 181 genes were up-regulated, including homeodomain-like protein, cytochrome P450 like, heat shock cognate 70 kDa protein and inositol-3-phosphate synthase from flower to fruit to 30 days post-harvest period. Similar differences in gene expression were observed at other stages of development, e.g., in 90 DAP-skin, 54 transcripts were up-regulated including isoflavone reductase, transcription and translation regulatory proteins, hydrolases, methyl transferase, etc. while four transcripts related to membrane and cytoplasm were down-regulated. Deshpande et al. (2017) also elaborated DEGs from other pathways like terpenes biosynthesis pathway and reported mono-terpene synthases, sesque-terpene synthases and di-terpene synthase genes which were dominant in the flowering and 30 DAP samples. Further, two genes of 9-LOX (lipoxygenase) and six genes of 13-LOX from the flavor related pathways were reported showing differential expression. In addition, other transcripts involved in cell wall hydrolysis and cysteine proteinase inhibition were also reported.

In another study, Dautt-Castro et al. (2015) reported DEGs of different gene families related to fruit ripening. The transcriptome study was performed on mesocarp of mature green and ripe mango cv. 'Kent', using 6.7 million reads generated using Illumina sequencing. The differential expression analysis showed that between mature green and ripe tissues total of 1,178 and 1,128 genes were up- and down-regulated, respectively. Several genes of carbohydrate catabolism, sucrose, and ethylene biosynthesis among others related to fruit ripening were up-regulated. In addition, 327 genes involved in metabolic pathways of fruit ripening were reported, namely hormone signal transduction, starch and sucrose metabolism, galactose metabolism, terpenoid backbone, and carotenoid biosynthesis. Similarly, Bajpai et al. (2018) has reported DEGs involved in anthocyanin biosynthesis pathway in mango cv. Amrapali. In another transcriptome study on mango cv. Zill, Wu et al. (2014) reported transcriptome profiling at four fruit development stages (50 days after flowering: young fruit, 80 days after flowering: young fruit, 100 days after flowering: immature stage, and fully ripe fruit), using pooled RNA sequencing strategy to generate total of 6.8 million reads and assembled these into 54,207 transcripts of which 23,741 were mapped to 128 different pathways. They reported different genes like malic enzyme family, aspartate aminotransferase-like, and phosphoglycerate kinase, which were up-regulated during ripening.

10.9.3 Transcriptome Analysis During Post-harvest and Chilling Injury

Mango is an important tropical fruit rich in nutritional constituents with high commercial value, however due to short shelf life and quick softening, mangos are highly susceptible to postharvest disease-enabling pathogens to infect and degrade the harvested fruit (Prusky et al. 2001). Post-harvest hot water treatment is widely adopted for cleaning and disinfecting the freshly harvested fruits, it enhances the storage of fruits for 3–4 weeks at 12 °C. Luria et al. 2014, conducted transcriptome analysis on the effect of hot water treatment on mango fruit cultivar 'Shelly'. They sequenced total of eight libraries of freshly harvested and hot water treated mango samples at different time intervals and sequence reads were generated using Illumina Hiseq 2000 platform. The DEG analysis showed that during 0 h of hot water treatment total of 827 genes were expressed, while after 48 h of heat treatment only 87 genes were expressed. A significant finding of the hot water brushing (HWB) treatment was that color index of the fruit skin was comparatively increased and the expression of chlorophyll (*LHCIIb, Oxepch, PIRC, and Thl1ch*) and anthocyanin biosynthesis-related genes (85A2 and Anthocyanin 5) was decreased. Apart from this, genes related to sugar and flavonoid metabolism were also highly up-regulated.

10.9.4 Differential Expression of Genes Associated with Alternate Bearing in Mango

Alternate bearing (AB) is a problem of irregular bearing in the fruit trees where fruit yield is high in one year (on year) and low in the next year (off year). Alternate bearing or biennial bearing habit is prominent in perennial fruits like mango, pistachio, olives, apple, pear, apricot, orange, jamun, and others (Barranco et al. 1998; Shalom et al. 2012; Guitton et al. 2012; Khezri et al. 2020). Several parameters have been proposed for the quantitative evaluation of AB, e.g., biennial bearing index (BBI), which is based on the mass of fruit on individual branch of the tree, however, the physiological mechanism of alternate bearing is largely unknown (Wilcox 1944; Bangerth 2009). Recent advancements in molecular and genetic studies on synchronization of flowering time and seed production have involved transcriptome analysis to explore the physiological processes associated with the AB in mango trees. Recently, Sharma et al. (2020) have reported a transcriptome-based comparative gene expression analysis of regular bearing cv. 'Neelam' and alternate bearing cv. 'Dashehari', and found that among 42,397 assembled transcripts, 6453 and 6,104 genes were up- and down-regulated, respectively. Transcriptome analysis between the two cultivars revealed that total of 26 DEGs were specifically related to AB, including SPL-like, and GA-20-oxidase genes, suggesting the hormonal regulation of AB. The reported study emphasized the role of potential genes like gibberellic acid (GA3) which initiates floral induction from the vegetative buds in mango. Similar role of Gibberellin has been reported in citrus and other fruits (Goldschmidt et al. 1997) (Table 10.3).

10.10 Conclusions and Way Forward

To conclude, attempts have been made over the last five years to assemble the genome sequence of five different varieties of mango, namely Amrapali (India), Tomy Atkins (USA), Kensington Pride (Australia), Hong Xiang Ya, and Alphonso (China). However, it has been difficult to assemble a high-quality reference genome of mango due to high level of heterozygosity and structural variations in the genome. Recently published mango genome assemblies have poor genome coverage even after using various software tools to merge the maternal and paternal chromosome sequences into a chromosomal-level mosaic genome assembly. Total of 36,000 to 46,000 protein-coding genes have been predicted using ab initio gene perdition software tools. There is a need to develop bin-separated maternal and paternal genome assemblies of the mango genome using the recently devised Trio-Binning (Koren et al. 2018) approach to separate the maternal and patterns sequence reads based on the genome sequence of the two parents and the progeny. A fully annotated bin-separated diploid assembly and pan genome of mango

Table 10.3 Chronological list of Bio-Projects of mango (*Mangifera indica*) registered with the NCBI/EBI/DDJB public repositories for genome and transcriptome sequence data (- indicates information not readily available)

Sr. no.	Accession no. (Registration date)	Lead institution	Objectives	Genotype	Sequencing technology	Genomic data availability	References
1.	PRJNA1935911 (Mar 2013)	ICAR-NIPB, IARI, New Delhi, India	Transcriptome (Leaf)	Dashehari	AB SOLiD 4 System	SRA,TSA	Mahato et al. (2016), Sharma et al. (2020)
2.	PRJNA193588 (March 21, 2013)	ICAR-NIPB, IARI, New Delhi, India	Transcriptome (Leaf)	Neelum	AB SOLiD 4 System	SRA,TSA	Mahato et al. (2016), Sharma et al. (2020)
3.	PRJNA2272431 (November 2013)	ARO, Israel	Transcriptome (Fruit peel, hot water treatment)	Shelly	Illumina	SRA,TSA	Luria et al. (2014)
4.	PRJNA1926768 (March 2013)	University of Karachi, Pakistan	Transcriptome (Leaf)	Langra	Illumina	SRA,TSA	Azim et al. (2014)
5.	PRJNA258477 (August 19, 2014)	UNAM, Mexico	Transcriptome (Ripening and shelf life)	Kent	Illumina	SRA	Dautt-Castro et al. (2015)
6.	PRJNA2547719 (July 2014)	ARO, Israel	Transcriptome (SNP, germplasm diversity)	Tommy Atkins and Keitt	Illumina	SRA,TSA	Sherman et al. (2015)
7.	PRJNA2532722 (June 3, 2014)	Cornell University, USA	Transcriptome (fruit peel, cuticle)	Keitt	Illumina	SRA	Tafella-Arellano et al. 2017
8.	PRJNA2352471 (January 6, 2014)	SSCRI, CATAS, China	Transcriptome (fruit peel and pulp)	Zill	Illumina	SRA	Wu et al. (2014)
9.	PRJNA300605 (October 30, 2015)	ICAR-NIPB, New Delhi, India	Genome (Bin-separated genome assembly)	Amrapali	Illumina PacBio RSII	SRA, AGP	Singh et al. (2016b, 2017, 2018, 2019)
10.	PRJNA3040932 (November 5, 2015)	ARO-VC, Israel	Transcriptome (cold storage chilling stress response)	Keitt and Shelly	Illumina	SRA	Sivankalyani et al. (2016)
11.	PRJNA2887972 (July 2015)	Indiana University, USA	Transcriptome (genetic diversity, SNP markers)	Tommy Atkins	Illumina	SRA	Kuhn et al (2016)
12.	PRJNA286253 (June 2015)	CIAD, Sonora, Mexico	Transcriptome (ripening due to hot water disinfection)	Ataulfo	Illumina	SRA	Daut-Castro et al. (2018)

(continued)

Table 10.3 (continued)

Sr. no.	Accession no. (Registration date)	Lead institution	Objectives	Genotype	Sequencing technology	Genomic data availability	References
13.	PRJNA2798293 (March 2015)	ICAR-NIPB, New Delhi, India	Leaf Transcriptome (SSR, SNP, Heterozygosity)	Amrapali	Illumina	SRA, TSA	Mahato et al. (2016)
14.	PRJNA274728 (February 6, 2015)	ICAR-IIHR, Bengaluru, India	Transcriptome (variants, mutational studies)	Alphonso	Illumina	SRA	–
15.	PRJNA325581 (June 14, 2016)	ICAR-IIHR, Bengaluru, India	Transcriptome (Peel color)	–	–	SRA	–
16.	PRJNA3133402 (February 7, 2016)	ICAR-CISH, Lucknow, India	Transcriptome (Flowering pathway)	Dashehari	Illumina	SRA	CISH Annual Report (2018)
17.	PRJNA3177128 (April 2016)	CSIR-NBRI, Lucknow, India	Transcriptome (Ripening and aroma)	Dashehari	Illumina 454 GSFLX	SRA	Srivastava et al. (2016)
18.	PRJNA3170441 (April 2016)	ICAR-CISH, Lucknow, India	Transcriptome (transcription regulation)	Chausa	Illumina	SRA	Bajpai et al. (2018)
19.	PRJNA3204584 (May 2016)	STCRI, CATAS, China	Transcriptome	F1 population	Illumina	SRA	–
20.	PRJNA3157912 (March 1, 2016)	SATCRI, CATAS, China	Transcriptomes (Mango/Fusarium interaction)	–	Illumina	SRA	–
21.	PRJNA3101932 (January 9, 2016)	SSCRI, China	Transcriptome	–	Illumina	SRA	–
22.	PRJNA3944881 (July 4, 2017)	King Abdul Aziz University, Riad, Saudi Arabia	Transcriptome Chloroplast genome	–	Illumina	SRA	–
23.	PRJNA4501431 (April 3, 2018)	USDA-ARS HSRS, Florida, USA	Genome (Mosaic assembly)	Tommy-Atkins	Illumina	SRA	Baruch et al. (2017) Kuhn et al. (2018)
24.	PRJNA4311252 (January 3, 2018)	ICAR-CISH, Lucknow, India	Transcriptome, Jelly seed development	Dashehari	Illumina	SRA	–

(continued)

Table 10.3 (continued)

Sr. no.	Accession no. (Registration date)	Lead institution	Objectives	Genotype	Sequencing technology	Genomic data availability	References
25.	PRJDB4532 (January 10, 2018)	NARO Institute of Fruit Tree Science	Genome sequencing DNA markers.	Irwin	454 GS FLX	SRA	Nashima et al. (2017)
26.	PRJNA517351SRA Acc. SRR8521837– SRR8521844. (January 7, 2019)	Florida International University, USA	Origin, Diversity	106 genotypes	Illuimna ddRAD	SRA	Warschefski et al. (2019)
27.	PRJNA533518 (April 8, 2019)	UQ, Brisbane, Australia	Gene discovery	cv. 1234	Illumina	SRA,TSA	Chabikwa et al. (2020)
28.	PRJNA5155641 (January 7, 2019)	University of Nottingham, Malaysia Campus	Transcriptome (Fruit ripening)	Chokanan and Golden Phoenix	Illumina	SRA	Chin et al. (2019)
29.	PRJCA002248 (March 2020)	SCAU, Guangzhou, China	Genome (Mosaic assembly)	Hong ZiangYa	Pacbio Sequel II Illumina	SRA, AGP	Li et al. (2020)
30.	PRJNA4871549 (March 2020)	TCGRI, CATAS, Haikou, China	Genome (Mosaic assembly)	Alphonso	PacBio Sequel II Illumina	SRA, AGP	Wang et al. (2020)

together with high-quality genome annotations will be helpful in genome-assisted gene discovery and marker-assisted breeding in mango.

Acknowledgements This study was carried out under the Network Project on Functional Genomics and Genetic Modification (FGGM) in Crops and Extra-Mural project with funding support from Indian Council of Agricultural Research (ICAR), Ministry of Agriculture and Farmers Welfare, Government of India.

References

Amorim LLB, da Fonseca Dos Santos R, Neto JPB, Guida-Santos M, Crovella S, Benko-Iseppon AM (2017) Transcription factors involved in plant resistance to pathogens. Curr Protein Pept Sci 18(4):335–351

Argout X, Salse J, Aury J-M, Guiltinan MJ, Droc G, Gouzy J, Allegre M et al (2011) The genome of *Theobroma cacao*. Nat Genet 43:101–108

Arumuganathan K, Earle ED (1991) Nuclear DNA content of some important plant species. Plant Mol Biol 9:208–218

Azim MK, Khan IA, Zhang Y (2014) Characterization of mango (*Mangifera indica* L.) transcriptome and chloroplast genome. Plant Mol Biol 85:193–208

Bajpai A, Khan K, Muthukumar M, Rajan S, Singh NK (2018) Molecular analysis of anthocyanin biosynthesis pathway genes and their differential expression in mango peel. Genome 61(3):157–166

Bangerth F (2009) Floral induction in mature, perennial angiosperm fruit trees: similarities and discrepancies with annual/biennial plants and the involvement of plant hormones. Sci Hort 122:153–163

Barranco D, Fernández-Escobar R, Rallo L (1997) El cultivo del olivo (Olive cultivation), 1st[a] edn. Mundi-PrensaLibros, Madrid, Spain

Baruch K (2017) De novo assembly of the Tommy Atkins mango genome. In: Plant and animal genome conference XXV, San Diego, CA, W597, January 16, 2017

Ben AB, Wirth S, Merchan F et al (2009) Novel long non-protein coding RNAs involved in Arabidopsis differentiation and stress responses. Genome Res 19(1):57–69

Bhandari A, Jayaswal PK, Yadav N, Singh R, Singh Y et al. (2019). Genomics-assisted backcross breeding for infusing climate resilience in high-yielding green revolution varieties of rice. Indian Soc Genet Plant Breed 29. doi:10.31742/IJGPB.79S.1.5

Budak H, Kaya SB, Cagirici HB (2020) Long non-coding RNA in plants in the era of reference sequences. Front Plant Sci 11:276

Coelho EM, de Souza MEAO, Corrêa LC, Viana AC, de Azevêdo LC, Dos Santos Lima M (2019) Bioactive compounds and antioxidant activity of mango peel liqueurs (*Mangifera indica* L.) produced by different methods of maceration. Antioxidants 8(4):102

Chabikwa TG, Barbier FF, Tanurdzic M, Beveridge CA (2020) *De novo* transcriptome assembly and annotation for gene discovery in avocado, macadamia and mango. Sci Data 7:9

Chakrabarti DK (2011) Mango malformation. Springer, Berlin

Chin CF, Teoh FY, Chee MJY, Al-Obaidi JR et al (2019) Comparative Proteomic analysis on fruit ripening processes in two varieties of tropical mango (*Mangifera indica*). Protein J. https://doi.org/10.1007/s10930-019-09868-x

Dautt-Castro M, Ochoa-Leyva A, Contreras-Vergara CA et al (2015) Mango (*Mangifera indica* L.) cv. Kent fruit mesocarp de novo transcriptome assembly identifies gene families important for ripening. Front Plant Sci 6:62

Dautt-Castro M, Ochoa-Leyva A, Contreras-Vergara CA et al (2018) Mesocarp RNA-Seq analysis of mango (*Mangifera indica* L.) identify quarantine postharvest treatment effects on gene expression. Sci Hort 227:146–153

Deshpande AB, Anamika K, Jha V et al (2017) Transcriptional transitions in Alphonso mango (*Mangifera indica* L.) during fruit development and ripening explain its distinct aroma and shelf life characteristics. Sci Rep 7(1):8711

Dillon N, Innes D, Ming Y, Fang X, Li X, Gao X (2016) Drafting the Kensington pride mango genome. In: Plant and animal genome conference XXIV, San Diego, CA, W821, 10 Jan 2016

Duval MF, Bunel J, Sitbon C, Risterucci AM (2005) Development of microsatellite markers of mango. Mol Ecol Notes 5:824–826

Iquebal MA, Sarika J, Mahato AK, Jayaswal PK et al (2017) MiSNPDb: a web-based genomic resources of tropical ecology fruit mango (*Mangifera indica* L.) for phylogeography and varietal differentiation. Sci Rep 7 (1):1–9

Franco-Zorrilla JM, Valli A, Todesco M et al (2007) Target mimicry provides a new mechanism for regulation of microRNA activity. Nat Genet 39 (8):1033–1037

Goldschmidt EE, Tamim M, Goren R (1997) Gibberellins and flowering in Citrus and other fruit trees: a critical analysis. Acta Hort 463:201–208

Guitton B, Kelner JJ, Velasco R, Gardiner SE, Chagné D, Costes E (2012) Genetic control of biennial bearing in apple. J Exp Bot 63(1):131–149

Honsho C, Nishiyama K, Eiadthong W, Yonemori K (2005) Isolation and characterization of new microsatellites markers in mango. Mol Ecol Notes 5:152–154

Huang S, Li R, Zhang Z, Li L, Gu X, Fan W, Lucas WJ et al (2009) The genome of the cucumber, *Cucumis sativus* L. Nat Genet 141:1275–1283

Huang X, Yan H, Wu Z, Yi Y (2020) Putative Allergens Identified in Mango (*Mangifera indica* L) Leaf and fruit with transcriptome analysis. J Food Sci Technol 5 (2):98–110

Jahurul MHA (2014) Supercritical carbon dioxide extraction and studies of mango seed kernel for cocoa butter analogy fats. J Food 12:97–103

Jayaswal PK, Mahato AK, Singh S, Rai V et al (2019) Characterization of transposable elements (TEs) in the reference genome of mango (*Mangifera indica* L. cv. Amrapali). In: XIV Agriculture Science Congress, NAAS, New Delhi, India

Jayaswal PK, Srivastava M, Bajpai A, Ravishankar KV, Singh NK (2020) Mango fruit transcriptome. Reference Module in Food Science. Elsevier https://doi.org/10.1016/B978-0-08-100596-5.22742-X

Khezri M, Heerema R, Brar G et al (2020) Alternate bearing in pistachio (*Pistacia vera* L.): a review. Trees https://doi.org/10.1007/s00468-020-01967-y

Kim DS, Hwang BK (2014) An important role of the pepper phenylalanine ammonia-lyase gene (*PAL1*) in salicylic acid-dependent signaling of the defense response to microbial pathogens. J Exp Bot 65:2295–2306

Koren S, Rhie A, Walenz BP, Dilthey AT, Bickhart DM et al (2018) *De novo* assembly of haplotype-resolved genomes with trio binning. Nat Biotech 36:1174–1182

Kuhn DN, Dillon NL, Innes DJ, Wu L-S, Mockaitis K (2016) Development of single nucleotide polymorphism (SNP) markers from the mango (*Mangifera indica*) transcriptome for mapping and estimation of genetic diversity. Acta Hort 1111:315–322

Kuhn DN, Bally ISE, Dillon NL et al (2017) Genetic map of mango: a tool for mango breeding. Front Plant Sci 8:577

Kuhn DN, Campbell M, Bally I, Dillon N, Innes D, Rahman J et al (2018) Assembly and annotation of the mango cultivar 'Tommy Atkins'. In: Plant and animal genome conference XXVI, San Diego, California, p W672

Lal S, Singh SK, Singh AK, Srivastav M, Singh A, Singh NK (2017) Character association and path analysis for fruit chromatic, physicochemical and yield attributes in mango (*Mangifera indica*). Indian J Agric Sci 87(1):122–131

Lai S, Singh SK, Singh AK, Singh NK (2019) Assessment of genetic divergence of mango genotypes using multivariate techniques. J Environ Biol 40(1):17–28

Landis JB, Soltis DE, Li Z et al (2018) Impact of whole-genome duplication events on diversification rates in angiosperms. Am J Bot 105(3):348–363

Lerat E (2010) Identifying repeats and transposable elements in sequenced genomes: how to find your way through the dense forest of programs. Heredity 104:520–533

Li H, Peng Z, Yang X, Wang W, Fu J, Wang J et al (2013) Genome-wide association study dissects the genetic architecture of oil biosynthesis in maize kernels. Nat Genet 45:43–50

Li W, Zhu X, Zhang Q, Li K, Zhang D, Shi C et al (2020) SMRT sequencing generates the chromosome-scale reference genome of tropical fruit mango, *Mangifera indica*. bioRxiv 2(20):960880

Liu F, Wu JB, Zhan RL, Ou XC (2016) Transcription profiling analysis of mango-*Fusarium mangiferae* interaction. Front Microbiol 7:1443

Luo C, Shu B, Yao Q, Wu H, Xu W, Wang S (2016) Construction of a high-density genetic map based on large-scale marker development in mango using specific-locus amplified fragment sequencing (SLAF-Seq). Front Plant Sci 7:1310

Liu TT, Zhu D, Chen W et al (2013) A global identification and analysis of small nucleolar RNAs and possible intermediate-sized non-coding RNAs in *Oryza sativa*. Mol Plant 6(3):830–846

Luria N, Sela N, Yaari M, Feygenberg O et al (2014) De-novo assembly of mango fruit peel transcriptome reveals mechanisms of mango response to hot water treatment. BMC Genom 15:957

Mahato AK, Sharma N, Singh A et al (2016) Leaf transcriptome sequencing for identifying genic-SSR markers and SNP heterozygosity in crossbred mango variety 'Amrapali' (*Mangifera indica* L.). PLoS One 11(10):e0164325

Maldonado-Celis ME, Yahia EM, Bedoya R et al (2019) Chemical composition of mango (*Mangifera indica* L.) fruit: nutritional and phytochemical compounds. Front Plant Sci 10:1073

Martinelli F, Sandra L, Uratsu SL, Albrecht U, Reagan RL et al (2012) Transcriptome profiling of citrus fruit in response to Huanglongbing disease. PLoS ONE 7:1

Mehrotra RC, Dilcher DL, Awasthi N (1998) A Paleocene *Mangifera*-like leaf fossil from India. Phytomorphology 48:91–100

Ming R, Hou S, Feng Y, Yu Q, Dionne-Laporte A, Saw JH, Senin P, Wang W et al (2008) The draft genome of the transgenic tropical fruit tree papaya (*Carica papaya* Linnaeus). Nature 452:981–996

Mukherjee SK (1950) Cytological investigation of the mango and the allied Indian species. Proc Nat Inst Sci India 16:281–303

Mukherjee SK (1951) Origin of mango. Indian J Genet 11:49–56

Mukherjee SK (1957) Cytology of some Malayan species of *Mangifera*. Cytologia 22:239–241

Mukherjee SK (1963) Cytology and breeding of mango. Punjab Hort J 3:107–115

Mukherjee SK, Singh RN, Majumder PK, Sharma DK (1968) Present position regarding breeding of mango (*Mangifera indca* L.) in India. Euphytica 17:462–467

Nascimento FX, Rossi MJ, Glick BR (2018) Ethylene and 1-Aminocyclopropane- 1-carboxylate (ACC) in plant-bacterial interactions. Front Plant Sci 9:114

Nashima K, Terakami S, Kunihisa M, Nishitani C, Shoda M et al (2017) Retrotransposon-based insertion polymorphism markers in mango. Tree Genet Genomics 13:110

Nor M, Nik MI, Salleh B, Leslie F (2013) Fusarium species associated with mango malformation in peninsular Malaysia. J Phytopathol 161:617–624

Pandey SP, Somssich IE (2009) The role of WRKY transcription factors in plant immunity. Plant Physiol 150:1648–1655

Paterson AH, Bowers JE, Chapman BA (2004) Ancient polyploidization predating divergence of the cereals, and its consequences for comparative genomics. Proc Natl Acad Sci USA 101:9903–9908

Prusky D, Eshel D, Kobiler I, Yakoby N, Beno-Moualem D et al (2001) Postharvest chlorine treatments for the control of the persimmon black spot disease caused by *Alternaria alternata*. Postharvest Biol Technol 22 (3):271–277

Ravishankar KV, Anand L, Dinesh MR (2000) Assessment of genetic relatedness among mango cultivars of India using RAPD markers. J Hort Sci Biotechnol 75:198–201

Ravishankar KV, Mani BHR, Anand L, Dinesh HR (2011) Development of new microsatellite markers from mango (*Mangifera indica*) and cross-species amplification. Am J Bot e6-e99

Karanjalker GR, Ravishankar KV, Shivashankara KS, Dinesh MR, Roy TK, Sudhakar Rao DV (1917) A study on the expression of genes involved in carotenoids and anthocyanins during ripening in fruit peel of green, yellow, and red colored mango cultivars. Appl Biochem Biotechnol. https://doi.org/10.1007/s12010-017-2529-x

Roy B, Visweswariya SS (1951) Cytogenetics of mango and banana (Review of work done, 1948-1951). Maharashtra Association for Cultivation of Science, Poona, India

Schieber A, Berardini N, Carle R (2003) Identification of flavonol and xanthone glycosides from mango (*Mangifera indica* L. cv. 'Tommy Atkins') peels by high-performance liquid chromatography-electrospray ionization mass spectrometry. J Agric Food Chem 51:5006–5011

Shalom L, Samuels S, Zur N et al (2012) Alternate bearing in citrus: changes in the expression of flowering control genes and in global gene expression in ON- versus. OFF-crop trees. PLoS One 7(10): e46930

Sharma N, Singh AK, Singh SK et al (2020) Comparative RNA sequencing based transcriptome profiling of regular bearing and alternate bearing mango (*Mangifera indica* L.) varieties reveals novel insights into the regulatory mechanisms underlying alternate bearing. Biotechnol Lett 42:1035–1050

Sherman A, Rubinstein M, Eshed R, Benita M, Ish-Shalom M et al (2015) Mango (*Mangifera indica* L.) germplasm diversity based on single nucleotide polymorphisms derived from the transcriptome. BMC Plant Biol 15:277

Singh NK, Dalal V, Batra K, Singh BK, Chitra G et al (2007) Single-copy genes define a conserved order between rice and wheat for understanding differences caused by duplication, deletion, and transposition of genes. Funct Integr Genom 7(1):17–35

Singh NK, Mahato AK, Sharma N, Gaikwad K, Srivastava M et al (2014) A draft genome of the king of fruit, mango (*Mangifera indica* L). In: Plant and animal genome XXII conference, San Diego, CA, P045, 13 Jan 2014

Singh R, Singh Y, Xalaxo S, Verulkar S, Yadav N et al (2016a) From QTL to variety-harnessing the benefits of QTLs for drought, flood and salt tolerance in mega rice varieties of India through a multi-institutional network. Plant Sci 242:278–287

Singh NK, Mahato AK, Jayaswal PK, Singh A, Singh S et al (2016b) Origin, diversity and genome sequence of mango (*Mangifera indica* L.). Indian J Hist Sci 51:355–368

Singh NK, Mahato AK, Jayaswal PK, Singh S, Singh A et al (2017) A reference genome of mango (*Mangifera indica*L. cv Amrapali). In: Plant and Animal Genome Conference XXV, San Diego, CA, W598, 16 Jan 2017

Singh NK, Mahato AK, Jayaswal PK, Singh S, Singh N et al (2018) A Reference genome assembly of the mango variety Amrapali (*Mangifera indica L.*). In: Plant and animal genome conference XXVI, San Diego, CA. W673, 15 Jan 2018

Singh NK, Mahato AK, Jayaswal PK, Singh S, Singh N et al (2019) A Reference assembly of the highly heterozygous genome of mango (*Mangifera indica* L. cv. Amrapali) In: Plant and animal genome conference XXVII, San Diego, CA, 13 Jan 2019

Singh RN, Majumder PK, Sharma DK, Mukherjee SK (1972) Some promising mango hybrids. Acta Hort 24:117–119

Sivankalyani V, Sela N, Feygenberg O, Zemach H, Maurer D, Alkan N (2016) Transcriptome dynamics in mango fruit peel reveals mechanisms of chilling stress. Front Plant Sci 7:1579

Sogi DS, Siddiq M, Greiby I, Dolan KD (2013) Total phenolics, antioxidant activity, and functional properties of 'Tommy Atkins' mango peel and kernel as affected by drying methods. Food Chem 141(3):2649–2655

Srivastava S, Rajesh K. Singh RK, Pathak G, Ridhi Goel R et al (2016) Comparative transcriptome analysis of unripe and mid-ripe fruit of *Mangifera indica*var."Dashehari" unravels ripening associated genes. Sci Rep 6:32556

Tafolla-Arellano JC et al (2017) Transcriptome analysis of mango (*Mangifera indica* L.) fruit epidermal peel to identify putative cuticle-associated genes. Sci Rep 7:46163

Velasco R, Zharkikh A, Troggio M, Cartwright DA, Cestaro A et al (2007) A high quality draft consensus sequence of the genome of a heterozygous grapevine variety. PLoS ONE 12:e1326

Velasco R, Velasco R, Zharkikh A, Affourtit J, Dhingra A, Cestaro A, Kalyanaraman A et al (2010) The genome of the domesticated apple (*Malus* × *domestica*Borkh.). Nat Genet 42:833

Wang P, Luo Y, Huang J et al (2020) The genome evolution and domestication of tropical fruit mango. Genome Biol 21:60

Warschefsky EJ, von Wettberg EJB (2019) Population genomic analysis of mango (*Mangifera indica*)

suggests a complex history of domestication. New Phytol 222:2023–2037

Wilcox J (1944) Some factors affecting apple yields in the Okanagan Valley: tree size, tree vigor, biennial bearing, and distance of planting. Sci Agric 25:189

Wood TE, Takebayashi N, Barker MS, Mayrose I et al (2009) The frequency of polyploid speciation in vascular plants. Proc Natl Acad Sci USA 106:13875–13879

Wu HX, Jia HM, Ma XW et al (2014) Transcriptome and proteomic analysis of mango (*Mangifera indica* L) fruits. J Proteomics 105:19

Xu Q, Chen L-L, Xiaoan RX, Chen D, Zhu A et al (2010) The draft genome of sweet orange (*Citrus sinensis*). Nat Genet 45:59–68

Yadav A, Jayaswal PK, Venkat Raman K et al (2020) Transcriptome analysis of flowering genes in mango (*Mangifera indica* L.) in relation to floral malformation. J Plant Biochem Biotechnol 29:193–212

Zhan RL, Yang SJ, Liu F, Zhao YL, Chang JM, He YB (2012) First report of *Fusarium mangiferae* causing mango malformation in China. Plant Dis 96:762

Zhang Y, Ou K, Huang G, Lu Y, Yang G, Pang X (2020) The complete chloroplast genome sequence of *Mangifera sylvatica* Roxb. (Anacardiaceae) and its phylogenetic analysis. Mitochondrial DNA Part B 5:738–739

The Mango Chloroplast Genome

11

Manish Srivastav, Sanjay K. Singh, and Nimisha Sharma

Abstract

Chloroplast is an important cell organelle that performs photosynsthesis and helps in several cellular processes. Metabolic compounds synthesized in chloroplasts help in signal transfer and play a seminal role in alleviating biotic and abiotic stresses. Although chloroplast is smaller in size and gene number compared to nuclear genome, it is equally important for evolution and molecular ecological studies. NGS sequencing technologies made it easier to characterize chloroplast genome in mango. Chloroplast genome of 'Langra' cultivar was first sequenced by using Sanger and GS-FLX systems which provided insight in new functional genes in mango. Thereafter, chloroplast genome of 'GuiFei' cultivar was sequenced using Illumina Hiseq-2000 platform, while in wild mango (*Mangifera sylvatica* Roxb) it was sequenced using Illumina's Hiseq 2500 which are available in the public domain. Genes identified from chloroplast genome are having significant role in photosynthesis, cell regulatory mechanisms, resistance for herbicides, biotic and abiotic stresses. The highly conserved nature of chloroplast genome sequences opened new avenues in phylogenetic and evolutionary studies on mango, which thus offers enormous opportunities for refining our insight into its origin, diversity, evolutionary relationship, and species identification.

11.1 Introduction

Mango (*Mangifera indica* L.) (Family, Anacardiaceae; (n)genome, ~ 439 Mbp; (cp)genome, 1,57,837 bp) is considered as difficult plant species to handle in improvement programs due to certain inherent characteristics including long juvenile phase, high level of heterozygosity, heavy fruit drop, and lack of knowledge about the genetics of important agronomic traits. Breeders have been relying on traditional breeding approaches for understanding the genetics of mango and utilization of that knowledge for further improvement of mango. Despite, its huge economic significance, the researchers paid very late and also little attention to the augmentation of genome resources in mango as compared to other fruit crops of temperate origin. This may be due to the fact that sequencing technologies and their cost was very high initially. Next generation sequencing (NGS) technologies offered rapid and cost-effective generation of sequencing data by an

M. Srivastav (✉) · S. K. Singh · N. Sharma
Division of Fruits and Horticultural Technology, ICAR- Indian Agricultural Research Institute, New Delhi 110012, India
e-mail: msrivastav@iari.res.in

S. K. Singh
e-mail: sanjaydr2@gmail.com

N. Sharma
e-mail: nims17sharma@gmail.com

C. Kole (ed.), *The Mango Genome*, Compendium of Plant Genomes,
https://doi.org/10.1007/978-3-030-47829-2_11

order of magnitude for a variety of applications. Indeed, sequencing of majority of fruit species has involved utilization of NGS techniques. Apart from using NGS for whole genome sequencing (Lam et al. 2012), these technologies were also used for whole transcriptome shotgun sequencing also known as RNA sequencing (Wang et al. 2009), whole-exome sequencing (Rabbini et al. 2014), targeted or candidate gene sequencing (Mardis et al. 2009; Kulski et al. 2014; Leo et al. 2015), and methylation sequencing (Pelizzola and Ecker 2011).

Cell organelles genome sequencing was once a tedious job; however, it is now a very commonly adopted procedure (Smith 2016). Accumulation of more number of sequences means more information for relevant studies and better understanding of genome evolution. Although chloroplast and mitochondrial genomes are smaller in size and gene number, They are equally important for evolution and molecular ecology studies. Chloroplast is the site of photosynthesis and mitochondria is powerhouse of cell, and both have their own DNA. Fruit plants cell commonly has one copy of nuclear and multiple copies of organelle genomes (Wang et al. 2018). Hence, deciphering the DNA sequences from mitochondria and chloroplast serve as primary source of data for molecular systematic, phylogeographic, and population genetic studies on crops like mango to unravel the basic information which is still not clear. It is more pertinent since out of the 70 plus *Mangifera* species hardly one dozen is under cultivation globally, while many are under the constant threat of extinction in their regions of origin due to the wrath of natural calamities, climate change, etc. before they are actually collected, studied, documented, and conserved for immediate or long term use by the mankind.

With the advent of cost and time-efficient NGS techniques and bioinformatics (Metzker 2010; Mariac et al. 2014; Tang et al. 2014), mango genome sequencing got momentum and the first report of draft nuclear mango genome appeared in 2014 during Animal and Plant Genomics Conference at Sandiago, USA (Singh et al. 2014). Similarly, genome sequencing of mango chloroplast was also accomplished almost during the same period (Azim et al. 2014). Whereas, complete mitochondrial and ribosomal genome sequences have not been attempted in mango. In mango, nuclear and chloroplast genomes have been sequenced (Table 11.1 and Fig. 11.1). The quantum of genome data generated in the recent past would be highly useful in genomic assisted breeding and deciphering the genetics of important agronomic traits. In this chapter, information pertaining to mango chloroplast genome has been discussed.

11.2 Chloroplast Genome

Chloroplast is the site of photosynthesis process in which light energy is converted into chemical energy for the purpose of synthesis of organic compounds in plants. Chloroplast being the most important organelle of cell also helps in several biochemical processes (Daniell et al. 2016). It is also suggested by several researchers that metabolic compounds synthesized in chloroplasts are helping in signal transfer between different parts

Table 11.1 Details of mango chloroplast genome

Cp genome	Size (bp)/Gene	Sequencing method	Reference
Mangifera sylvatica Roxb.	Size-158,063 bp Total genes-112	llumina sequencing	Zhang et al. (2020)
Mangifera indica variety 'GuiFei'	Size-157,837 bp Total genes- 171	llumina Hiseq 2000 Platform	Zhao et al. (2019)
Mangifera indica L.	Size-157,780 bp Total genes-112	Illumina HiSeq 2000 system	Jo et al. (2017)
Mangifera indica variety 'Langra'	Size-151,173 bp Total genes-139	Sanger and next generation sequencing	Azim et al. (2014)

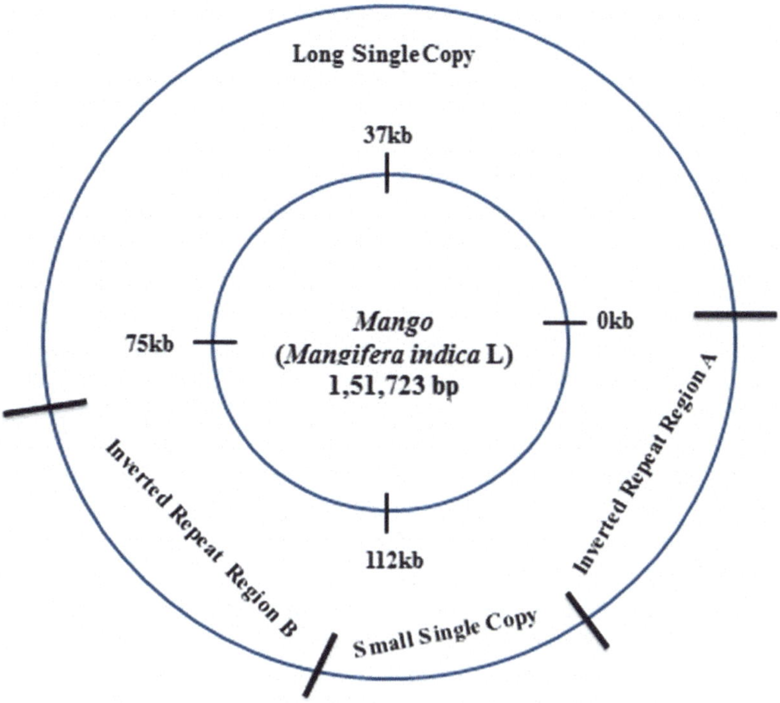

Fig. 11.1 Chloroplast genome map of *Mangifera indica*. (*Source* Azim et al. 2014)

of plant and play a role in stresses caused by abiotic and biotic factors (Daniell et al. 2016). Chloroplast evolved from endosymbiosis of a cyanobacterium and during evolution genes of the ancestral chloroplasts have been transferred into the cell nucleus. However, proteins essential for the photosynthesis phenomenon remained with the chloroplast genome. The chloroplast genome confers several advantages over nuclear genome in phylogenetic and population studies. These includes (i) highly conserved genes and low mutation rate, (ii) small size, (iii) haploid number, (iv) high copy number, (v) variation in intergenic regions, (vi) lacks recombination, and (vi) principally maternal inheritance (Schaal et al. 1998). Tobacco (*Nicotiana tabacum*) (Ohyama et al. 1986) and liverwort (*Marchantia polymorpha*) were the first plants of which chloroplast genome was sequenced (Shinozaki et al. 1986). Since then, the chloroplasts DNA of more than 3,000 plant species have been deciphered. Approximately 1,193 chloroplast genomes of land plants are available in the National Center for Biotechnology Information (NCBI) (http://www.ncbi.nlm.nih.gov/genome/browse/) (Wang et al. 2018).

Complete chloroplast genome information has enriched our knowledge of plant cell biology and significantly contributed in phylogenetic studies. Information pertaining to variation in chloroplast genome within a species and between species in terms of sequences, structure, and organization proved highly valuable for understanding the molecular mechanism involved in adaptation of economically important species to different climates (Wambugu et al. 2015; Brozynska et al. 2016). Variations in chloroplast genomes have also helped in understanding the specific gene transfer from chloroplast to mitochondrial or nuclear genomes, and deciphered the relationship among three valuable genomes in plants. Since, the chloroplast gene order is comparatively conserved, small in size, non-recombinant type, maternally inherited, and haploid. Thus, it can overcome the disadvantages encountered with nuclear DNA markers analysis. Conservative nature of chloroplast genome offers advantage in parentage and taxonomic studies (Cheng et al. 2005).

The first draft chloroplast genome sequence of mango cultivar 'Langra' was obtained using a combination of Sanger and GS-FLX NGS System (Roche Inc., USA) and provided insight in the identification of new functional genes in mango (Azim et al. 2014). The draft mango chloroplast genome size was 1,51,173 bp with inverted repeats of 27,093 bp. Total 139 chloroplast genes have been detected of which 119 were single-copy genes and 20 were duplicated genes. There were 29 distinct *tRNA* genes, four *rRNA* genes, and seven of which were duplicated in the inverted repeats. Azim et al. (2014) described different genes found in mango cpDNA and their involvement in different cellular processes. Sequence analysis of chloroplast genomes confirmed *Citrus sinensis* (160,129 bp) as the closest fruit species to mango.

Interestingly, *M. indica* had *infA* gene coding for translation initiation factor was absent in *Citrus* (Bausher et al. 2006). Chloroplast genome map was constructed and a phylogenetic tree of complete cpDNA sequences was accomplished. However, complete sequence could not be obtained as mango cpDNA sequence of 151,173 bp contained 30 gaps, and it was thought that only 95% of genome sequences have been finally obtained. It was also noted that inverted repeats of mango chloroplast was bigger than several previously reported species. The length of mango cpDNA inverted region was 97 bp larger than *C. sinensis* (Bausher et al. 2006). This supported the fact that expansion and contraction of inverted region of chloroplast DNA as source of length variation. The phylogenetic tree indicated grouping of mango chloroplast sequences with *C. sinensis*, *G. hirsutum* and *V. vinifera*. Along with two large inverted repeats, i.e., IRA and IRB, large number of small repeats have also been observed and these repeats occur as part of intergenic spacers as well as genes (Azim et al. 2014). This draft sequence of mango chloroplast genome with accession number FJ212316 was submitted in GenBank.

In plant evolution, intracellular gene transfer is very common and continuous process. Chloroplast genome generally has resistance for foreign DNA incorporation, while nuclear and mitochondrial genomes are known to integrate foreign DNA. Jo et al. (2017) completed chloroplast sequences of plant species including mango, fig, guava, pomegranate, cashew, litchi, cinnamon, okra, quinoa, and basil. They observed that most chloroplast have a set of conserved coding sequences with size stability and configuration. The phylogenetic tree indicated grouping of mango with *Anacardium excelsum*, *A. carymbosum*, *A. occidentale*, *A. humile* and *A. nanum*.

Chloroplast genome of mango cultivar 'Gui-Fei' was reported by Chinese researchers using NGS platform Illumina Hiseq-2000 (Zhao et al. 2019). The complete genome sequence was 1,57,837 bp long comprising of inverted repeats pair with small single-copy region of 43,323 bp and large single-copy region of 59,717 bp. Genes (171) were predicted having protein coding (118), *rRNA* (8) and *tRNA* (45) genes. Notably, the organization of genome, genes, and their relative positions were different compared to other fruit species. Interestingly, 32 genes had more copies, 18 genes were found to be duplicated in inverted region and 22 genes in single-copy region. Phylogenetic analysis of 41 (cp)genomes sequences of different species clearly revealed close relationship of mango with *C. sinensis*, *C. paltvmamma* and *C. aurantifolia*. This confirmed the fact that mango (cp)genome is ideal resource for generation of DNA markers for taxonomic and evolutionary research and provides valuable information on highly variable markers for population studies (Zhao et al. 2019).

In continuation, another research group from China reported complete (cp)genome of wild mango (*Mangifera sylvatica* Roxb) for the first time (Zhang et al. 2020). *Mangifera sylvatica*, popularly known as *Ban-am* or *Laksmi-am* in Assam is having close affinity with common mango (*Mangifera indica*). *Mangifera sylvatica* is having medicinal and nutrition values (Akhter et al. 2016). Chloroplast genome was sequenced using Illumina's Hiseq 2500. The chloroplast genome of *M. sylvatica* was found to be comparably smaller (158,063 bp) in length than what was reported by the same group in 'GuiFei' cultivar. The large single copy was of 87,008 bp and had 82 genes. However, small single copy was of 18,340 bp and contained 13 genes. The two inverted repeats were of 26,379 bp.

The annotation of 78 functional genes ascertained its similarity with *M. indica* chloroplast (Azim et al. 2014; Zhao et al. 2019).

Among the genes coding for protein, atpF, PetB, *PetD*, *ndhA*, *ndhB*, *ropC1*, *rpl2*, rpl16, and *rps16* had one intron, however, *clpP*, rps12, and *ycf3* had two introns. *rRNAs* were only restricted to inverted region while *tRNA* genes were located in all the four regions. Comparing chloroplast genome of *M. sylvatica* and 29 other species from 10 different families revealed the closest relationship of this wild mango with common mango (*M. indica*) and other species of Anacardiaceae family, i.e., *Spondias bahiensis*, *Spondias tuberose,* and *Rhus chinensis*.

11.3 Use of Chloroplast Genome Sequences

The recognition that chloroplast has its own unique DNA has led to intensive investigation about organization of chloroplast genomes, sequence properties, and modes of expression of constituent genes. Genes having role in photosynthesis, cell regulatory mechanism, resistance for herbicides and biotic, and abiotic stresses have been decoded. The highly conserved cpDNA made it possible to use them for phylogenetic, evolutionary, and cultivar identification studies in mango. Xing-Hua et al. (2007) reported high degree of chloroplast DNA polymorphism and suggested cp base DNA markers highly useful in taxonomic studies, genetic diversity analysis, hybridity testing, and analysis of cytoplasm inheritance in mango (Xin-hua et al. 2007). Similarly, variations in *rpl20-rps12* intergenic spacer of chloroplast genome have been successfully used for taxonomic studies and identification of cultivars in mango (Khan and Azim 2011). They compared variations in *atp-rbcL* and *rpl20-rps12* spacers of 12 mango cultivars and found that *atp-rbcL* region had no variation, while *rpl20-rps12* region variation at both inter-species and intra-varietal levels. Noncoding regions which are the hot spots for mutations proved to be an important resource for analyzing intra-and-inter specific variations in sequences (Intrieri et al. 2007).

The usefulness of cpDNA intergenic spacer *trnL-F* sequences in phylogenetic relationships among *Mangifera* species was demonstrated by Fitmawati and Hartana (2010). They compared *trnL-F* sequences of *M. laurina* with related species and found *Mangifera* sp. Hiku (*Mangga hiku*) as the main cultivar in the clade as progenitor of *M. laurina*. Similarly, *MaturaseK (matK)* gene of chloroplast DNA has served as appropriate candidate for DNA barcoding. Using this DNA marker, phylogenetic relationship among 19 *Mangifera* species collected from Indonesia and Thailand were analyzed. This study demonstrated that the *matK* as a barcode classified the *Mangifera* into three major groups. Furthermore, *matK* barcode identified species originated in Thailand, and it was found that *matK* gene in *M. laurina* and *M. macrocarpa* were different in specimens collected from Indonesia and Thailand (Hidyat et al. 2011). Identification of mango cultivar 'Totapuri' at molecular level was attempted using chloroplast *trnL-F* region by Jagarlamudi et al. (2011). They found that these gene sequences are highly conserved in nature and remain same in any part of the plant. Based on sequence of dominant DGGE band in tested leaves, cent percent similarity was noted with chloroplast ITS sequences of *M. indica.*

Dinesh et al. (2015) studied phylogenetic relationships among *M. andamanica, M. camptosperma*, *M. indica*, *M.* griffithii, and *M. odorata* using markers designed from chloroplast genome, viz., *petB-petD*, *rps16*, *trnL-trnF* and nuclear markers. The chloroplast and nuclear markers-based phylogenetic analysis revealed close affinity of common mango *M. indica* L. with *M. camptosperma* and *M. griffithii* of *Mangifera* sub-genus. However, *M. odorata* of *Limus* sub-genus separately grouped with *M. andamanica.*

Harahap et al. (2017) studied phylogenic relationships among *Mangifera* species in Northern Sumatra using *trnLF* intergenic spacer. They categorized *Mangifera* species into two clades, first consisting of *M. odorata* (Aceh Tengah, Aceh), *M. odorata* (Rantau Prapat,

Table 11.2 Important chloroplast genes and their functions

Gene(s)	Function	Reference(s)
psaA, psaB, psaC, psaJ, psaI, psbA, psbB, psbC, psbD, psbE, psbF, psbH, psbI, psbJ, psbL, psbM, psbN, psbT, cytb6f, ATP synthases, rbcL, NAD(P)H, Ycf3a, ycf4	Photosynthetic function	Bansal and Saha (2012) Azim et al. (2014)
tRNA genes: trnaH, trnK *rRNA genes:rrn16, rrn5, rpoA, rpoB*	Regulatory role in cell metabolism	Bansal and Saha (2012)
rps2, rps3, rps4, rps7, rps8, rps11, rps12, rps14, rps15, rps16, rps18, rps19, rpl14, rpl16, rpl20, rpl22, rpl23, rpl32, rpl33, rpl36	Ribosomal proteins	Azim et al. (2014)
trnH-psbA, rps16-trnQ, rpoB-trnC, rps4-trnT-trnL, ndhF, ndhF-rpl32-trnL, ycf1a, ycf1b	Hypervariable chloroplast DNA markers	Li et al. (2018)
infA, rpl22, ndh	Phylogenetic analyses and evolutionary studies	Jansen et al. (2007)
matK, ycfs	Phylogenetic analyses and evolutionary studies	Carbonell-Caballero et al. (2015)
infA	Initiates translation	Millen et al. (2001)
ndhF	Stress acclimation	Peng et al. (2011)
PTA (Pinellia ternata agglutinin)	Multiple resistances against bacteria, insects, and virus pathogens	Jin et al. (2012)
CP4EPSPS	Glyphosate resistance	Ye et al. (2001)
rpl20-rps12	Cultivar identification	Khan and Azim (2011)
ndhA, clpP	Cultivar identification	Chung et al. (2019)
*trnL, trn*F	Hybridity testing	Muthukumar et al. (2018)
cemA	Gene for inorganic carbon uptake	Azim et al. (2014)

Labuhanbatu, N. Sumatra), *M. laurina* (Takengon, Aceh Tengah, Aceh), *M. laurina* (Medan, N. Sumatra), *M. indica*, *M. quadrifida*, *M. zeylanica*, and second clade consisted of *M. foetida* (Aceh Barat, Aceh) and *M. foetida* (Medan, N. Sumatra). The intergenic spacer *trnL-F* with less variation in sequences ascertained conserved nature and low rate of evolution in *Mangifera*. Muthukumar et al. (2018) affirmed hybridity of mango cultivars using chloroplast genes *trn*L and *trn*F localized in large single-copy region. Mango is known to have diverse varieties for different traits owing to their evolution by open pollination and natural selection and phylogenetic relationship among cultivars and ascertainment of hybridity of open-pollinated progenies of mango is very important for trait-specific breeding. They also postulated that inheritance of cp genes was not strictly maternal but could be paternal or bi-parental in nature (Muthukumar et al. 2018). A list of chloroplast genes and their functions are given in Table 11.2.

11.4 Conclusion

Advancements in sequencing technologies and simultaneous development of bioinformatics tools facilitated rapid and cost-effective characterization of complex cell organelles genomes in different plant species in short span of time and in a cost-effective way. The complete chloroplast DNA in mango and its wild relatives have been sequenced. Complete chloroplast genome of *Mangifera indica* varieties 'Langra' and 'GuiFei', and wild mango *Mangifera sylvatica* are now available in public domain. In fruit tree like mango, cp genome sequencing would offer unprecedented opportunities for refining our insights into its origin, diversity, evolutionary studies, cultivar and species identification, multigene engineering, transgenic expression and genome engineering for conferring biotic and abiotic stress tolerance, biosynthesis of bioactive compounds and enhancing nutritional value of mango.

References

Akhter S, McDonald MA, Marriott R (2016) *Mangifera sylvatica* (Wild Mango): a new cocoa butter alternative. Sci Rep 6(1): 32050. https://doi.org/10.1038/srep32050

Azim MK, Khan IA, Zhang Y (2014) Characterization of mango (*Mangifera indica* L.) transcriptome and chloroplast genome. Plant Mol Biol 85(1–2):193–208. https://doi.org/10.1007/s11103-014-0179-8

Bansal KC, Saha D (2012) Chloroplast genomics and genetic engineering for crop improvement. Agric Res 1 (1):53–66. https://doi.org/10.1007/s40003-011-0010-6

Bausher MG, Singh ND, Lee SB, Jansen RK, Daniell H (2006) The complete chloroplast genome sequence of *Citrus sinensis* (L.) Osbeck var 'Ridge Pineapple': organization and phylogenetic relationships to other angiosperms. BMC Plant Biol 6:21. https://doi.org/10.1186/1471-2229-6-21

Brozynska M, Furtado A, Henry RJ (2016) Genomics of crop wild relatives: expanding the gene pool for crop improvement. Plant Biotech J 14:1070–1085. https://doi.org/10.1111/pbi.12454

Carbonell-Caballero J, Alonso R, Ibañez V, Terol J, Talon M, Dopazo J (2015) A phylogenetic analysis of 34 chloroplast genomes elucidates the relationships between wild and domestic species within the genus *Citrus*. Mol Biol Evol 32:2015–2035

Cheng YJ, Carmen-de-Vicente M, Meng HJ, Guo WW, Tao NG, Deng XX (2005) A set of primers for analyzing chloroplast DNA diversity in Citrus and related genera. Tree Physiol 25:661–672

Chung HY, Won SY, Kim YK, Kim JS (2019) Development of the chloroplast genome based InDel markers in Niitaka (*Pyrus pyrifolia*) and its application. Plant Biotechnol Rep 13:51–61. https://doi.org/10.1007/s11816-018-00513-0

Daniell H, Lin C, Yu M, Chang W (2016) Chloroplast genomes: diversity, evolution, and applications in genetic engineering. Genome Biol 17:134. https://doi.org/10.1186/s13059-016-1004-2

Dinesh MR, Ravishankar KV, Nischita P, Sandya BS, Padmakar B, Ganeshan S, Chithiraichelvan R, Sharma TVRS (2015) Exploration, characterization and phylogenetic studies in wild *Mangifera indica* relatives. Am J Plant Sci 6:2151–2160. https://doi.org/10.4236/ajps.2015.613217

Fitmawati F, Hartana A (2010) Phylogenetic study of *Mangifera laurina* and its related species using cpDNA trnL-F spacer markers. Hayati J Biosci 17 (1):9–14. https://doi.org/10.4308/hjb.17.1.9

Harahap SP, Fitmawati Sofiyanti N (2017) Phylogenetic analysis of mango (*Mangifera*) in Northern Sumatra based on gene sequences of cpDNA *trnL-F* intergenic spacer. Biodiversitas 18:715–719. https://doi.org/10.13057/biodiv/d180239

Hidayat T, Pancoro A, Kusumawaty D, Eiadthong W (2011) Development *matK* gene as DNA barcode to assess evolutionary relationship of important tropical forest tree genus *Mangifera* (Anacardiaceae) in Indonesia and Thailand. J Teknologi 59:17–20

Intrieri MC, Muleo R, Buiatti M (2007) Chloroplast DNA polymorphisms as molecular markers to identify cultivars of *Olea europaea* L. J Hort Sci Biotechnol 82:109–113. https://doi.org/10.1080/14620316.2007.11512206

Jagarlamudi S, Rosaiah G, Kurapati RK, Pinnamaneni R (2011) Molecular identification of Mango, *Mangifera indica* L. *var. Totupura*. Bioinformation 5(10):405–409. https://doi.org/10.6026/97320630005405

Jansen RK, Cai Z, Raubeson LA, Daniell H, Leebens-Mack J, Müller KF (2007) Analysis of 81 genes from 64 plastid genomes resolves relationships in angiosperms and identifies genome scale evolutionary patterns. Proc Natl Acad Sci USA 104:19369–19374

Jin S, Zhang X, Daniell H (2012) Pinellia ternata agglutinin expression in chloroplasts confers broad spectrum resistance against aphid, whitefly, Lepidopteran insects, bacterial and viral pathogens. Plant Biotechnol J 10:313–327

Jo S, Kim HW, Kim YK, Sohn JY, Cheon SH, Kim KJ (2017) The complete plastome sequences of *Mangifera indica* L. (Anacardiaceae). Mitochondrial DNA Part B 2(2):698–700

Khan IA, Azim MK (2011) Variations in intergenic spacer *rpl20-rps12* of mango (*Mangifera indica*) chloroplast DNA: implications for cultivar identification and phylogenetic analysis. Plant Sys Evol 292:249–255. https://doi.org/10.1007/s00606-011-0424-4

Kulski JK, Suzuki S, Ozaki Y, Mitsunaga S, Inoko H, Shiina T (2014) Phase HLA genotyping by next generation sequencing—A comparison between two massively parallel sequencing bench-top systems, the Roche GS Junior and Ion Torrent PGM. In: Xi Y (ed) HLA and Associated Important Diseases. Croatia: Intech, pp 141–81. https://doi.org/10.5772/57556

Lam HY, Clark MJ, Chen R (2012) Performance comparison of whole-genome sequencing platforms. Nat Biotechnol 30:78–83. https://doi.org/10.1038/nbt.2065

Leo VC, Morgan NV, Bern D (2015) Use of next-generation sequencing and candidate gene analysis to identify underlying defects in patients with inherited platelet function disorders. J Thromb Haemost 13:643–650. https://doi.org/10.1111/jth.12836

Li W, Liu Y, Yang Y, Xie X, Lu Y, Yang Z, Jin X, Dong W, Suo Z (2018) Interspecific chloroplast genome sequence diversity and genomic resources in *Diospyros*. BMC Plant Biol 18:210. https://doi.org/10.1186/s12870-018-1421-3

Mardis ER, Wilson RK (2009) Cancer genome sequencing: A review. Hum Mol Genet 18(R2):R163–R168. https://doi.org/10.1093/hmg/ddp396

Mariac C, Scarcelli N, Pouzadou J, Barnaud A, Billot C, Faye A, Kougbeadjo A, Maillol V, Martin G, Sabot F, Santoni S, Vigouroux Y, Couvreur TLP (2014) Cost-effective enrichment hybridization capture of chloroplast genomes at deep multiplexing levels for population genetics and phylogeography studies. Mol Ecol Resour 14:1103–1113

Metzker ML (2010) Sequencing technologies—The next generation. Nat Rev Genet 11:31–46. https://doi.org/10.1038/nrg2626

Millen RS, Olmstead RG, Adams KL, Palmer JD, Lao NT, Heggie L (2001) Many parallel losses of *inf*A from chloroplast DNA during angiosperm evolution with multiple independent transfers to the nucleus. Plant Cell 13:645–658

Muthukumar M, Bajpai A, Rajan S (2018) Chloroplast genes reveal hybridity in mango (*Mangifera indica* L.). J Appl Hort 20(1):55–59

Ohyama K, Fukuzawa H, Kohchi T, Shirai H, Sano T, Sano S, Umesono K, Shiki Y, Takeuchi M, Chang Z, Aota S, Inokuchi H, Ozeki H (1986) Chloroplast gene organization deduced from complete sequence of liverwort *Marchantia polymorpha* chloroplast DNA. Nature 322:572–574

Pelizzola M, Ecker JR (2011) The DNA methylome. FEBS Lett 585:1994–2000. https://doi.org/10.1016/j.febslet.2010.10.061

Peng L, Yamamoto H, Shikanai T (2011) Structure and biogenesis of the chloroplast NAD(P)H dehydrogenase complex. Biochem Biophys Acta 1807:945–953

Rabbini B, Tekin M, Mahdieh N (2014) The promise of whole-exome sequencing in medical genetics. J Hum Genet 59:5–15. https://doi.org/10.1038/jhg.2013.114

Schaal BA, Hayworth DA, Olsen KM, Rauscher JT, Smith WA (1998) Phylogeographic studies in plants: problems and prospects. Mol Ecol 7(4):465–474. https://doi.org/10.1046/j.1365-294x.1998.00318.x

Shinozaki K, Ohme M, Tanaka M, Wakasugi T, Hayashida N, Matsubayashi T (1986) The complete nucleotide sequence of the tobacco chloroplast genome: its gene organization and expression. EMBO J 5:2043–2049

Singh NK, Mahato AK, Sharma N, Gaikwad K, Srivastava M, Tiwari K, Dogra V, Rawal HC, Jayaswal P, Singh A, Rai V, Mithra SVA, Bajpai A, Dinesh MR, Ravishankar KV, Rajan S, Rai A, Singh AK, Sharma TR (2014) A draft genome of the king of fruit, mango (Mangifera indica L.). Paper presented in Plant and Animal Genomic XXII conference January 11–15, 2014, San Diego, USA

Smith DR (2016) The past, present and future of mitochondrial genomics: have we sequenced enough mtDNAs? Brief Funct Genomics 15:47–54

Tang M, Tan M, Meng G, Yang S, Su X, Liu S, Song W, Li Y, Wu Q, Zhang A, Zhou X (2014) Multiplex sequencing of pooled mitochondrial genomes- a crucial step toward biodiversity analysis using mito-metagenomics. Nucleic Acids Res 42:e166

Wambugu P, Brozynska M, Furtado A, Waters D, Henry R (2015) Relationships of wild and domesticated rices (*Oryza* AA genome species) based upon whole chloroplast genome sequences. Sci Rep 5:13957

Wang X, Cheng F, Rohlsen D, Bi C, Wang C, Xu Y, Wei S, Ye Q, Yin T, Ye N (2018) Organellar genome assembly methods and comparative analysis of horticultural plants Hort Res 5:3. https://doi.org/10.1038/s41438-017-0002-1

Wang Z, Gerstein M, Snyder M (2009) RNA-Seq: a revolutionary tool for transcriptomics. Nat Rev Genet 10:57–63. https://doi.org/10.1038/nrg2484

Xin-hua HE, Yong-ze G, Yang-mi L, Sh-jin O (2007) Assessment of the genetic relationship and diversity of mango and its relatives by cpISSR marker. Agril Sci China 6(2):137–142

Ye GN, Hajdukiewicz PT, Broyles D, Rodriguez D, Xu CW, Nehra N, Staub JM (2001) Plastid- expressed 5-enolpyruvyl shikimate-3-phosphate synthase genes provide high level glyphosate tolerance in tobacco. Plant J 25:261–270

Zhang Y, Ou K, Huang G, Lu Y, Yang G, Pang X (2020) The complete chloroplast genome sequence of *Mangifera sylvatica* Roxb. (Anacardiaceae) and its phylogenetic analysis, Mitochondrial DNA Part B 5(1):738–739. https://doi.org/10.1080/23802359.2020.171 5286

Zhao Z, Gao A, Huang J, Luo R (2019) The complete sequence of chloroplast genome from mango (*Mangifera indica* var 'GuiFei'). Mitochondrial DNA Part B 4(1):1916–1917. https://doi.org/10.1080/23802359.2019.1606678

Mango Functional Genomics

12

Anju Bajpai, M. Muthukumar,
and Sandeep Kumar

Abstract

Mango, one of the most relished fruits globally, has attracted the attention of the global research community for breeding new varieties and hybrids utilizing the naturally existing diverse genetic resources. Functional genomics approaches are useful in mango for expediting the trait-specific varietal improvement through precision breeding programs. This chapter presents the status of functional genomics resources in mango that are available in public domain and their applications in trait mapping as well as in breeding for specific target traits such as enhanced shelf-life, attractive peel color and fruit quality traits, regularity in bearing, malformation and abiotic stress tolerance. The methods, tools, and techniques used in functional genomics, such as transcriptomics, proteomics, and metabolomics, are detailed in this chapter with examples. The traditional or conventional approaches in transcriptional and translational levels of expression profiling are discussed in this chapter with illustrations. Till date, 19 transcriptome data sets based on RNA sequencing have been deposited in the NCBI sequence repository which includes whole transcriptome analysis data of different tissues such as flowers, panicles, leaves, fruit pulp, peel, and roots pertaining to different commercial cultivars from various mango functional genomics working groups. Despite non-availability of mango genome sequence in public domain, transcriptomics have provided leads in marker development and also for augmentation of genomic resources. Classical examples and illustrations on the functional genomics studies in trait mapping have also been discussed in detail. Various aspects of functional analysis of genes in terms of full-length cDNAs, EST resources, identification of key candidate genes for specific traits from transcriptome, proteome and metabolic approaches, their cloning and expression analysis, and genetic transformation have also been summarized. This chapter also highlights on the challenges, possible solutions, and future prospects of functional genomics in mango.

12.1 Introduction

Mango is an important fruit for the Indian subcontinent supporting livelihood and nutrition for many small and marginal farmers besides being a part of cultural heritage. Unfortunately, mango breeding and crop improvement are impeded by

A. Bajpai (✉) · M. Muthukumar · S. Kumar
Division of Crop Improvement and Biotechnology, ICAR-Central Institute for Subtropical Horticulture, Lucknow 226101, India
e-mail: anju.bajpai@gmail.com

C. Kole (ed.), *The Mango Genome*, Compendium of Plant Genomes,
https://doi.org/10.1007/978-3-030-47829-2_12

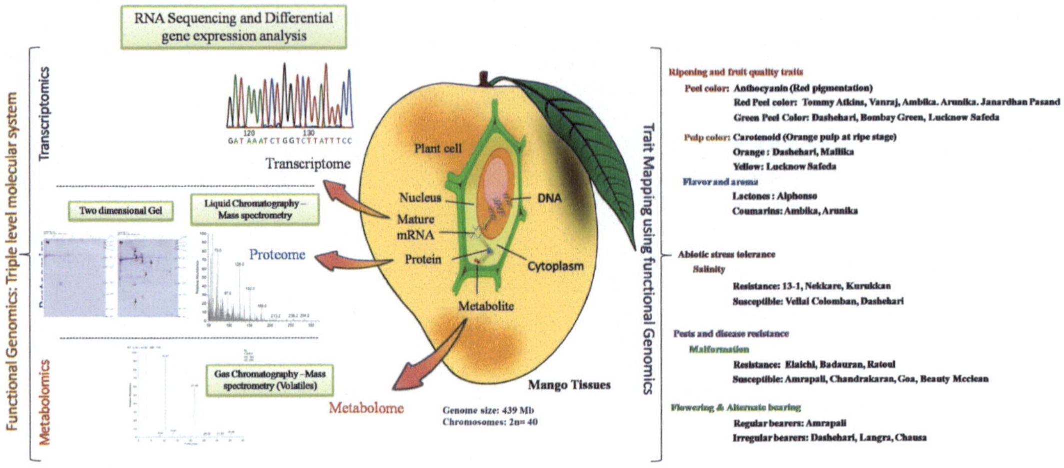

Fig. 12.1 Overview of functional genomics based approaches and important target traits of mango

long juvenility, high heterozygosity and poorly understood genetics for many important traits (Bajpai and Muthukumar 2019). Mango functional genomics offers scope for unraveling genetic network regulating complex traits, viz, attractive fruit with delayed ripening, canopy characteristics, improved rootstock, biotic and abiotic stress resistance in important cultivars. The molecular information of mango genome would help improvise breeding strategies toward the development of elite varieties with improved agronomic attributes through supplementing traditional methods of genetic improvement. Functional genomics approach encompasses whole genome gene expression profiling through transcriptomes to assess transcriptional level changes and whole genome proteome profiling to assess the translational level changes associated with a specific trait. In mango, transcriptome analysis especially RNA sequencing has been widely explored for understanding several complex traits. Limited work has been reported with regard to proteomics and metabolomics in mango with scarce reports on characterization for ripening related enzymes and flavor, and aroma compounds only. The major research efforts by various working groups are focused on understanding the genetic and molecular basis of important horticultural traits such as ripening and fruit quality traits, pulp and peel color, flowering and alternate bearing, malformation resistance, salinity and chilling injury (Fig. 12.1).

12.2 Functional Genomics Resources

Functional genomics resources pertain to the complimentary cDNA sequence information (related to a specific mature mRNA) or protein sequences that are submitted and stored in the public domain database either as a partial or complete sequence and are ascribed to genes with specific functions. These include either gene coding sequences often called expressed sequence tags (ESTs) or amino acid sequences that relate to specific peptides or protein sequences. All these sequences should be annotated and featured with specific exon predictions for mRNA or motif/domain predictions for a protein sequence. In mango, as of April 2020 NCBI data base holds 88,412 nucleotide sequences with 1,231 molecular types representing either genomic DNA or RNA and 2,035 sequences accounting for mRNA contributed by 88.411 gene bank source and one from reference sequence. Out of 88,412 sequences, about 86,722 sequences are nucleotide sequences and 1,690 are EST sequences. Although the draft genome has been reported, unavailability on the public domain limits the utilization of this genomic resource for functional genomics. Yet, considerable amount of SRA submissions pertaining to RNA sequencing datasets of different tissues from various commercial and popular cultivars or varieties such as Tommy Atkins, Shelly, Zill, Kent, Keitt, Zill, Amrapali, Chausa, and Dashehari are available in NCBI database under 19 biosamples as given in Table 12.1. These transcriptome resources which are available in the public domain have allowed gene discovery and gene expression studies associated with complex traits of economic importance like ripening, shelf-life, pulp traits, etc. These resources also have potentials for utilization in mining genes associated with specific traits in crop improvement programs. Similarly with respect to protein resources, Uniprot database houses information on 1,197 protein sequences of *Mangifera indica*, while the structural

Table 12.1 Transcriptome resources in public domain NCBI database (Information up to April 2020)

S. no.	Bio samples	Sequencing platform	SRA accession number (in relative order)	Institute
1	Mango-Leaves (young:1 monthand mature: 5 years)	Illumina HiSeq 2500	SRX5770298 SRX5770299	The University of Queensland, Brisbane, QLD, Australia
2	Mango—A mixture of roots, young and mature leaves, stems, axillary buds, flowers, and fruits	Illumina NextSeq 550	SRX5707082 SRX5707081 SRX5707080 SRX5707079	School of Biological Sciences, The University of Queensland, Brisbane, QLD, Australia
3	Dashehari—Pulp	Illumina MiSeq	SRX3598947	ICAR-CISH, Lucknow, India
4	Dashehari—Pulp	Illumina HiSeq 2000	SRX1687174	CSIR-NBRI, Lucknow, India
5	Chausa—Leaves	Illumina NextSeq 500	SRX1672529	ICAR-CISH, Lucknow, India
6	Amrapali—Leaves	Illumina NextSeq 500	SRX1603794	ICAR-CISH, Lucknow, India
7	Renong and Uno-bagged and unbagged fruits	Illumina HiSeq 2000	SRX1645280 SRX1641640 SRX1641636 SRX1641632	South Subtropical Crops Research Institute, China
8	Keitt peel Fruit stored—for 14 days at 12 and 5 °C —for 7 days at 12 and 5 °C —for 2 days at 12 and 5 °C At time of harvest	Illumina HiSeq 2000	SRX1452237 SRX1452217 SRX1452216 SRX1452194 SRX1452149 SRX1452112 SRX1452110	ARO, Volcani Center, Israel
9	Tommy Atkins—mixed organs, seed, mesocarp, Leaf, flower, exocarp, seed coat	Illumina HiSeq 2000	SRX1153795 SRX1149975 SRX1149974 SRX1149973 SRX1149972 SRX1149971 SRX1146011	Monoisolate Indiana University, USA

(continued)

Table 12.1 (continued)

S. no.	Bio samples	Sequencing platform	SRA accession number (in relative order)	Institute
10	Amin Abrahimpur, Burma, M. Casturi "Purple", Neelum, Thai Evergreen, Turpentine	Illumina HiSeq 2000	SRX1149425 SRX1149469 SRX1149478 SRX1149493 SRX1149504 SRX1149515	Monoisolate Indiana University, USA
11	Ataulfo—Hot water disinfection control and treated fruits' mesocarp	Illumina Genome Analyzer IIx	SRX1058207 SRX1058206 SRX1058205 SRX1058204 SRX1058203 SRX1058201 SRX1058199 SRX1058169	CIAD Centro de InvestigacionenAlimentacion y Desarrollo A.C.
12	Amrapali—Malformed flower, fruit, and leaf	Illumina MiSeq	SRX3123804 SRX3123782 SRX982233	ICAR-NRCPB, New Delhi, India
13	Kent—Mesocarp	Illumina HiSeq 2000	SRX690486 SRX690439 SRX690236 SRX689551	Universidad Nacional Autonoma de Mexico, Mexico
14	Keitt—Young leaves, young inflorescences, young fruit, flesh and peels of mature fruit	454 GS FLX Titanium	SRX651793	Agricultural Research Organisation, (ARO), Israel
15	Keitt—Peel of overripe and ripe fruits	Illumina HiSeq 2000	SRX620268 SRX620267 SRX620257	Cornell University, USA
16	Zill—Fruits	Illumina HiSeq 2000	SRX434022	South Subtropical Crops Research Institute, Chinese Academy of Tropical Agricultural Sciences, Hongxia Wu
17	Shelly—Control and hot Water Brush treatment at 0, 4,17, and 48 h	Illumina HiSeq 2000	SRX375390	ARO, Israel
18	Dashehari-Leaves	ABISOLiD 4 System	SRX553039	ICAR-NRCPB, New Delhi, India
19	Neelam-Leaves	ABISOLiD 4 System	SRX551448	ICAR-NRCPB, New Delhi, India

database of protein, i.e., protein data bank (PDB) database has merely one well-predicted structure of *Mi*GST encoding Glutathione S-Transferase protein reported through X-ray diffraction studies (Chavira et al. 2017). These resources are limited in comparison with other fruit crops such as apple, citrus, and grapes wherein such information are available in plenty because of their well-annotated genomes and huge information on resequencing. Over the years, the functional genomic resources flowing into the public domain is increasing with the advancements made in next-generation sequencing (NGS) platforms expediting RNA sequencing in terms of cost, time as well as information generation.

RNA-Seq has been widely used by several researchers which not only allows differential gene expression analysis for understanding molecular mechanism of complex traits but also identification of simple sequence repeat (SSR) and single nucleotide polymorphism (SNP) markers. For instance, Iquebal et al. (2017) have used a panel of 84 mango varieties, and generated 28.6 Gb data by ddRAD-Seq approach on Illumina HiSeq 2000 platform from which 1.25 million SNPs were discovered. A web genomic resource *MiSNPDb*, based on 3-tier architecture, was developed using PHP, MySQL, and Javascript, which could be exploited for high-density linkage mapping in mango. Another database of mango genome called as Mango Genome Database (MGDB) has been reported by Qamar-ul-Islam et al. (2018). Similarly, mango genomics resource has been developed at ICAR-Central Institute of Subtropical Horticulture (CISH), Lucknow (www.mangifera.res.in/primer.php) which houses information on genomic and genic SSR markers. A web resource on cuticle-associated epidermal peel transcriptome data sets generated by RNA sequencing of fruit peel has been documented and is accessible from url: http://bioinfo.bti.cornell.edu/cgi-bin/mango/index.cgi (Tafallo-Aranello et al. 2017).

12.3 Functional Genomics Strategies, Methods, and Approaches

Functional genomics strategies of transcriptomics, proteomics, and metabolomics have been extensively used for elucidating the molecular mechanisms underlying complex traits in mango. The methods and approaches used in functional genomics could be categorized into three generations (Table 12.2) based on the technique used, robustness, and efficiency in understanding gained in a particular trait. Details of the techniques and their successful examples have been elaborated in the subsequent sections.

12.3.1 First Generation Functional Genomics

First generation functional genomics involves use of tools like northern blotting and polyacrylamide gel electrophoresis (PAGE) for individual transcript and protein characterization studies. Limited work has been done in case of mango and much focus was on characterization of full-length cDNA related to gene homologs of ripening. The first report of cDNA encoding alternative oxidase associated with climateric nature, respiration, and ripening processes was isolated and characterized in mango by Cruz et al. (1995). The gene of 623-

Table 12.2 Classification of different tools and techniques used in mango functional genomics

Time line	Techniques for quantification of gene expression and metabolite analysis		
	Transcriptome	Proteome	Metabolome
First Generation	Northern blot (Specific RNA detection)	SDS-PAGE, Native PAGE, Western blot, N-terminus Edman sequencing (Specific protein isolation and analysis)	–
Second Generation	RT-PCR (individual transcript assays), Differential Display (gene discovery)	Isoform analysis, Peptide mass fingerprinting and Protein sequencing (Specific protein isolation and characterization)	–
Third/Next Generation	RNA sequencing (all genes) and Microarrays (known genes)	2D-GE-MS (few differential proteins), LC-MS (whole proteome)	GC-MS (only volatiles) LC-MS (volatile and non-volatile metabolites)

Note The content inside the parenthesis indicates the robustness/limitations of the technique bp fragment was used as a probe to screen a ripe mango mesocarp cDNA library, from which a full-length cDNA clone (pAOMI.1), encoding of mango homolog of the alternative oxidase was obtained with open reading frame (ORF) yielding a polypeptide of 318 amino acids. Gel blot hybridization revealed that pAOMI.1 was likely to be encoded by a single-copy gene. Western blot analysis using antibodies recognized three bands of 27, 33, and 36 kDa sizes representing mitochondrial proteins. The latter two bands were detectable before ripening and the former band (27 kDa) was accumulated during ripening. Similarly, Bojorquez and his coworkers (1995) characterized a cDNA pTHMF1 encoding for a peroxisomal mango thiolase using differential screening of mesocarp cDNA libraries of cv. Manila and Northern hybridization proved the role of fatty acid metabolism in fruit ripening. In another study, Zainal et al. (1996) have isolated a full-length cDNA clone from fruit having homology to rab11/YPT3 class of small GTPases, involved in trafficking cell-wall modifying enzymes through the trans-Golgi network, which was found to be expressed in the fruits during ripening only.

12.3.2 Second Generation Functional Genomics Approaches

The second generation gene expression tools include differential display and real-time PCR techniques for studying expression at transcript level and protein sequencing at translational level. Luo et al. (2014) have used differential display technique for analyzing differences in gene expression of abiotic stress (cold, drought, salinity, and heavy metal) treatments in mango. Oligo-dT anchored cDNA-SCoT technique was exploited to identify differentially expressed genes during several stress treatments in mango. Ninety-two transcript-derived fragments (TDFs) were characterized and their expression patterns were confirmed by quantitative reverse transcription–polymerase chain reaction (qRT-PCR). In a study by Yunhe et al. (2016), a gene encoding ethyelene response factor (*MiETR1b*) was isolated from the cotyledon of mango cv. Zihua using RT-PCR, and a full-length cDNA (2,530 bp) was generated using the 5′ and 3′ rapid amplification of cDNA ends (RACE). Genomic sequence of *MiETR1b* was 4,116 bp with ORF representing 3,305 bp sequence from the start to terminator codon and harboring 6 exons and 5 introns while the deduced amino acids showed conserved domains of the GAF and H-ATPase C super-families. *MiETR1b* was found to be expressed in the proximal and distal cut surfaces throughout the adventitious root formation period which was evident from qRT-PCR wherein significant up-regulation in roots was ascertained. Unfortunately, proteomics studies are not much advanced in mango and no reports on protein sequencing were reported which could be because of poor specificity in detection while using homologous proteins and antibody of relative homologs.

12.3.3 Third/Next Generation Functional Genomics

The third generation functional genomics approach includes the latest tools and techniques that are used in gene expression analysis. This can be grouped as transcriptome analysis techniques such as microarrays and RNA sequencing, proteomics techniques such as 2D-gel electrophoresis and liquid chromatography coupled with mass spectrometry (LC-MS), and metabolomics techniques such as gas chromatography coupled with mass spectrometry (GC-MS) and LC-MS. Development of microarray chip for gene expression studies is in infancy stage excepting for the SNP chips developed for genotyping (Iquebal et al. 2017). A defined set of well-characterized gene probes is essential for developing gene expression microarrays and without the whole genome information this is highly impractical. Moreover, microarray-based gene expression profiling would be biased because of its inherent limitation of characterization of known genes only. With advancements made in NGS platforms, using RNA sequencing for gene expression profiling is more rapid, easier, and efficient owing to the advantages it offers, i.e., studying whole genome gene expression profiling, understanding minor genes contributing to cumulative function, etc. Thus, transcriptome profiling using RNA sequencing has been widely used for gene expression profiling in mango to understand different traits.

12.4 Transcriptome Analysis Using RNA Sequencing

Till date, 14 transcriptome data have been well characterized in mango (Table 12.3) which indicates their potentials in understanding complex traits. Perhaps, unavailability of mango genome data in the public domain is a critical limitation that compulsorily forces to opt only for de novo assembly and genomic annotations with related genera or family. But if the whole genome sequence is made available in open source it would assist in reference mapping of the transcripts to the well-annotated genomic regions that facilitates precise gene predictions. Whole genome transcriptome profiling can be performed using experimentation and sampling based on (i) case-control that helps in elucidation of mechanisms related to abiotic and biotic stress tolerance (salinity, drought, cold, malformation, anthracnose, wilt) and/or (ii) hypothesis-mechanism model that allows undermining mechanisms behind a specific phenomena such as flowering, alternate bearing, peel and pulp coloration. The transcriptomics work flow is presented in Fig. 12.2 which describes the details of the steps involved in RNA-Seq and bioinformatics pipelines. Majority of the transcriptome studies pertain to understanding the molecular intricacies of ripening so as to develop varieties with enhanced shelf-life. Most recently, efforts have been made in assessing the gene expression profiles related to other complex traits of mango such as flowering, anthocyanin pigmentation in peel, chilling resilience in peel and hot water treatment for eliminating pathogens as detailed in Table 12.3.

12.4.1 Small RNA Sequencing and Identification of MiRNAs for Key Traits

Small RNA sequencing is a technique for characterization of miRNAs and is also often quoted as MicroRNA-Seq which utilizes the NGS or massively parallel high-throughput sequencing technologies. This technique differs from RNA-Seq technique because it allows sequencing of enriched small RNAs only and is advantageous in terms of sequence independence and coverage. Yet, it has also demerits such as huge costs, infrastructure, run lengths, and potential artifacts. Alternately, small RNAs can also be mined from RNA sequencing transcriptome raw data without filtering quality reads. The major idea behind the small RNA sequencing is mining of the small non-coding RNAs and prediction of the targets gives insights into regulation of complex traits which is possible by correlating the target gene expression profiles from transcriptome date with miRNA expression profiles of small RNA sequencing data. As a thumb rule, if miRNA is upregulated, then its corresponding target is expected to be downregulated and vice versa and thus, it is presumed to play a crucial role in regulating downstream genes of a signal transduction pathway associated with a complex trait. Ahmad et al. (2018) used an in silico analysis of transcriptome data from NCBI database and identified 18 conserved miRNAs belonging to 15 families. Some important miRNAs identified were min-miRNA159, min-miRNA319, min-miRNA397, min-miRNA160, and min-miRNA393. Similarly, miRNAs have been characterized through small RNA sequencing in floral and vegetative buds of mango cv. Dashehari at ICAR-CISH, Lucknow (unpublished), wherein several known and novel miRNAs were identified. A list of known miRNAs and their target transcription factors involved in flowering that were identified in the study are presented in Table 12.4. Novel miRNAs which are upregulated resulted in its corresponding

Table 12.3 RNA Sequencing based transcriptomics and their utilization in mango functional genomics

S. No.	Cultivar	Tissue	Platform	Pathway	Prominent genes	RT validation	Ref
1	Neelam and Dashehari	leaves	Sequenced-RNA-sequence reads and transcript assembly from Mahato et al. 2016 SOLiD total RNA Seq	Regular bearing and alternate bearing	*SPL*-like gene (GBVX01015803.1), *Rumani GA-20-oxidase-like gene* (GBVX01019650.1), *LOC103420644* (GBVX01016070.1), and GBVW01004309.1	Yes	Sharma et al. (2020)
2	Zill	Fruits	Raw transcriptome reads from Hong et al. (2016). llumina paired-end sequencing	Resistance gene analogs (RGAs)	*NBS*, *NBS-LRR*, *CNL*, *TNL*, *CN*, *TN*, *RLP*, *RLK*, *TM-CC*, and *NBS*-encoding proteins with other domains.	No	Lantican et al. (2020)
3	Amrapali	Healthy and malformed floral buds	Illumina NextSeq	Flowering genes in relation to floral malformation	*MYB30*, *TPL*, *bHLH*, *FTIP1*, *CDKC2*, *CPK33*, *ATH1*, *UBP12*, *EFS*, *AGL8*, *AGL14*, *AGL20*, *AGL24*, *KIN10*, *MYB30*, *SUS2*, *FTIP1*, *CCT*, *LDL2*, *TIL1*, *TIC*, *DCL3*, *GA20OX3*, *CCT*, *AP1*, *AGL6*, *AGL8*, *MYB30*, *AGL8*, *GCT*, and *GA3OX1*	Yes	Yadav et al. (2019)
4	Amrapali	Leaf	Illumina NextSeq 500/MiSeq	Anthocyanin biosynthesis pathway related to fruit peel color	Alternate spliced variance of *CHS*, *CHI*, and *F3H*	Yes	Bajpai et al. (2018a, b)
5	Ataulfo	Mesocarp	Genome analyzer GAIIx II	Effect of hot water treatment on Cv. Ataulfo	*PME*, *PL*, *PG*, *GLB-1*, and *RG1*	Yes	Dautt-Castro et al. (2018)
6	Keitt	Peel	Illumina HiSeq 2000	Chilling stress on fruit peel	*NADPH Oxidase*, *MAP Kinase*, *WRKYs*.	No	Sivankalyani et al. (2016)
7	Zihua	–	Illumina Hiseq 2500	Adventitious root formation from cotyledon segment	G1/S-specific cyclin D, CDK inhibitor, Pectate Lyase, Expansin, Xyloglucan endotransglycosylase, Endo-1,4-beta-glucanase	Yes	Li et al. (2017b)

(continued)

Table 12.3 (continued)

S. No.	Cultivar	Tissue	Platform	Pathway	Prominent genes	RT validation	Ref
8	Alphonso	Flowers and fruits	Illumina HiSeq 1000	Fruit Development and ripening with respect to aroma and shelf-life	*CS1*, *CS2*, *CTN1*, *CTN2*, *CTN3*, *PL1*, *PL2*, *PL3*, *O3FAd*, *O6FAD*, *CFACL*, *G3PAT1*, *G3PAT2*, *G3PAT3*, *ADH1*, *ADH2*, *ADH3*, *ERF1*, *ERF2*, *ERF3*, *ERF4*, *ERF5*, *DRP1*, *DRP2*, *MTPS1*, *MTPS2*, *MTPS3*, *MTPS4*, *MTPS5*, *MTPS6*,*STPS1*, *STPS2*, *STPS3*, *STPS4*, *STPS5*, *DTPS1*, *DTPS2*,*DTPS3*	Yes	Deshpande et al. (2017)
9	Keitt	Fruit	Illumina HiSeq 2500 system	Identification of cuticle-associated genes	*CD2*. *SHN1*, *CER2*, *CER3*, *CS2*, *CS6*, *LTP3*, *LTP1*, *WBC11*, *LTP2*, *LTPG1*, *CUS1*, *CUS2*, *PEL1*, *NAC*, *C2C2-GATA*, *G2-like*, *CRPK1* Like Kinase, *HSF*, *MYB* and *bHLH*	Yes	Tafolla-Arellano et al. (2017)
10	Dashehari	Fruit	454-GS-FLX	Ripening associated genes	*MYB*, *bHLH*, *WRKY*, *HSF*, *NAC*, *bZIP*, *ERF*, *SAM Synthase*, *ACS*, *ACO*, *CTR ETR*, *EIN3*, Endo-beta-1,4-glucanase, Expansin, Pectate Lyase, Polygalacturonase	Yes	Srivastava et al. (2016)
11	Zill	Fruits	Illumina paired-end sequencing	Host responses of the mango fruit against *C. gloeosporioides*	*ERFs*, *NBS-LRR*, *NPR1* , and *PR*	Yes	Hong et al. (2016)
12	Kent	Mesocarp	Genome Analyzer GAIIx II	Fruit ripening	*ACO*, *PME*, *PG*, *PL*, *GLB-1*, *GUN*, *GAL-A*, *EXP*, *RGL*, *IDI*, *FPPS*, *GGPS*, *SuSy*, *CRTISO*, *CHYB*, *BAM*, etc.	Yes	Dautt-Castro et al. (2015)
13	Alphonso	Fruit	Megabase 100 DNA sequencer	Fruit development and ripening	*IPPI*, *GPPS*, *MTPS*, *GGPPS*, *FPPS*, *SqTPS*, *IsoCH*, *GT*, *MDHAR*, *14-3-3*, *MT*, *MeTR*, *sHSP*, *CHI*, *CysPI*, *ERF*, *UbqPL*	Yes	Pandit et al. (2010)
14	Dashehari	Pulp	ABI 373A	Ethylene induced post-harvest softening	Pectate Lyase	Normal PCR/semi-quantitative PCR	Chourasia et al. (2006)

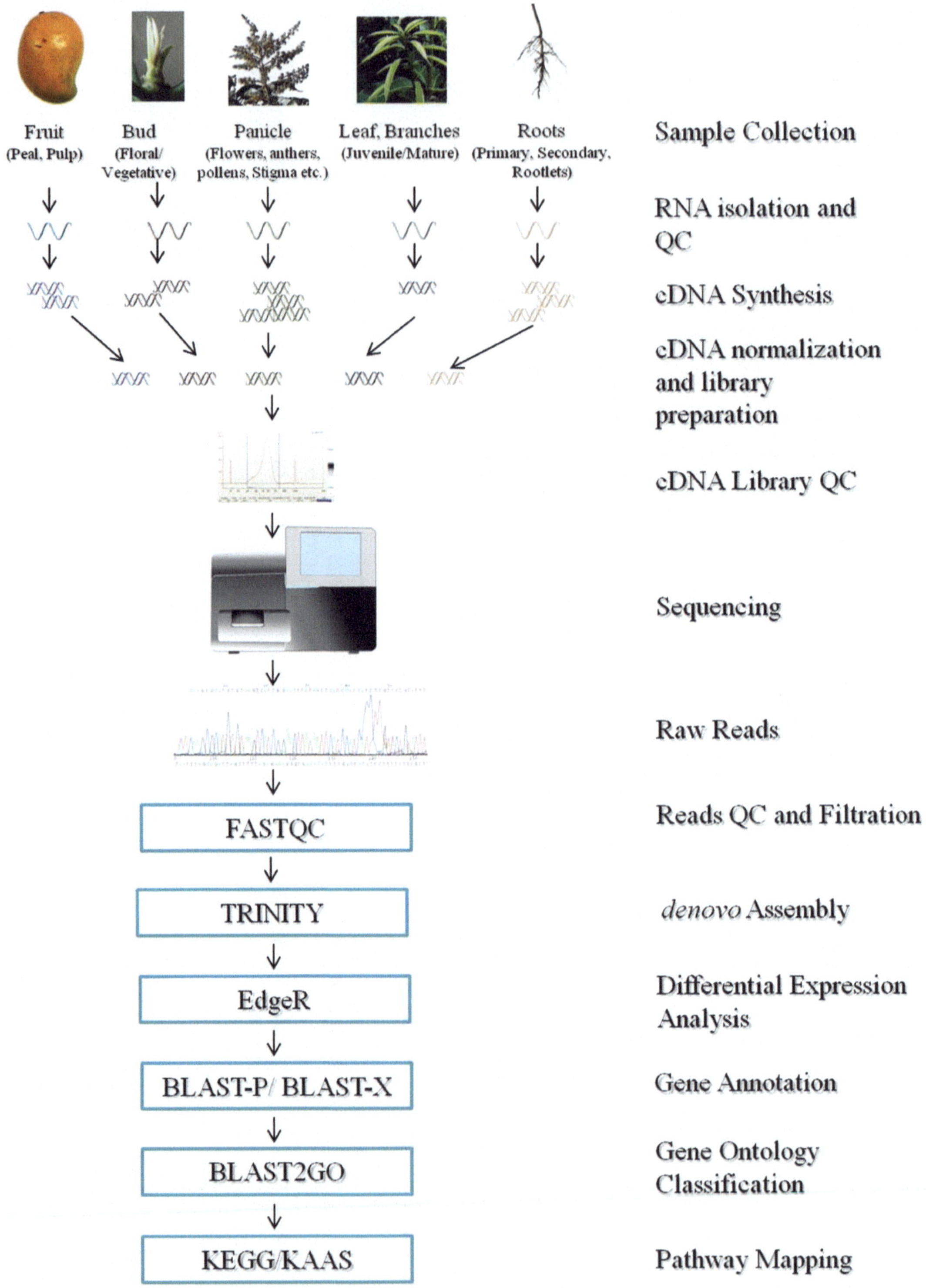

Fig. 12.2 The transcriptomics work flow describing the details of RNA-Seq and bioinformatics pipelines

Table 12.4 Flowering related known miRNAs identified in flower bud of mango cv. Dashehari

Known miRNAs	Gene target	Significance and function
miR156	*SPL*	Phase transition V-R, promotes flowering
miR157	*SPL*	Phase transition V-R, promotes flowering
miR172	*AP2*	Repressor in photoperiod induction
miR159	*MYB*	Promotes flowering
miR171	*SCL/LM*	Promotes/delays flowering
miR319	*TCP*	Promotes flowering
miR390	*tasi RNA*	Promotes juvenility
miR393	*TIR*	Repressor
miR399	*PHT2*	Temperature, phosphate starvation responsive, Repressor
miR5200	*FT*	Photoperiod induction and promotes flowering
miR169	*NF-Y*	Repress/induce flowering in response to abiotic stress
miR824	*AGL*	Repressor based on environment/genetic background

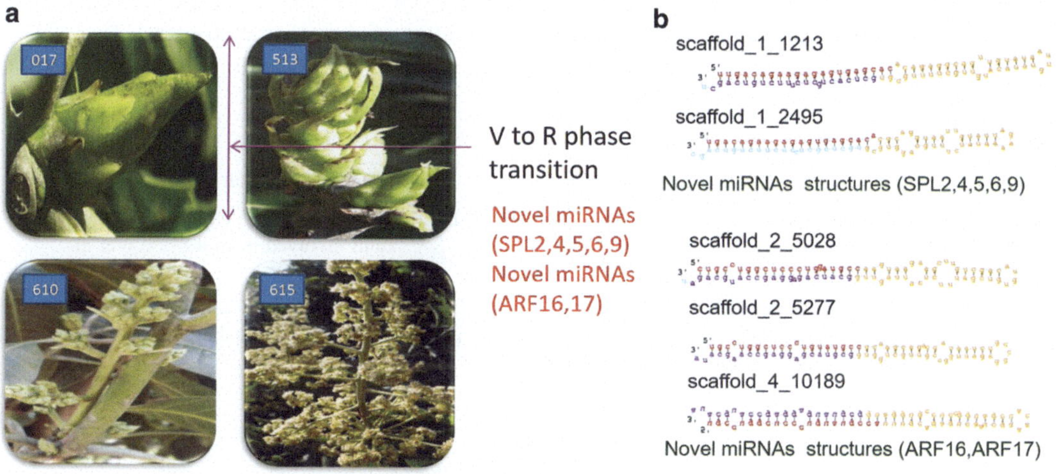

Fig. 12.3 Novel miRNAs involved in GA signalling in flowering

downregulation of target GRAS family transcription factors involved in GA signaling (GA inhibition-based floral induction) in flowering (Fig. 12.3). Other novel miRNAs identified targeted SPL and ARF transcription factors. Unique miRNAs targeting NBS-LRR/Phosphate 2, S-locus protein kinase were identified in floral bud.

12.5 Proteomics Approaches in Understanding Complex Traits

Proteomics is a study of total proteome configuration of the cell or organelle which comes from all the expressed genes at a given point of time and tissue. This is a powerful tool that can be used to elucidate the complex traits at the cellular and molecular levels. An overall status of protein-based studies is presented in Table 12.5. The modern proteomics tools include 2D-MS, LC-MS, and HPLC-MS as explained briefly in the earlier classification section. Andrade et al. (2012) have reported on the first comparative proteomic investigation between the pre-climacteric and climacteric fruits of mango cv. Keitt to identify protein species with variable abundance during ripening. In their study, proteins were extracted from fruits through phenol-extraction protocol followed by cyanine-dye-labeling and separation on 2D gels at pH 4–7. Around 373 proteins spots were identified from which 47 were consistently differentially regulated and were subjected to LC-MS/MS analysis. This study helped in the identification of proteins associated with ripening, color development, and pulp softening. Fasoli (2013) has developed combinatorial peptide ligand libraries (CPLLs) for investigating the proteomes of mango peel and pulp as well their peptidome content (the latter as captured with a C18 resin) and used this technique to assess the potential presence of allergens and of peptides endowed with biological activities. Similarly, Chin et al. (2019) studied the comparative proteome profiles of mesocarps of ripe and unripe fruits of mango cvs. Chokanan and Golden Phoenix using 2-dimensional gel electrophoresis (2D-GE) coupled with MALDI-TOF/TOF. Differentially abundant proteins identified were found to be involved in ethylene synthesis and aromatic volatiles, cell wall degradation, and stress-response proteins. Six differential proteins were validated using qRT-PCR in which genes encoding glutathione S-transferase (GST) and alpha-1,4 glucan phosphorylase (AGP) were found to express in concordant with differential proteome profiles.

Table 12.5 Brief overview on protein isolation and characterization studies in mango

Time line	Approach and technique	Protein class	Function and trait	References
First generation	Individual protein analysis-SDS-PAGE	Protein kinase II (PK-II)	Phosphorylating casein and involved in fruit ripening	Frylinck et al. (1998)
Second generation	Differential protein regulation 2D-DIGE	Rubisco large subunit 1, phosphoglycerate kinase, triose–phosphate isomerase, NADB Rossmann superfamily peptides, malate dehydrogenase, phosphoglucomutase, enolase, aldehyde dehydrogenase, phosphoglycerate kinase, F1 ATPase family, transaldolase, electron transfer flavoprotein–ubiquinone oxidoreductase, adenylate kinase	carbon fixation, hormone biosynthesis, catabolism, stress-response	Andrade et al. (2012)
Second generation	SDS-PAGE and Western blot analysis	Lewis type α1,3/α1,4-fucosyltransferase	Biosynthesis of Le(a) containing glycoconjugates	Okada et al. (2017)
Second generation	SDS-PAGE and Zymography	Tau class glutathione S-transferase	Cellular detoxification	Valenzuela-Chavira et al. (2017)

(continued)

Table 12.5 (continued)

Time line	Approach and technique	Protein class	Function and trait	References
Third Generation	Protein purification using affinity chromatography, Analysis by SDS-PAGE, Zymography, and LC-MS/MS	α-amylase iso-enzymes	Fruit ripening	Ubonbal et al. (2017)
Third Generation	2-dimensional gel electrophoresis (2D-GE) coupled with MALDI-TOF/TOF	glutathione S-transferase (GST) and alpha-1,4 glucan phosphorylase (AGP)	Fruit ripening	Chin et al. (2019)
Third Generation	2-dimensional gel electrophoresis (2D-GE) coupled with MALDI-TOF/TOF	Cytosolic class II small heat-shock protein HSP17.5, Superoxide dismutase, Glutamine synthetase, Vitamin-b12 independent methionine synthase, NADP-dependent malic enzyme, NAD-malate dehydrogenase	Immunity and defense, oxidation and reduction, and amino acid and carbohydrate metabolism	Liao et al. (2016)
Third Generation	Transcriptome, SDS-PAGE, LC-MS/MS	14-3-3, Small heat-shock protein, Chloroplast chlorophyll A/B binding protein, Succinate dehydrogenase, 3-ketoacyl-CoA thiolase B: acetyl-CoA C-acyltransferase, Glyceraldehyde-3-phosphate dehydrogenase, Glycolate oxidase	Signal transduction, apoptosis, cell cycle control, Protein folding, Photosynthesis, Citric acid cycle and electron transport chain, Fatty acid beta-oxidation, Glycolysis, Oxidoreduction	Renuse et al. (2012)

12.6 Metabolomics Approaches in Understanding Complex Traits

Metabolomics is a study of metabolome of an organism's specific tissue at a given developmental stage or pooled stages. It involves studying the total metabolite profile which includes both primary and secondary metabolites; under a given situation or a developmental stage that defines its temporal, spatial, and developmentally controlled production. The modern tools for metabolomics approach include gas chromatography and liquid chromatography coupled with mass spectrometry commonly called GC-MS and LC-MS, respectively. Metabolomics studies in mango are focussed on the characterization of flavor and aroma compounds in different mango varieties (Table 12.6) and mainly using GC-MS technique. Major flavor and aroma compounds identified were mainly falling in the groups of coumarins or lactones, terpenes, and terpene hydrocarbons. The metabolite studies could also be confirmed by transcriptome studies or real-time gene expression profiling pertaining to the genes involved in specific lactone or terpene or flavonoid biosynthetic pathways.

Table 12.6 Status of metabolite characterization in different mango varieties using metabolomics approach

S. no.	Technique	Cultivar and tissue	Metabolite Profile			References
			Total metabolites	Major metabolites	Unique metabolites	
1	GC-MS	Haden, Irwin, Manila and Tommy Atkins, Hilacha, Vallenato, Van Dyke, Yulima (Fruits)	145	Terpene hydrocarbons (major volatiles)	δ-3-carene (Haden, Irwin, Manila and Tommy Atkins), α-pinene (Hilacha and Vallenato), α-phellandrene (Van Dyke), and terpinolene (Yulima)	Quijano et al. (2007)
2	GC-Olfactometry	Haden, White Alfonso, PrayaSowoy, Royal Special, and Malindi (Tree ripe fruits)	54	27 aroma volatiles	4-hydroxy-2,5-dimethyl-3(2H)-furanone	Munafo et al. (2014)
3	GC-MS	Shindri, Langra (Peel)	34	Bicyclo [4.1.0] hept-3-ene (49%), 3, 7, 7-trimethyl-, (1S) (47%) (Major essential oils)	–	Baloch et al. (2017)
4	GC-MS	Twenty-five mango varieties of China, America, Thailand, India, Cuba, Indonesia, and the Philippines	127	Monoterpenes (Major aroma and aroma volatile)	β-Myrcene (Xiaofei, Philippines); γ-Octanoic lactone (Guiya, China); Hexamethyl cyclotrisiloxane (Magovar/Mulgoa, India)	Li et al. (2017a)
5	GC-MS	Ambika, Arunika, Amrapali, Vanraj, and Janardhan Pasand	60	Coumarins (Major aroma volatiles)	Furo-coumarin (Arunika, both parents Amrapali and Vanraj) dimethoxy-flavones (Ambika and Amrapali female parent and absent in Janardhan Pasand)	Unpublished

12.7 Trait Mapping, Identification, and Functional Characterization of Genes Using Functional Genomics

The transcriptome of mango mesocarp cv. Kent identified gene families important for ripening and captured most of the gene space in mango (Dautt-Castro et al. 2015). The expression data obtained by RNA-seq was favorably validated with the data obtained by qRT-PCR. Limited shelf-life and perishability of mango fruits limit marketing and global transport, due to postharvest desiccation and senescence. Recent studies utilizing RNA-Seq analysis (Tafolla-Arellano et al. 2017) suggested that the pathway leading to biosynthesis of the cuticle component, cutin, is upregulated during overripening. RNAseq and de novo assembly of cv. Langra leaf transcriptomic sequences yielded 13,500 unigenes that could be assigned to 293 KEGG pathways with emphasis on an array of biochemical pathways involved in (a) biosynthesis of bioactive flavonoids, flavones and flavonols, (b) biosynthesis of terpenoids and lignins, and (c) plant hormone signal transduction (Azim et al. 2014). Mango fruit transcriptomes of cv. *Zill* (Wu et al. 2014), cv. *Shelly* (Luria et al. 2014), cv. *Kent* (Dautt-Castro et al. 2015), cv. *Dashehari* (Srivastava et al. 2016), cv. *Keitt* (Tafolla-Arellano et al. 2017) were utilized to divulge insights on genetic resources responsible for flavor and color of mango fruit. Sequence diversity and differential expression of major phenylpropanoid-flavonoid biosynthetic genes among three mango varieties elucidated characteristic patterns that differed between mango peel and flesh of important germplasm (Hoang et al. 2015). The leaf transcriptome repository was utilized to mine transcript sequences from phenylpropanoid flavonoid pathway for elucidating red peel development, and anthocyanin genes with MYB, bHLH transcription factors were upregulated in red peel cultivars (Fig. 12.4; Bajpai et al. 2018a).

Genes involved in salinity stress tolerance were shortlisted from transcriptome data of salinity tolerant mango cv. 13-1 and these mined genes such as *OLP*, *ADH*, *APN*, *YABBY*, *FLP*, *DREB*, *DNaseI*, *WRKY72*, *LOG*, *GT* were validated using RT-PCR assay (Bajpai et al. 2018b). Muthukumar et al. (2018) mined leaf transcriptomes (SRA Nos. SRP070908, SRP072727) and shortlisted 6 key genes, viz., *STPK*, *MAPK6*, *ERF2*, *MYC2*, *MDH*, and *TPS* related to salinity stress responses and used for expression analysis

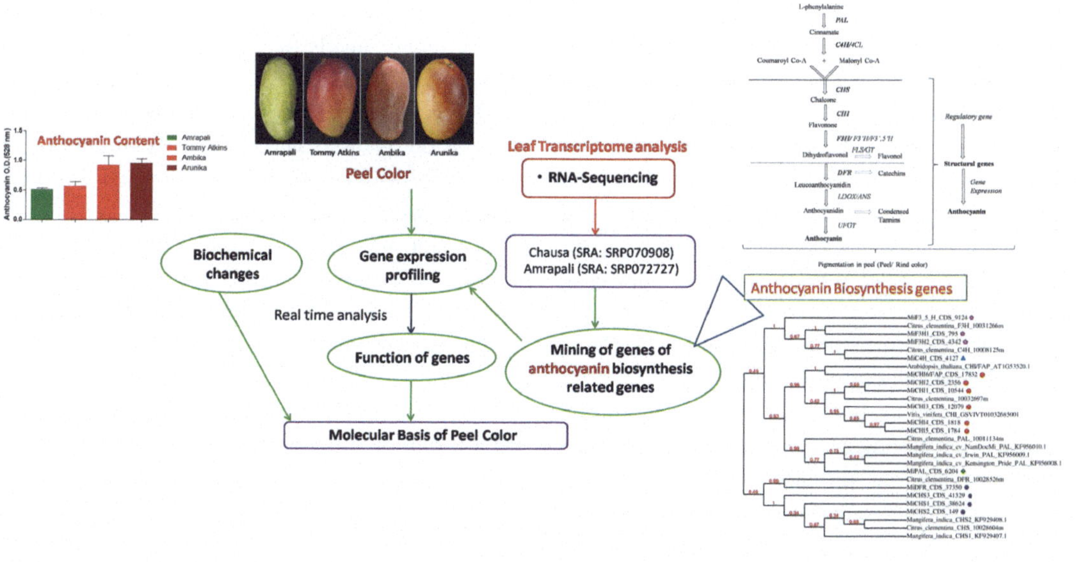

Fig. 12.4 Utilization of leaf transcriptome repository for mining anthocyanin pathway genes

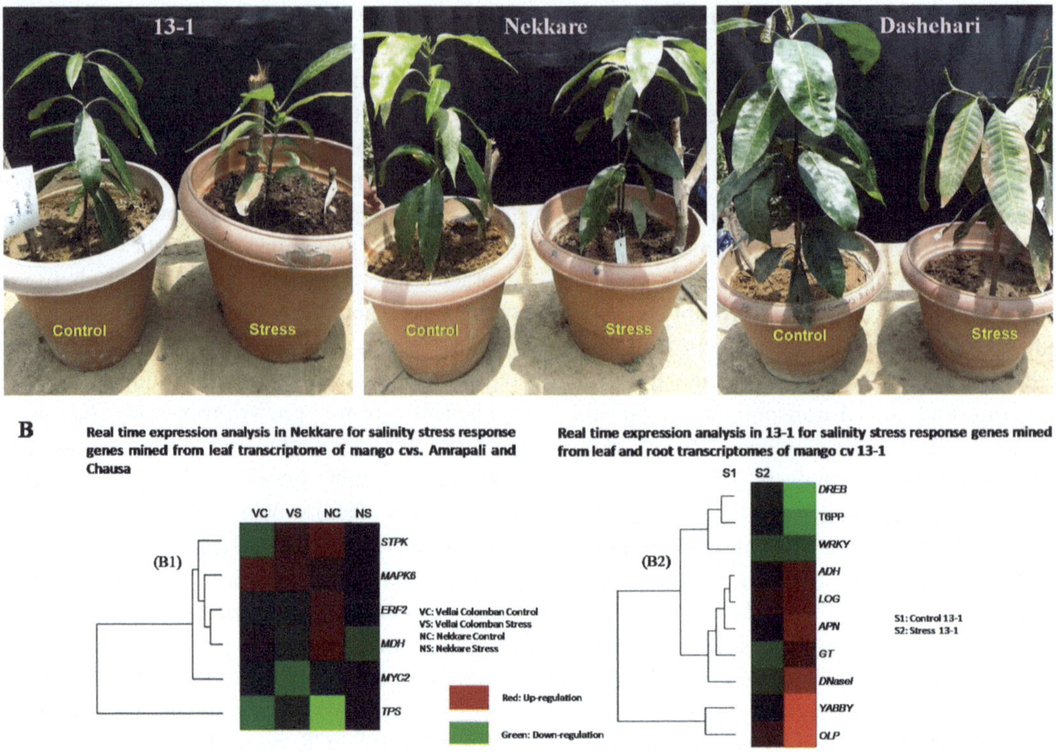

Fig. 12.5 Molecular mechanism of salinity tolerance in mango cvs

using salinity tolerant cv. Nekkare and salinity susceptible cv. Vellai Colomban. The overall strategy used for elucidation of molecular mechanism of salinity tolerance in mango cvs 13-1 and Nekkare is presented in Fig. 12.5. Chilling stress in fruit peel of "Keitt" mango was monitored by transcriptomic, physiological, and microscopic analyses. Symptoms of chilling injury include red and black spots that start from discolored lenticels and develop into pitting. Transcriptomic changes in the mango fruit peel were evaluated during optimal (12 °C) and sub-optimal (5 °C) cold storage (Sivankalyani et al. 2016).

Mango malformation disease (MMD) is characterized by symptoms manifested in vegetative and floral tissues which includes abnormal, compact development of shoots, flowers, and leaves. Metabolite profiles of malformed and non-malformed (normal) panicles of mango were obtained by GC-MS that reveal distinct chemistries (Fig. 12.6). Role of *F. mangiferae* in stimulating stress ethylene production and over load of "stress ethylene" causing malformed tissue development suggest that cyanide can be an etiological agent (Ansari et al. 2015). Significantly higher levels of cyanide, a by-product of ethylene biosynthesis, and corresponding lowered β-cyanoalanine synthase gene (β-CAS) expression is associated with accumulation of unmetabolized cyanide in malformed inflorescence (Ansari et al. 2019).

Annual transition of apical meristem from vegetative to floral is the key feature that determines regular bearing behavior of mango cultivars under subtropical ecosystems. Characterization of *Constans like* gene (Zhang et al. (2005) and role of *Flowering locus T* in inductive conditions were established by different workers (Nakagawa et al. 2012; Santos-Villalobos et al. 2012). Expression analysis of mango *Flowering Locus T (FT)* and *Terminal Flower 1 (TFL1)* genes was done in

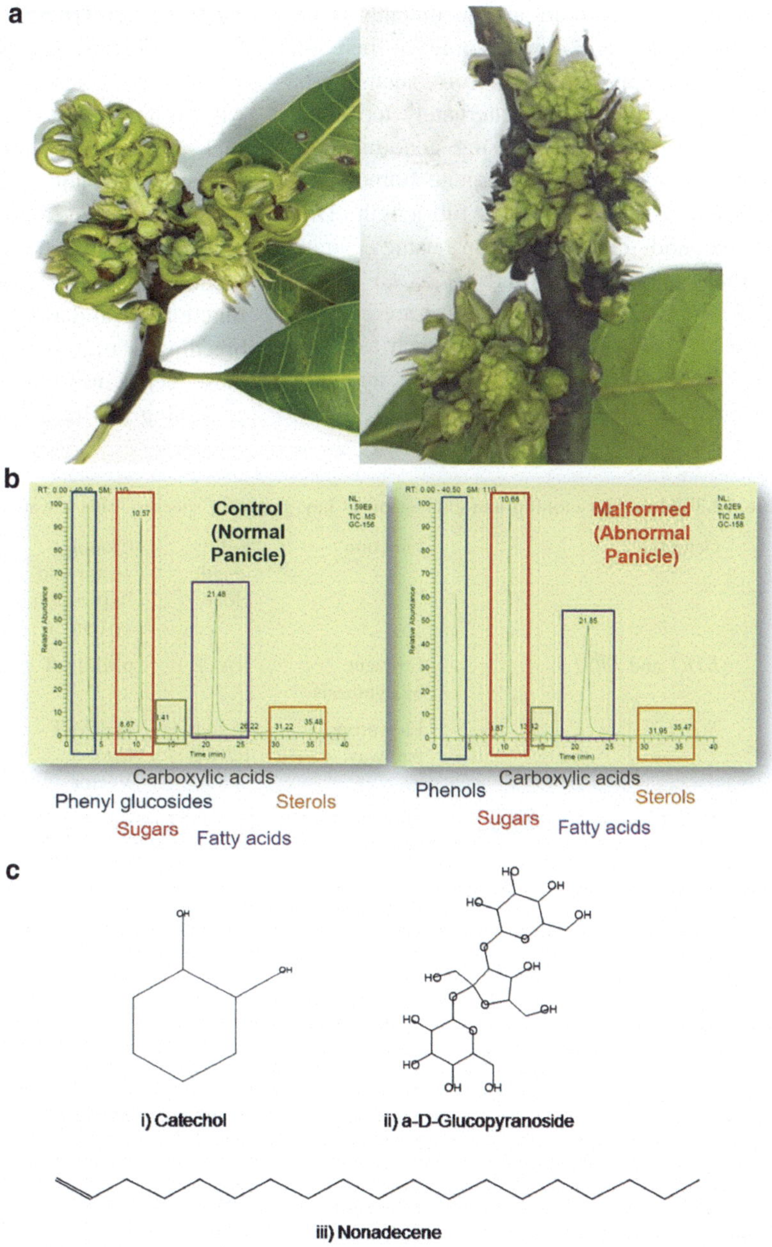

Fig. 12.6 Metabolite profiles of malformed and non-malformed panicles

mango cultivars Alphonso and Ratna in Inida (Vyavahare et al. 2017). RNA-seq data has also allowed identification of key differentially regulated genes (DEGs) coding for transcription factors related to vegetative growth, ABA and auxin hormonal changes, sugar metabolism and dormancy related genes (Bajpai et al. 2019), that indirectly influence flower induction.

12.8 Discovery, Tagging, Isolation, and Characterization of Genes for Specific Traits

Gene tagging methods are traditional approaches for mapping traits which are done through activation tagging or transposon tagging methods.

Using mapping populations and mutants in tagging the genes are cumbersome in perennial systems like mango and therefore, such approaches have limited scope. Alternately techniques for gene discovery are i. using genomic information and identification of gene homologs in wild species of *Mangifera*, ii. full-length cDNA isolation and characterization using 3′-rapid amplification of cDNA ends (RACE) and 5′-RACE. Some successful examples are available related to ripening related genes such as ETR1, ERS1, Rab5, anthocyanin synthesis, and flowering pathway genes (Table 12.7).

12.9 Functional Validation of Key Genes for Specific Traits

All 3-omics components of functional genomics such as transcriptomics, proteomics, and metabolomics in mango have helped in short-listing several key genes associated with the specific traits as discussed earlier in the trait mapping section. These key genes are ideal candidates for functional validation and utilization in mango breeding or trait improvement programs. At first, these genes need to be confirmed if they are governing the trait. This could be ascertained by

Table 12.7 Molecular cloning and expression of key genes for specific traits toward functional validation

S. No.	Gene (s)	Function	Full-length clone	Cloning vector and Expression system	Trait	Reference
1	*ETR1* and *ERS1*	Ethylene biosynthesis	RACE	pMD18-T	Ripening	Li et al. (2019)
2	*SOC1*	Flowering pathway	3′ and 5′ RACE	pGEM-T Easy Vector	Flowering	Wei et al. (2016)
3	*F3'5'H*	Flavonoid pathway	3′ and 5′ RACE	–	Peel coloration	Zhao et al. (2016)
4	*Rab5*	GTP-binding Protein	5′ RACE	pET-30a	Fruit ripening and stress response	Liu et al. (2014)
5	*ANS*	Anthocyanin biosynthesis	3′ and 5′ RACE	–	Peel coloration	Zhichang et al. (2014)
6	*ETR1b*	Ethylene biosynthesis	3′ and 5′ RACE	pMD18-T vector	Ripening	Yunhe et al. (2013)
7	*ERS1*	Ethylene biosynthesis	DNA walking	–	Ripening	Contreras-Vergara et al. (2012)
8	*ACC*, *ACO1*, *ACS1*, *ACS3*, *AMY*, AOX, *APX*, *BGAL*, *CHS1*, *CHS2*, *EXP*, *FFT*, *FGT*, UDP glucose: flavonoid 3-O-glucosyltransferase; *PG*, *PME*, *PHO*, Alpha 1,4-glucan phosphorylase; R1, Starch debranching enzyme	Ethylene enzymes, Starch mobilization, Ascorbate metabolism	From NCBI database	–	Fruit softening, cell wall integrity, peel and pulp color	de Godoy et al. (2009)

either using knock-out mutant analysis or genetic transformation using *Agrobacterium*-mediated transformation system using the gene associated with the trait. Functional validation is usually done through knock down mutant screens in Arabidopsis models or through induced mutation created by site-directed mutagenesis followed by screening the mutants in subsequent generations. This need to be further confirmed in native host plant, i.e., mango through transgenics approach. But owing to the reasons such as perenniality, heterozygosity, and heterogeneity in mango, the mutant screening seems to be highly imperative and thus mutant screening studies are not advisable. Next best choice for functional validation is molecular cloning of the candidate genes and expression in model organisms such as tobacco, tomato, or Arabidopsis system through *Agrobacterium*-mediated transformation. Both for mutant screen assays and genetic transformation, the first and foremost requisite is an efficient protocol for regeneration and generation of large number of tissue culture plants for functional characterization and validation. Unfortunately, limited success has been reported in mango regeneration systems and only few sporadic reports are available on successful regeneration of tissue culture mango plants (Mishra et al. 2010; Busaidi et al. 2016). Li et al. (2019) have first reported successful functional validation of eukaryotic translation initiation factors (eIFs) associated with abiotic stress, viz., *MieIF1A-a*, *MieIF5*, and *MieIF3sB* that were most strongly expressed under salinity, osmotic, and low temperature (2 °C) stress, respectively. Overexpression of *MieIF1A-a* in Arabidopsis resulted in enhanced seedling growth and survival rates, higher chlorophyll content, increased ROS scavenging enzyme activities (superoxide dismutase, peroxidase, and catalase), and decreased malondialdehyde accumulation under salt treatment compared to the wild-type Arabidopsis. This study provides new insights into the *MieIF* genes that could be applied to the breeding of salt-tolerant mango varieties. With all these challenges to be resolved, presently only resort could be *Agrobacterium*-mediated *in-planta* transformation and thus efforts are underway in different laboratories. There are so many constraints in developing transgenic mangoes and in the present scenario, functional genomics approaches would aid in smart breeding through using functional genomics for mining functional markers, *i.e.*, SSR or SNP markers co-located within the genes for specific traits and using these markers for genome-wide association studies would help in molecular breeding (Fig. 12.7).

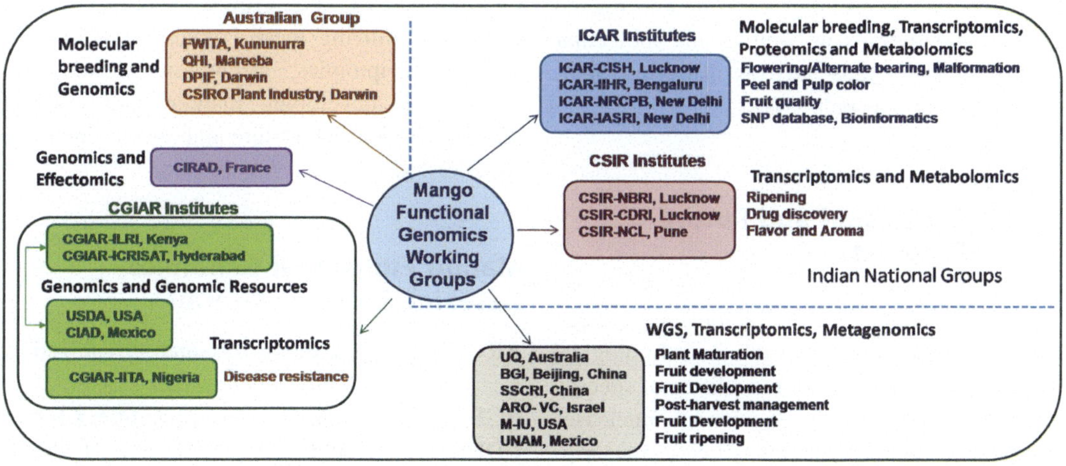

Fig. 12.7 Global research organizations and their research priorities

12.10 Global Mango Functional Genomics' Working Groups

Globally several research organizations have research priorities (Fig. 12.7) that address mango genetic resources management, genomic resources augmentation and their utilization through functional genomics tools such as transcriptomics, proteomics, and metabolomics. An Australian group comprising of Frank Wise Institute of Tropical Agriculture, Agriculture Western Australia, Kununurra Queensland Horticulture Institute Mareeba, Queensland Northern Territory Department of Primary Industries and Fisheries and Darwin CSIRO Plant Industry, Darwin have a major focus on molecular breeding and genomics. CIRAD, France has emphasis on "genomics and effectomics". CGIAR institutes such as ILRI, Kenya; ICRISAT, Hyderabad in collaboration with USDA, USA and CIAD, Mexico has been concentrating efforts on genomics and genomic resources. IITA, Nigeria another CGIAR institute has been working on disease resistance and functional genomics aspects related to disease resistance. In India, CSIR and ICAR institutes are major contributors of functional genomics in mango. ICAR-NRCPB, New Delhi, is a global leader in sequencing mango genome in collaboration with ICAR-CISH, Lucknow; ICAR-IARI, New Delhi; and ICAR-IIHR, Bengaluru. The same team is involved in functional genomics with specific focus expression profiling of different mango varieties and hybrids as well as the molecular genetics of key straits such as flowering, malformation, peel and pulp color.

12.11 Conclusion and Future Prospects

Precision functional genomics tools have been developed in the recent past due to technological advancements made in tailored editing of specific genomic regions using engineered nucleases such as zinc finger nucleases (ZFNs), transcription activator-like effector nucleases (TALENs), and the Clustered Regularly Interspaced Short Palindromic Repeats (CRISPR-Cas/) systems, called as genome editing techniques. Unlike the *Agrobacterium*-mediated transformation which integrates the gene into any location, the genome editing tools offer scope for engineering the plant genome directly through targeting the location of the gene to specific genomic locations and making minimum alternations in the genome. As of now, genome editing approaches have not been initiated in mango yet, it has ample scope in near future. Similarly, a new area of genomics tool called genetical genomics has evolved which integrates the genetics and gene expression analysis. The identification of the causal genes and markers underlying quantitative trait loci (QTLs) would help in effective marker-assisted breeding in mango. So far, paucity of markers and genotyping were presumed to be a major challenge in mango improvement. But with the latest developments in NGS and genotyping by sequencing (GBS) platforms, genotyping costs have been curtailed to a large extent and the available mapping populations at ICAR-CISH, ICAR-IIHR, and ICAR-IARI could be genotyped and used for genome-wide association studies. ICAR-CISH, Lucknow, is a national germplasm active repository of mango with more than 800 mango germplasm accessions and several hybrid progenies and thus has enormous core and mapping populations for selection of association panels. Efforts are made in this direction by different working groups including ICAR-CISH for utilizing these "omics" technologies, viz., transcriptomics, metabolomics, and proteomics for developing markers and utilizing them in molecular breeding and developing more superior hybrids.

Mango Functional Genomics Research Working Groups

ARO VC	Agricultural Research Organization, Volcani Center, Israel
BGI	Beijing Genomics Institute, China
CDRI	Central Drug Research Institute, Lucknow, India

CGIAR	Consultative Group on International Agricultural Research
CIRAD	The French Agricultural Research Centre for International Development. Montpellier, France.
CISH	Central Institute for Subtropical Horticulture, Lucknow, India
CSIR	Council of Scientific and Industrial Research
CSIRO	Plant Industry, Darwin, Australia
FWITA	Frank Wise Institute of Tropical Agriculture, Agriculture Western Australia, Kununurra
IARI	Indian Agriculture Research Institute, New Delhi, India
IASRI	Indian Agricultural Statistics Research Institute, New Delhi, India
ICRISAT	International Crops Research Institute for the Semi-Arid Tropics, Patancheru, Hyderabad, India
CIAD	Research Center for Food and Development, Mexico
IIHR	Indian Institute of Horticulture Research, Bangalore, India
IITA	International Institute of Tropical Agriculture, Nigeria
ILRI	International Livestock Research Institute (ILRI), Nairobi, Kenya
MIU	Marconi International University, USA
NBRI	National Botanical research Institute, Lucknow India
NCL	National Chemical Laboratory, Pune, India
NRCPB	National Research Centre for Plant Biotechnology, New Delhi, India
QHI	Queensland Horticulture Institute Mareeba, Queensland Northern Territory Department of Primary Industries and Fisheries and Darwin, Australia
SSCRI	South Subtropical Crops Research Institute, Chinese Academy of Tropical Agricultural Sciences, Zhanjiang, China
UNAM	National Autonomous University of Mexico, Mexico
USDA	US Department of Agriculture, Washington, USA

References

Ahmad I, Muthukumar M, Hudedamani U, Rajan S (2018) In silico identification of mi-RNAs from Mango (Mangifera indica L.) fruit transcriptome data. Res J Biotechnol 13(10):98–104

Al-Busaidi KT, Shukla M, Al-Burashdi AH, Al-Blushi GS, Al-Jabri MH, Al-Kalbani BS, Al-Hasani HD (2016) *In vitro* regeneration of Mango (*Mangifera indica* L.) cv. Baramasi through nucellar embryogenesis. J Hort For 8:37–43

Andrade J de M, Toledo TT, Nogueira SB, Cordenunsi BR, Lajolo FM, do Nascimento JRO (2012) 2D-DIGE analysis of mango (*Mangifera indica* L.) fruit reveals major proteomic changes associated with ripening. J Proteom 75(11):3331–3341

Ansari MW, Kaushik S, Bains G, Tula S, Joshi B, Rani V et al (2019) Cyanide produced with ethylene by ACS and its incomplete detoxification by β-CAS in mango inflorescence leads to malformation. Sci Rep 9:18361

Ansari MW, Rani V, Shukla A, Bains G, Pant RC, Tuteja N (2015) Mango (*Mangifera indica* L.) malformation: a malady of stress ethylene origin. Physiol Mol Biol Plants 21(1):1–8. https://doi.org/10.1007/s12298-014-0258-y

Azim MK, Khan IA, Zhang Y (2014) Characterization of mango (*Mangifera indica* L.) transcriptome and chloroplast genome. Plant Mol Biol 85(1–2):193–208

Bajpai A, Khan K, Muthukumar M, Rajan S, Singh NK (2018a) Molecular analysis of anthocyanin biosynthesis pathway genes and their differential expression in mango peel. Genome 61(3):157–166

Bajpai A, Muthukumar M, Bajpai Y, Trivedi AK (2018b) RNA-Seq analysis of salinity stress response in mango. In: e-book of Abstracts in national conference on arid horticulture for enhancing productivity and economic empowerment, Bikaner, India, 27–29 Oct, 2018, p 74

Bajpai A, Muthukumar M (2019) Advances in genomics for fruit improvement. J Ecofriend Agric 14(2):38–40

Bajpai A, Muthukumar M, Bajpai Y, Rajan S (2019) Novel insights into the regulatory mechanisms underlying floral induction in mango. In: 2nd International conference on recent advances in agricultural, environmental and applied sciences for global development, 27–29 Sept, 2019, Solan, HP, India, p 30

Baloch FS, Tahir SS, Sherazi STH, Jilani NS, Khokhar AL, Rajput MT (2017) Evaluation of essential oil components from the fruit peelings of sindhri and langra varieties of mango (*Mangifera indica* L.). Pak J Bot 49:1479–1484

Bojorquez G, Gómez-Lim MA (1995) Peroxisomal thiolase mRNA is induced during mango fruit ripening. Plant Mol Biol 28:811–820

Chin CF, Teoh EY, Chee MJY, Al-Obaidi JR, Rahmad N, Lawson T (2019) Comparative proteomic analysis on fruit ripening processes in two varieties of tropical mango (*Mangifera indica*). Protein J 38(6):704–715

Chourasia A, Sane VA, Nath P (2006) Differential expression of pectate lyase during ethylene-induced postharvest softening of mango (*Mangifera indica* var. Dashehari). Physiol Plant 128(3):546–555

Cruz-Hernández A, Gómez-Lim MA (1995) Alternative oxidase from mango (*Mangifera indica* L.) is differentially regulated during fruit ripening. Planta 197:569–576

Dautt-Castro M, Ochoa-Leyva A, Contreras-Vergara CA, Muhlia-Almazán A, Rivera-Domínguez M, Casas-Flores S et al (2018) Mesocarp RNA-Seq analysis of mango (*Mangifera indica* L.) identify quarantine postharvest treatment effects on gene expression. Sci Hort 227:146–153

Dautt-Castro M, Ochoa-Leyva A, Contreras-Vergara CA, Pacheco-Sanchez MA, Casas-Flores S, Sanchez-Flores A et al. (2015) Mango (*Mangifera indica* L.) cv. Kent fruit mesocarp de novo transcriptome assembly identifies gene families important for ripening. Front Plant Sci 6(FEB):1–12

de Godoy A, Morita RJ, Cordenunsi BR, Lajolo FM, do Nascimento JRO (2009) Expression analysis of a set of genes related to the ripening of bananas and mangoes. Braz J Plant Physiol 21(4):251–259

Santos-Villalobos S, Parra-Cota FI, de-Folter S, Pena-Cabriales JJ (2012) Primers to amplify flowering locus T (FT) transcript in mango (*Mangifera indica*) and their potential use in other angiosperms. Plant Omics 5 (5):453–457

Deshpande AB, Anamika K, Jha V, Chidley HG, Oak PS, Kadoo NY et al (2017) Transcriptional transitions in Alphonso mango (*Mangifera indica* L.) during fruit development and ripening explain its distinct aroma and shelf life characteristics. Sci Rep 7(1):1–19

Fasoli E, Righetti PG (2013) The peel and pulp of mango fruit: A proteomic samba. Biochim Biophys Acta-Proteins Proteom 1834(12):2539–2545

Frylinck L, Dubery IA (1998) Protein kinase activities in ripening mango, *Mangifera indica* L., fruit tissue: I: Purification and characterization of a calcium-stimulated casein kinase-I. Biochim Biophys Acta-Protein Struct Mol Enzymol 1382(1):65–79

Hoang VL, Innes DJ, Shaw PN, Monteith GR, Gidley MJ, Dietzgen RG (2015) Sequence diversity and differential expression of major phenylpropanoid-flavonoid biosynthetic genes among three mango varieties. BMC Genom 16:561

Hong K, Gong D, Zhang L, Hu H, Jia Z, Gu H, Song K (2016) Transcriptome characterization and expression profiles of the related defense genes in postharvest mango fruit against *Colletotrichum gloeosporioides*. Gene 156(1):275–283

Iquebal MA, Jaiswal S, Mahato AK, Jayaswal PK, Angadi UB, Kumar N, et al. (2017) MiSNPDb: a web-based genomic resources of tropical ecology fruit mango (*Mangifera indica* L.) for phylogeography and varietal differentiation. Sci Rep 7:14968

Lantican DV, Cortaga CQ, Manohar ANC, dela Cueva FM, Sison MLJ (2020) Transcriptome-wide analysis of expressed resistance gene analogs (RGAs) in mango. bioRxiv 2020.02.08.939736; https://doi.org/10.1101/2020.02.08.939736

Li L, Ma XW, Zhan RL, Wu HX, Yao QS, Xu WT, Luo C et al (2017a) Profiling of volatile fragrant components in a mini-core collection of mango germplasms from seven countries. PLoS ONE 12 (12):e0187487

Li L, Shuai L, Sun J, Li C, Yi P, Zhou Z et al. (2019) The role of 1-methylcyclopropene in the regulation of ethylene biosynthesis and ethylene receptor gene expression in *Mangifera indica* L. (mango fruit). Food Sci Nutr 8(2):1284–1294

Li YH, Zhang HN, Wu QS, Muday GK (2017b) Transcriptional sequencing and analysis of major genes involved in the adventitious root formation of mango cotyledon segments. Planta 245(6):1193–1213

Liao DJ, Lu XP, Chen HS, Lu Y, Mo ZY (2016) Evaluation of four protein extraction methods for proteomic analysis of mango peel. Genet Mol Res 15 (3):1–9. https://doi.org/10.4238/gmr.15039006

Liu Z-liang, Luo C, Dong L, Toan-C Van, P-Xiao Wei, X-hua He (2014) Molecular characterization and expression analysis of a GTP-binding protein (*MiRab5*) in *Mangifera indica*. Gene 540(1):86–91

Luo C, He XH, Hu Y, Yu HX, JinOu S, Fang ZB (2014) Oligo-dT anchored cDNA–SCoT: a novel differential display method for analyzing differential gene expression in response to several stress treatments in mango (*Mangifera indica* L.). Gene 548(2):182–189. ISSN 0378-1119

Luria N, Sela N, Yaari M, Feygenberg O, Kobiler I, Lers A et al (2014) De-novo assembly of mango fruit peel transcriptome reveals mechanisms of mango response to hot water treatment. BMC Genom 15:957

Mahato AK, Sharma N, Singh A, Srivastav M, Jaiprakash, Singh SK, et al. (2016) Leaf transcriptome sequencing for identifying genic-SSR markers and SNP heterozygosity in crossbred mango variety "Amrapali" (*Mangifera indica* L.). PLoS One 11 (10):1–16

Mishra M, Shree Y, Pati R, Seal S, Shukla N, Kamle M (2010) Micropropagation of *Mangifera indica* L. cv. Kurakkan through somatic embryogenesis. Indian J Genet 70(1): 85–89

Munafo JP Jr, Didzbalis J, Schnell RJ, Schieberle P, Steinhaus M (2014) Characterization of the major aroma-active compounds in mango (*Mangifera Indica* L.) Cultivars Haden, White Alfonso, Praya Sowoy, Royal Special, and Malindi by application of a comparative aroma extract dilution analysis. J Agric Food Chem 62(20):4544–4551

Muthukumar M, Bajpai A, Bajpai Y, Singh VK and Rajan S (2018). Differential gene expression responses induced by salinity stress in polyembryonic mango cultivars. In: Souvenir cum book of abstracts. National Conference on Strategies & Challenges in doubling farmers' Income through Horticultural Technologies in Subtropics, held at ICAR-CISH, Lucknow during 21–22 June, 2018. pp 161–162.

Nakagawa M, Honsho C, Kanzaki S, Shimizu K, Utsunomiya N (2012) Isolation and expression analysis of FLOWERING LOCUS T-like and gibberellin metabolism genes in biennial-bearing mango trees. Sci Hort 139:108–117

Okada T, Ihara H, Ito R, Ikeda Y (2017) Molecular cloning and functional expression of lewis type α1,3/α1,4-fucosyltransferase cDNAs from *Mangifera indica* L. Phytochemistry 144:98–105

Pandit SS, Kulkarni RS, Giri AP, Köllner TG, Degenhardt J, Gershenzon J, Gupta VS (2010) Expression profiling of various genes during the fruit development and ripening of mango. Plant Physiol Biochem 48 (6):426–433

Qamar-ul-Islam T, Khan MA, Faizan R, Mahmood U (2018) MGDb: An analyzed database and a genomic resource of mango (*Mangifera indica* L.) cultivars for mango research. bioRxiv; 301358

Quijano C, Salamanca GG, Pino, Jorge J (2007) Aroma volatile constituents of Colombian varieties of mango (*Mangifera indica* L.). Flav Frag J 22. https://doi.org/10.1002/ffj.1812

Renuse S, Harsha HC, Kumar P, Acharya PK, Sharma J, Goel R et al (2012) Proteomic analysis of an unsequenced plant—*Mangifera indica*. J Proteom 75 (18):5793–5796

Sharma N, Singh AK, Singh SK, Mahato AK, Srivastav, Singh NK (2020) Comparative RNA sequencing based transcriptome profiling of regular bearing and alternate bearing mango (*Mangifera indica* L.) varieties reveals novel insights into the regulatory mechanisms underlying alternate bearing. Biotechnol Lett 8. https://doi.org/10.1007/s10529-020-02863-8

Sivankalyani V, Sela N, Feygenberg O, Zemach H, Maurer D, Alkan N (2016) Transcriptome dynamics in mango fruit peel reveals mechanisms of chilling stress. Front Plant Sci 7:1–17

Srivastava S, Singh RK, Pathak G, Goel R, Asif MH, Sane AP, Sane VA (2016) Comparative transcriptome analysis of unripe and mid-ripe fruit of *Mangifera indica* (var. "Dashehari") unravels ripening associated genes. Sci Rep 6(8):1–13

Tafolla-Arellano JC, Zheng Y, Sun H, Jiao C, Ruiz-May E, et al. (2017) Transcriptome analysis of mango (*Mangifera indica* L.) fruit epidermal peel to identify putative cuticle-associated genes. Sci Rep 7(4):1–13

Ubonbal R, Porsoongnoen S, Daduang J, Klaynongsruang S, Daduang S (2017) Purification and characterization of two isoforms of native α amylase from Ok-Rong mango (*Mangifera indica* Linn. cv. Ok-Rong). Turk J Biochem 42(6):624–632

Valenzuela-Chavira I, Contreras-Vergara CA, Arvizu-Flores AA, Serrano-Posada H, Lopez-Zavala AA, García-Orozco KD et al (2017) Insights into ligand binding to a glutathione S-transferase from mango: Structure, thermodynamics and kinetics. Biochimie 135:35–45

Vyavahare SN, Krishna B, Joshi SS, Chaudhari RS, Subramaniam VR, Sane PV (2017) Characterization of mango Flowering Locus T (*FT*) and Terminal Flower 1 (*TFL1*) genes. Acta Hort 1183:113–124

Wei J, Liu D, Liu G, Tang J, Chen Y (2016) Molecular cloning, characterization, and expression of MiSOC1: A homolog of the flowering gene suppressor of overexpression of constans1 from mango (*Mangifera indica* L). Front Plant Sci 7(11):1–13

Wu H, Jia H, Ma X, Wang S, Yao Q et al (2014) Transcriptome and proteomic analysis of mango (*Mangifera indica* Linn) fruits. J Proteom 105:19–30

Yadav A, Jayaswal PK, Venkat Raman K, Singh B, Singh NK, Usha K (2019) Transcriptome analysis of flowering genes in mango (*Mangifera indica* L.) in relation to floral malformation. J Plant Biochem Biotechnol https://doi.org/10.1007/s13562-019-00541-z

Zainal ZB, Tucker GA, Lycett GW (1996) A rab11-like gene is developmentally regulated in ripening mango (*Mangifera indica* L.) fruit. Biochim Biophys Acta—Mol Cell Res 1314(3):187–190

Zhao Z, Chen Y, Gao A, Huang J, Dang Z (2014) Cloning and expression of anthocyanidin synthase (ANS) gene from peel of mango (*Mangifera indica* Linn). Afr J Plant Sci 8(3):147–152

Zhao Z, Gao A, Chen Y, Huang J, Dang Z (2016) Cloning and expression analysis of F3'5'H gene from mango (*Mangifera indica* L.). Hans J Comput Biol 6(1):1–7

Zhang T, Ying Z, Davenport TL (2005) Isolation and characterization of a CONSTANS-like gene in mango. Paper presented at: Plant Biology 2005—Annual Meeting of the American Society of Plant Biologists. American Society of Plant Biologists, Seattle, Washington, p 151

Mango Genomic Resources and Databases

13

Sarika Jaiswal, Mir Asif Iquebal, UB Angadi, Sunil Kumar, Anil Rai, Nagendra K. Singh, and Dinesh Kumar

Abstract

Web resources are the archival repository for data and also a research accelerator tool for global community, economizing cost of data along with the advantage of rapidity, and ease of accessibility. Recently, the mango genome has been sequenced but very limited web genomic resources of this crop are available. This chapter covers mango web genomic resources under four sub-heads, viz., linkage map based, whole genome based, transcriptome based and SNP marker based. Linkage Map genomic resource describes how two distinct varieties, *i.e.*, Jin-Hwang and Irwin have been used to make the very first high-density genetic map of mango. This is based on 156,368 polymorphic specific-locus amplified fragments (SLAFs) with total length of 3,148.28 cM and 20 linkage groups with an average marker distance of < 0.50 cM. Whole genome-based genomic resources covers latest whole genome sequence (WGS) data accessibility at NCBI (BioProject: PRJNA487154). It also provides INSDC ID of all 12 chromosome for download. Transcriptome-based genomic resources describe existing genomic resource MGdb based on four mango varieties, viz., langra (Pakistan), zill (China), shelly (Israel), and kent (Mexico). This section also tabulates transcriptome resources available in GenBank with varieties, types/treatment of tissues, Bio-project number with country and organizational details. The last section of this chapter describes how such web genomic resources can be used for DNA-based varietal signature supplementing distinctness, uniformity and stability (DUS) test, quantitative trait locus (QTL) detection, gene discovery, variety improvement for better biotic and abiotic stresses management, traceability, adulteration issues in mango product or origin or protection in terms of intellectual property (IP) of geographical indication (GI) law. More such resources are warranted not only for

S. Jaiswal · M. A. Iquebal · U. Angadi · S. Kumar · A. Rai · D. Kumar (✉)
Centre for Agricultural Bioinformatics, ICAR-Indian Agricultural Statistics Research Institute, New Delhi, India
e-mail: dinesh.kumar@icar.gov.in

S. Jaiswal
e-mail: sarika@icar.gov.in

M. A. Iquebal
e-mail: ma.iquebal@icar.gov.in

U. Angadi
e-mail: ub.angadi@icar.gov.in

S. Kumar
e-mail: sunil.kumar7@icar.gov.in

A. Rai
e-mail: anil.rai@icar.gov.in

N. K. Singh
ICAR-National Institute for Plant Biotechnology, New Delhi, India
e-mail: nksingh4@gmail.com

C. Kole (ed.), *The Mango Genome*, Compendium of Plant Genomes,
https://doi.org/10.1007/978-3-030-47829-2_13

knowledge discovery but also for better productivity and profitability of mango growers with better consumer interest protection.

13.1 Introduction

Web resources or databases have become not only an archival space for data, but are also a tool for organizing voluminous biological data which can be used by global researchers to access and retrieve customized information. Users can easily access and share/use the data in minimum time and in a user-friendly manner, economizing cost of data and also with advantage of rapidity (Lathe et al. 2008). Though mango is the 'king of fruits' of tropical and subtropical ecological region of the world with relatively smaller allotetraploid genome size of < 450 Mb made of 20 pairs of chromosome, still genomic resources of mango are very limited. Very recently, in the year 2020, chromosome-wise whole genome sequence assembly of mango genome has become available (Wang et al. 2020). Unless more and more genomic resources are available, the germplasm improvement using genomics-based approach can never be able to see the light of the day.

The mango crop is highly sensitive to diversity of both, soil and tropical ecology. This has led mango crop to evolve with diverse varieties having unique appearance, taste, and aroma in different geographical regions of the world. Hence their genomic resources are equally diverse. The web genomic resources as a source of knowledge discovery can be used to understand the crop with better management of germplasm. The following mango genomic resources are available in the public domain for the global community:

1. Mango genomic resources as linkage map
2. Whole genome-based mango genomic resources
3. Transcriptome-based genomic resources
4. SNP marker-based genomic resources.

13.2 Mango Genome Resources as Linkage Map

Genetic maps are important resources used in mapping if simply inherited and quantitative trait loci (QTLs) and thereby in marker-assisted selection (MAS) for improvement in target traits. In mango, a genetic map has been prepared by the crossing of two distinct varieties, Jin-Hwang and Irwin, by researchers at the Key Laboratory of Tropical Fruit Biology, Ministry of Agriculture; South Subtropical Crops Research Institute, Chinese Academy of Tropical Agricultural Sciences, Zhanjiang, China. This was achieved by genomic and phenomic data of 173 F_1 plants and high-throughput sequencing and specific-locus amplified fragment (SLAF) library construction. This very first high-density genetic map of mango was constructed by 66.02 GB genomic data having 330.64 M reads, 156,368 polymorphic SLAFs (out of 318,414 SLAFs). This map has a total length of 3,148.28 cM, with an average marker distance of <0.50 cM by use of 6,594 SLAFs distributed on 20 linkage groups (Table 13.1). The single nucleotide polymorphism (SNP) distribution and transition/transversion numbers are shown in Table 13.2.

13.3 Whole Genome-Based Genomic Resources of Mango

Though cytogenetic, transcriptome and genetic map data had been available in mangoes, whole-genome resources were publically available only in the year 2020. The popular traditional Indian mango variety Alphonso was sequenced by China in collaboration with the USA. This was done by using the PacBio Sequel II platform along with paired-end, mate-paired short reads and Hi-C sequencing for scaffolding. De novo assembly having 20 chromosomes (pseudomolecules) were made by anchoring scaffolds over genetic maps having 20 linkage groups

Table 13.1 Basic characteristics of the 20 linkage groups of mango. (Luo Chun et al. 2016)

Linkage group	Total number of markers			Total distance (cM)			Average distance (cM)			Gap <5 (cM)			Max. gap (cM)			Number of markers with segregation distortion
	Female map	Male map	Integrated map	Female map	Male map	Integrated map	Female map	Male map	Integrated map	Female map (%)	Male map (%)	Integrated map (%)	Female map	Male map	Integrated map	Integrated map
LG1	230	145	362	161.96	113.72	168.04	0.71	0.79	0.47	96.51	92.36	98.06	22.148	22.188	9.612	17
LG2	170	111	272	180.35	165.48	173.21	1.07	1.50	0.64	94.67	93.64	97.05	21.295	23.033	12.019	8
LG3	245	67	299	198.74	104.87	170.13	0.81	1.59	0.57	96.31	90.91	98.32	13.030	14.758	10.450	28
LG4	390	162	526	188.39	137.92	184.04	0.48	0.86	0.35	99.23	95.65	99.43	13.548	31.201	10.715	1
LG5	209	80	262	144.75	88.97	150.23	0.70	1.13	0.58	95.67	93.67	97.70	10.940	17.055	10.940	22
LG6	207	76	255	146.29	108.31	119.44	0.71	1.44	0.47	98.06	92.00	96.85	16.249	38.316	13.690	3
LG7	195	49	224	125.31	116.33	121.40	0.65	2.42	0.54	96.39	91.67	97.76	12.290	39.192	10.311	9
LG8	191	74	219	167.78	105.03	175.09	0.88	1.44	0.80	97.37	91.78	96.79	28.000	45.865	12.075	15
LG9	202	171	368	182.83	258.06	198.19	0.91	1.52	0.54	95.02	93.53	97.28	20.624	45.959	18.734	3
LG10	192	130	287	119.61	122.70	122.33	0.63	0.95	0.43	96.86	96.12	99.30	9.610	34.369	5.144	0
LG11	392	235	480	178.46	189.62	184.04	0.46	0.81	0.38	98.21	96.58	99.37	9.523	19.466	5.388	12
LG12	160	108	231	136.15	106.60	126.40	0.86	1.00	0.55	93.71	93.46	98.70	12.401	21.295	5.245	9
LG13	264	160	371	160.10	141.03	150.57	0.61	0.89	0.41	97.72	93.08	98.92	16.249	21.295	6.088	17
LG14	219	86	272	116.84	109.00	123.21	0.54	1.28	0.45	96.33	96.47	98.89	6.143	46.874	8.047	14
LG15	183	98	262	140.72	85.10	141.19	0.77	0.88	0.54	95.05	92.78	96.17	9.926	16.249	6.801	4
LG16	477	223	591	206.09	169.13	205.69	0.43	0.76	0.35	98.74	96.40	99.49	11.665	31.035	6.967	1
LG17	174	80	228	132.75	116.64	133.93	0.77	1.48	0.59	95.38	86.08	95.15	45.959	24.032	13.799	0
LG18	158	160	266	140.96	192.00	170.57	0.90	1.21	0.64	93.63	96.23	98.11	21.026	33.952	12.696	11
LG19	359	294	520	193.20	258.51	206.90	0.54	0.88	0.40	98.04	95.22	98.27	38.932	27.908	17.809	0
LG20	249	76	299	122.95	58.87	123.68	0.50	0.78	0.42	97.18	88.00	97.99	10.227	6.705	8.827	0
Total	4866	2585	6594	3144.23	2747.89	3148.28	0.65	1.07	0.48	96.50	93.28	97.98	45.959	46.874	18.734	174

*"Gap < 5" indicates the percentages of gaps in which the distance between the adjacent markers was less than 5

Table 13.2 Distribution of SNP loci on the 20 linkage groups of mango (Luo et al. 2016)

Linkage group	SNP number	Transition/transversion number
LG1	750	487/263
LG2	575	371/204
LG3	691	439/252
LG4	1112	692/421
LG5	592	404/189
LG6	561	389/172
LG7	512	340/172
LG8	443	278/165
LG9	775	497/279
LG10	596	421/176
LG11	971	578/393
LG12	474	312/162
LG13	788	523/265
LG14	562	355/207
LG15	506	319/188
LG16	1139	726/414
LG17	515	351/164
LG18	549	329/221
LG19	1105	705/400
LG20	628	397/231

Table 13.3 Whole genome data of mango at NCBI (BioProject: PRJNA487154)

Total sequence length	:	392,983,452
Total ungapped length	:	392,900,608
Gaps between scaffolds	:	0
Number of scaffolds	:	252
Scaffold N50	:	17,652,500
Scaffold L50	:	10
Number of contigs	:	421
Contig N50	:	3,579,047
Contig L50	:	35
Total number of chromosomes and plasmids	:	22
Number of component sequences (WGS or clone)	:	252

(Wang et al. 2020). These genomic resources are freely accessible at NCBI (BioProject: PRJNA487154). The assembly statistics reflecting the quality of assembly are shown in Table 13.3. The entire 20 chromosomes can be downloaded using INSDC ID as shown in Table 13.4 along with their respective size and GC content.

Table 13.4 Mango chromosome-wise data (Wang et al. 2020)

Sl. no.	Type/name	*INSDC ID	Size (Mb)	GC%
1.	Chromosome 1	CM021837.1	29.46	32.5
2.	Chromosome 2	CM021838.1	24.4	32.5
3.	Chromosome 3	CM021839.1	23.14	32.9
4.	Chromosome 4	CM021840.1	21.51	32.7
5.	Chromosome 5	CM021841.1	21.08	33.1
6.	Chromosome 6	CM021842.1	18.81	32.4
7.	Chromosome 7	CM021843.1	20.62	32.5
8.	Chromosome 8	CM021844.1	18.24	33.5
9.	Chromosome 9	CM021845.1	18.23	32.5
10.	Chromosome 10	CM021846.1	17.65	32.4
11.	Chromosome 11	CM021847.1	17.14	32.9
12.	Chromosome 12	CM021848.1	16.03	32.8
13.	Chromosome 13	CM021849.1	15.46	32.5
14.	Chromosome 14	CM021850.1	14.81	32.7
15.	Chromosome 15	CM021851.1	14.77	32.3
16.	Chromosome 16	CM021852.1	13.82	32.4
17.	Chromosome 17	CM021853.1	13.51	32.4
18.	Chromosome 18	CM021854.1	13.37	32.6
19.	Chromosome 19	CM021855.1	13.09	32.5
20.	Chromosome 20	CM021856.1	12.29	32.3
21.	Mitochondria	CM021857.1	0.87	44.4
22.	Plasmids	CM021858.1	0.15	37.8
23.	Unplaced	–	34.52	33.5

*INSDC: International Nucleotide Sequence Database Collaboration

13.4 Transcriptome-Based Genomic Resources of Mango

Qamar-ul-Islam et al. (2018) have reported the mango web genomic resource MGdb, based on comparative transcriptome analysis of four mango varieties, viz., Langra (Pakistan), Zill (China), Shelly (Israel), and Kent (Mexico). This was based on limited public domain secondary transcriptome data which were already published. These transcriptome resources were pooled from mango varieties, viz., Langra (Azim et al. 2014), Zill (Wu et al. 2014), Shelly (Luria et al. 2014), and Kent (Dautt-Castro et al. 2015).

Though Mango Genome Atlas as web genomic resource for this particular plant is yet to be developed, considerable transcriptomic resources are available online in GenBank. In Table 13.5, such transcriptome resources are described with variety, type/treatment of tissue, Bioproject number of GenBank, and country with organizational details.

13.5 SNP Genomic Resources of Mango

World's first web-based genomic resources based on 84 distinct mango varieties was reported from India (Iquebal et al. 2017). This was based on 1.25 million SNPs, discovered from 28.6 Gb data, generated by ddRAD-Seq approach using Illumina HiSeq 2000 platform. The developed web genomic resources MiSNPDb is freely accessible at http://webtom.cabgrid.res.in/

Table 13.5 Transcriptomic resources of different mango varieties

Sl. no.	Variety	Tissue	BioProject number	Organization with country
1	Dashehari	Floral and vegetative buds of mango	PRJNA610491	ICAR-Central Institute for Subtropical Horticulture, Lucknow, India
2	Wild Mango	Six sets	PRJNA533518	University of Queensland, Australia
3	Dashehari	Jelly seed stage	PRJNA431125	ICAR-Central Institute for Subtropical Horticulture, Lucknow, India
4	Dashehari	Fruits with various stage of ripening and aroma	PRJNA317712	ICAR-Central Institute for Subtropical Horticulture, Lucknow, India
5	Chausa	Various stages of fruits	PRJNA317044	ICAR-Central Institute for Subtropical Horticulture, Lucknow, India
6	Zihua	Stages of root formation in cotyledon segments of mango	PRJNA316542	Chinese Academy of Tropical Agricultural Sciences, Zhanjiang, China.
7	Dashehari	Fruits stages for ripening and aroma	PRJNA317712	CSRI-National Botanical Research Institute, Lucknow, India
8	Keitt	Tissue bud with stages of infection with *Fusarium mangiferae*	PRJNA315791	South Asia Tropical Crop Research Institute, CATAS, China
9	Amrapali	various stages of flower	PRJNA313340	ICAR-Central Institute for Subtropical Horticulture, Lucknow India
10	Ono and Renong No.1	Pericarp	PRJNA310193	South Subtropical Crops Research Institute, Chinese Academy of Tropical Agricultural Sciences
11	Keitt	fruit peel with chilling stress	PRJNA304093	Agricultural Research Organization, Volcani Center, Israel
12	Tommy Atkins with other six varieties viz Turpentine' 'Thai Everbearing' 'neelum' M. casturi "Purple"' 'Burma' 'Amin Abrahimpur'	mixed organs, seed, seed coat, mesocarp, leaf, flower, exocarp for marker discovery	PRJNA288797	Keithanne Mockaitis, Indiana University, Bloomington
13	Ataulfo	hot water treated fruits for insect control	PRJNA286253	CIAD: Centro de Investigacion en Alimentacion y Desarrollo A.C. Mexico, Mexico
14	Amrapali	Malformed flower, Mango healthy fruit, Amrapali leaf	PRJNA27982	ICAR-Indian Agricultural Research Institute, New Delhi, India

(continued)

Table 13.5 (continued)

Sl. no.	Variety	Tissue	BioProject number	Organization with country
15	Kent	Mesocarp for fruit ripening	PRJNA258477	National Autonomous University of Mexico (Universidad Nacional Autonoma de Mexico), Mexico
16	Tomy Atkins and Keitt	Different fruit stages	PRJNA254771	Agricultural Research Organization, Volcani center, Israel
17	Keitt	Fruit peel	PRJNA253272	Cornell University, Ithaca, New York
18	Zill	Fruit development and ripening in mango	PRJNA235247	South Subtropical Crops Research Institute, Chinese Academy of Tropical Agricultural Sciences
19	Shelly	Fruit stress Hot water treatment	PRJNA227243	Agricultural Research Organization, Israel
20	Dushehari	leaves	PRJNA193591	ICAR-National Research Centre on Plant Biotechnology, Pusa Campus, New Delhi, India
21	Neelam	leaf	PRJNA193588	National Research Centre on Plant Biotechnology, IARI Pusa Campus, New Delhi, India
22	Langra	leaves	PRJNA192676	International Center for Chemical and Biological Sciences, University of Karachi, Karachi, Pakistan

mangosnps/. Table 13.6 describes type of SNP variation, their number and percentage of the SNP type in MiSNPdb. Table 13.7 describes number of SNP available in 84 different mango varieties in MiSNPdb.

13.6 Utility of Mango Web Genomic Resources

Mango genomic resources are required to improve the germplasm for resistance against various biotic and abiotic stresses. They are also required to accelerate the DNA-based varietal differentiation, phylogeographic studies, QTL detection, and gene discovery in germplasm management and improvement. It can be used in development of varietal DNA signature, traceability, genome finishing and SNP chip development for future genome-wide association studies (GWAS) in genomic selection programs (Iquebal et al. 2017). It can be used for other non-model species which are genetically close to this species. These SNPs can be mapped in latest genome assembly of mango and such location specific SNP can be of much greater value.

Markers of Mango SNP database can be used over present linkage map of mango. This can increase marker density further over chromosome. The database has markers from non-genic regions also thus ultra-dense genetic map is feasible covering entire chromosome region. It can be used for mango throughput SNP arrays (SNP chip) development where MAF

Table 13.6 Type of SNP variation, their number and percentage of the SNP type (Iquebal et al. 2017)

Type of variation	Number	Percentage of the SNP type
A/G	427968	34.00
C/T	429349	34.11
A/T	118403	9.40
A/C	98811	7.85
C/G	83914	6.66
G/T	100260	7.96
Total	1258705	100.00

Table 13.7 Number of SNPs detected in each of the 84 varieties of mango (Iquebal et al. 2017)

Variety	Number of SNPs	Variety	Number of SNPs	Variety	Number of SNPs
Afeam	17749	Fazri Kalam	11889	Mohanbhog	8980
Alphan	16439	Gilas	8132	Mombosa	18388
Alphanso	12305	Gola Bhadaiya	19057	Mulgoa	18270
Amin Prince	13929	Gourjeet	9891	Mundappa Black	8955
Amrapali	14002	Gulab Khas Green	11236	Neelum	15352
Arka Aruna	18636	Hardil Aziz	10045	Nekkare	16946
Arunika	8913	Heraswania	26619	Prabhashankar	5018
Baganapalli	19227	Himsagar	31479	Primor de Amoreira	9979
Banganpalli	16458	Hyb. 165	13375	Pusa Arunima	22087
Banglora	13851	Irwin	9889	Pusa Lalima	16964
Baramasi	19290	Iturba	18228	Pusa Peetamber	13154
Baramasi Ajholi	13286	Janardan Pasand	15850	Pusa Pratibha	11330
Bathui	11030	Kala	13867	Pusa Shersth	12922
Bhadaiya Sukul	15080	Kalapahar	8873	Pusa Surya	14123
Bhadauran	11122	Karishad	9704	Ramkela	16380
Bombay	7440	Kesar	19130	Rataul	17558
Bombay Green	13236	Khasulkhas	14202	Ratna	14419
Bombay Yellow	9793	Kothapalli Kobbari	13768	Rosari	16491
Bride of Russia	22978	Kurukkan	11571	Safdar Pasand	14382
Carabao	15576	Langra	30131	Samar Bahist Alibagh	7946
Chandrakaran	13538	Langra Gorakhpur	14260	Seipia	17643
Chinku	12245	Machhli	15548	Sensation	13110
Creeping II	15272	Malda	15530	Sonatol	9698

(continued)

Table 13.7 (continued)

Variety	Number of SNPs	Variety	Number of SNPs	Variety	Number of SNPs
Dushehari	15095	Malihabad Safeda	13653	Sukul	18788
Edward	13431	Mallika	19592	Suvarnarekha	37688
Elaichi	9243	Manipur dwarf	11064	Tatoul	3122
Extrema	15042	Manorajan	33248	Willard	26827
Fazri	13350	Mohammada Vikarabad	19862	Zardalu	4936

(minorallele frequency) information is critically required for selection of SNPs for array development to be used in GWAS and genomic selection programs.

Use of mango SNPs and mango genomic resources is very useful in DNA-based traceability in establishing the origin of mango products. In case of varietal dispute for intellectual property (IP) or sovereignty issues, few selected SNPs can be used as DNA signature of mango variety. SNPs offer advantages by their ability of high-resolution power and objectivity. It can easily be achieved by taking any sample in form of live germplasm or any tissue of processed products with rapidity and cost-effectiveness (Iquebal et al. 2017). Such DNA-based tests are legally admissible while establishing varietal identity during dispute resolution. Online availability of such resources to the global community can accelerate the development of a specific assay for varietal identification as and when required for group of mango varieties having adulteration issues in their product or origin or protection in terms of GI (geographical indication) protection. The phylogenetic tree using 749 common SNPs across 84 mango varieties has revealed three major lineages among the Indian mango varieties, which were superimposed over geographical map of India (Iquebal et al. 2017). It has clearly demonstrated the ability of such genomic resources to protect intellectual property given to mango growing community or association in terms of GI protection. Besides research and knowledge discovery such resources also play a critical role in terms of better profitability to mango growers and also ensuring consumer interests protection as envisaged in GI law of countries.

References

Azim MK, Khan, I, Zhang Y (2014) Characterization of mango (*Mangifera indica* L.) transcriptome and chloroplast genome. Plant Mol Biol 85(1–2):193–208. https://doi.org/10.1007/s11103-014-0179-8

Dautt-Castro M, Ochoa-Leyva A, Contreras-Vergaral CA, Pacheco-Sanchez MA, Casas-Flores S, Sanchez-Flores A, Kuhn DN, Islas-Osuna MA(2015) Mango (*Mangifera indica* L.) cv. Kent fruit mesocarp *de novo* transcriptome assembly identifies gene families important for ripening. Front Plant Sci 6:62. https://doi.org/10.3389/fpls.2015.00062

Iquebal MA, Jaiswal S, Mahato AK, Jaiswal PK, Angadi UB et al (2017). MiSNPDb: a web-based genomic resources of tropical ecology fruit mango (*Mangifera indica* L.) for phylogeography and varietal differentiation. Sci Rep 7:14968 (2017). https://doi.org/10.1038/s41598-017-14998-2

Lathe W, Williams J, Mangan M, Karolchik D (2008) Genomic data resources: challenges and promises. Nat Educ 1(3):2

Luo C, Shu B, Yao Q, Wu H, Xu W, Wang S (2016) Construction of a high-density genetic map based on large-scale marker development in mango using Specific-Locus Amplified Fragment Sequencing (SLAF-seq). Front Plant Sci 7:1310. https://doi.org/10.3389/fpls.2016.01310

Luria N, Sela N, Yaari M, Feygenberg O, Kobiler I, Lers A, Prusky D (2014) *De-novo* assembly of mango fruit peel transcriptome reveals mechanisms of mango response to hot water treatment. BMC Genom 15(1):957. https://doi.org/10.1186/1471-2164-15-957

Qamar-ul-Islam T, Khan MA, Faizan R, Mahmood U (2018) MGDb: An analysed database and a genomic resource of mango (*Mangifera Indica* L.) cultivars for mango research. bioRxiv. https://doi.org/10.1101/301358

Wang P, Luo Y, Huang J, Gao S, Zhu G et al (2020) The genome evolution and domestication of tropical fruit mango. Genome Biol 21(1):60. https://doi.org/10.1186/s13059-020-01959-8

Wu H, Jia H, Ma X, Wang S, Q Yao, Xu W, Gao Y (2014) Transcriptome and proteomic analysis of mango (*Mangifera indica* Linn) fruits. J Proteom 105:19–30. https://doi.org/10.1016/j.jprot.2014.03.030

The manufacturer's authorised representative in the EU is Springer Nature Customer Service Centre GmbH, Europaplatz 3, 69115 Heidelberg, Germany. If you have any concerns regarding our products, please contact ProductSafety@springernature.com

Printed and bound by CPI Group (UK) Ltd, Croydon, CR0 4YY
30/09/2026
02225691-0018